T0142394

Lecture Notes in Electrical Engineering

Volume 662

The book series *Lecture Notes in Electrical Engineering* (LNEE) publishes the latest developments in Electrical Engineering—quickly, informally and in high quality. While original research reported in proceedings and monographs has traditionally formed the core of LNEE, we also encourage authors to submit books devoted to supporting student education and professional training in the various fields and applications areas of electrical engineering. The series cover classical and emerging topics concerning:

- Communication Engineering, Information Theory and Networks
- Electronics Engineering and Microelectronics
- Signal, Image and Speech Processing
- Wireless and Mobile Communication
- Circuits and Systems
- Energy Systems, Power Electronics and Electrical Machines
- Electro-optical Engineering
- Instrumentation Engineering
- Avionics Engineering
- Control Systems
- Internet-of-Things and Cybersecurity
- Biomedical Devices, MEMS and NEMS

For general information about this book series, comments or suggestions, please contact leontina.dicecco@springer.com.

To submit a proposal or request further information, please contact the Publishing Editor in your country:

China

Jasmine Dou, Associate Editor (jasmine.dou@springer.com)

India, Japan, Rest of Asia

Swati Meherishi, Executive Editor (Swati.Meherishi@springer.com)

Southeast Asia, Australia, New Zealand

Ramesh Nath Premnath, Editor (ramesh.premnath@springernature.com)

USA, Canada:

Michael Luby, Senior Editor (michael.luby@springer.com)

All other Countries:

Leontina Di Cecco, Senior Editor (leontina.dicecco@springer.com)

** Indexing: The books of this series are submitted to ISI Proceedings, EI-Compendex, SCOPUS, MetaPress, Web of Science and Springerlink **

More information about this series at http://www.springer.com/series/7818

Rabindranath Bera · Prashant Chandra Pradhan ·
Chuan-Ming Liu · Sourav Dhar ·
Samarendra Nath Sur
Editors

Advances in Communication, Devices and Networking

Proceedings of ICCDN 2019

 Springer

Editors
Rabindranath Bera
Department of Electronics
and Communication Engineering
Sikkim Manipal Institute of Technology
Majitar, Sikkim, India

Prashant Chandra Pradhan
Department of Electronics
and Communication Engineering
Sikkim Manipal Institute of Technology
Majitar, Sikkim, India

Chuan-Ming Liu
Department of Computer Science
and Information Engineering
National Taipei University of Technology
Taipei, Taiwan

Sourav Dhar
Department of Electronics
and Communication Engineering
Sikkim Manipal Institute of Technology
Majitar, Sikkim, India

Samarendra Nath Sur
Department of Electronics
and Communication Engineering
Sikkim Manipal Institute of Technology
Majitar, Sikkim, India

ISSN 1876-1100   ISSN 1876-1119   (electronic)
Lecture Notes in Electrical Engineering
ISBN 978-981-15-4934-2   ISBN 978-981-15-4932-8   (eBook)
https://doi.org/10.1007/978-981-15-4932-8

This Springer imprint is published by the registered company Springer Nature Singapore Pte Ltd.
The registered company address is: 152 Beach Road, #21-01/04 Gateway East, Singapore 189721, Singapore

Organization

Chief Patron

Lt. Gen.(Dr.) Venkatesh, VC, SMU

Patron

Dr. K. S. Sherpa, Registrar, SMU
Dr. A. Sharma, Director, SMIT

General Chair

Dr. Sourav Dhar, ECE, SMIT

Program Chair

Dr. Rabindranath Bera, ECE, SMIT
Dr. P. C. Pradhan, ECE, SMIT

Convener

Dr. Samarendra Nath Sur, ECE, SMIT

Co-Convener

Mr. Debjyoti Ghosh, ECE, SMIT

International Advisory Committee

Dr. Hiroshi Iwai, Tokyo Institute of Technology, Kangawa, Japan
Dr. Durgamadhab Misra, New Jersey Institute of Technology (NJIT), USA
Dr. Valentina Emilia Balas, Romania
Dr. Efe Francis Orunmwense, CRART, South Africa
Dr. Rupesh Kumar, Technicolor, Rennes, France
Dr. Babu Sena Paul, University of Johannesburg, South Africa
Dr. Rajeev Kumar Kanth, University of Turku, Finland
Dr. Sanjeevikumar Padmanaban, University of Johannesburg, South Africa
Dr. Chuan-Ming Liu, National Taipei University of Technology (Taipei Tech), Taiwan
Dr. Supriyo Bandyopadhyay, Virginia Commonwealth University, USA

National Advisory Committee

Dr. Subir Kumar Sarkar, Jadavpur University, Kolkata
Dr. Mrinal Kanti Ghose, SMIT
Dr. Sanjay Dahal, SMIT
Dr. P. C. Pradhan, SMIT
Dr. Karma Sonam Sherpa, SMU
Dr. Kalpana Sharma, SMIT
Dr. B. B. Pradhan, SMIT
Dr. H. K. D. Sharma, SMIT
Dr. Sangeeta Jha, SMIT
Dr. Vinod Kumar Sayal, SMIT
Dr. Gobinda Chandra Mishra, SMIT
Dr. V. Sarala, SC-F, DRDO, Hyderabad
Dr. Monojit Mitra, IIEST, Shibpur
Mr. Arijit Mazumdar, SC-E, SAMEER, Kolkata
Dr. Bansibadan Maji, NIT, Durgapur
Dr. Samarjit Ghosh, Thapar University, Panjub
Dr. Debdatta Kandar, NEHU, Shillong
Dr. Rajeeb Dey, NIT, Silchar
Dr. Roy P. Paily, IIT, Guwahati
Dr. Abhilasha Mishra, Maharashtra Institute of Technology

Dr. Sumana Kumari, University Polytechnic, B.I.T. Mesra

Dr. Nagendra Pratap Singh, IIT-BHU

Dr. Sanjay Kumar Ghosh, Bose Institute, Kolkata

Dr. C. K. Sarkar, Jadavpur University

Dr. Sayan Chatterjee, Jadavpur University

Dr. Sujit Kumar Biswas, Jadavpur University

Dr. Kiran Shankar Hazra, Scientist, Institute of Nano Science and Technology

Dr. Rohit Sinha, IIT Guwahati

Dr. S. R. M. Prasanna, IIT Guwahati

Dr. P. K. Bora, IIT Guwahati

Dr. B. K. Rai, IIT Guwahati

Dr. Rajib Kumar Panigrahi, IIT Roorkee

Dr. Kuntal Deka, IIT Guwahati

Dr. Rafi Ahamed, IIT Guwahati

Dr. Rahul Shrestha, IIT Mandi

Dr. M. B. Meenavathi, Bangalore Institute of Technology

Dr. Sandeep Chakraborty, Indus University, Gujrat

Dr. Kabir Chakraborty, Tripura Institute of Technology

Dr. Aritra Acharya, Coochbehar Government Engineering College

Dr. Amitava Mukherjee, Adamas University, Barasat, Kolkata

Dr. Jawar Singh, IIT Patna

Dr. Rudra Sankar Dhar, NIT Mizoram

Dr. K. V. Srinivas, IIT BHU, Varanasi

Dr. Bijoy krishna mukherjee, BITS, Pilani, Rajasthan

Dr. Ujjwal Mondal, Applied Physics, University of Calcutta, Kolkata

Dr. Sunit Kumar Sen, Applied Physics, University of Calcutta

Dr. Ratna Ghosh, Jadavpur University

Dr. Soumen Das, School of Medical Science and Technology, IIT Kharagpur, WB, India

Dr. Rowdra Ghatak, Department of ECE, NIT Durgapur

Dr. Rajat Mahapatra, Department of ECE, NIT Durgapur

Dr. Durbadal Mandal, Department of ECE, NIT Durgapur

Dr. D. P . Chakraborty, BESU, West Bengal, India

Dr. C. K. Sarkar BESU, West Bengal, India

Dr. Arindam Biswas, Kazi Nazrul University (KNU), West Bengal

Dr. Sudipta Das, IMPS College of Engineering, Malda, West Bengal

Dr. Bhaswati Goswami, Jadavpur University

Dr. Avik Chattopadhyay, Department of Radio Physics and Electronics, University of Calcutta

Dr. Suchismita Tewari, Department of Radio Physics and Electronics, University of Calcutta

Dr. Ashish Kumar Ghunawat, Malaviya National Institute of Technology

Dr. Dharampal, Senior fellow, IETE

Dr. Debashis De, Professor and Director in the Department of Computer Science and Engineering Maulana Abul Kalam Azad University of Technology, India
Ms. Naiwrita Dey, Applied Electronics and Instrumentation Department, RCC Institute of Technology, Kolkata

Technical Program Committee

Dr. Rabindranath Bera, SMIT
Dr. Ashik Paul, IRPEL, CU
Dr. Angsuman Sarkar, Kalyani Government Engineering College, Kalyani
Dr. Himadri Sekhar Dutta, Kalyani Government Engineering College
Dr. Prolay Saha, Jadavpur University, Kolkata, India
Dr. Sanatan Chattopadhyay, University of Calcutta
Dr. Kaustavl, Physics Department, Jadavpur University
Dr. Nagendra Pratap Singh, Banaras Hindu University
Dr. Roy P. Paily, IITG
Dr. P. K. Banerjee, Jadavpur University, Kolkata
Dr. Sukla Bose, Kalyani Government Engineering College, Kalyani
Dr. Saurabh Das, Indian Statistical Institute, Kolkata
Dr. Anjan Kundu, IRPEL, CU
Dr. Dipanjan Bhattacharjee, SMIT
Dr. Md. Ruhul Islam, SMIT
Md. Nasir Ansari, SMIT
Mr. Amit Kumar Singh, SMIT
Dr. Mousumi Gupta, SMIT
Dr. Samarjeet Borah, SMIT
Dr. Utpal Deka, SMIT
Dr. Bibhu Prasad Swain, SMIT
Dr. Somenath Chatterjee, SMIT
Dr. Sourav Dhar, SMIT
Dr. Tanushree Bose Roy, SMIT
Dr. Akash Kumar Bhoi, SMIT
Dr. Amrita Biswas, SMIT
Dr. Swastika Chakraborty, SMIT
Dr. Hemanta Saikia, SMIT
Dr. Om Prakash Singh, SMIT
Dr. Prashant Chandra Pradhan, SMIT
Dr. Bikash Sharma, SMIT
Dr. Nitai Paitya, SMIT
Dr. Swarup Sarkar, SMIT
Dr. Barnali Dey, SMIT
Mr. Debjyoti Ghosh, SMIT
Ms. Soumyasree Bera, IIT, KGP

Mr. Arun Kumar Singh, SMIT
Dr. Samarendra Nath Sur, SMIT
Mr. Amit Agarwal, SMIT
Mr. Himangshu Pal, SMIT
Mr. Jayanta Kumar Baruah, SMIT
Mr. Jitendra Singh Tamang, SMIT
Mr. Kushal Pokhrel, SMIT
Mr. Nirmal Rai, SMIT
Mrs. Rijhi Dey, SMIT
Mr. Rochan Banstola, SMIT
Mr. Saumya Das, SMIT
Ms. Sayantani Roy, SMIT
Mr. Suman Das, SMIT

Preface

The Department of Electronics and Communication Engineering (ECE) of Sikkim Manipal Institute of Technology (SMIT), Sikkim, organized the 3rd International Conference on Communication, Device and Networking (ICCDN-2019) during December 9–10, 2019, technically co-sponsored by IEEE, Kolkata chapter, IEEE, GRSS society and publication sponsored by LNEE, Springer (SCOPUS Indexed).

The aim of the conference is to provide a platform for researchers, engineers, academicians, and industry professionals to present their recent research works and to explore future trends in various areas of engineering. The conference also brings together both novice and experienced scientists and developers, to explore newer scopes; collect new ideas; establish new cooperation between research groups; and exchange ideas, information, techniques, and applications in the field of Electronics, Communication, Devices, and Networking.

The ICCDN-2019 Committees rigorously invited submissions of manuscripts from researchers, scientists, engineers, students, and practitioners across the world related to the relevant themes and tracks of the conference. The call for papers of the conference was divided into six tracks as mentioned, Track-1: Electronics & Nano-technology, Track-2: Energy & Power, Track-3: Microwave, Track-4: Wireless Communication & Digital Signal Processing, Track-5: Control & Instrumentation, and Track-6: Data Communication and Networking.

The conference is enriched with five keynote speeches each of 1 hour duration by eminent Prof. Dr. Alejandro C. Frery from Universidade Federal de Alagoas, Brazil.; Prof. Dr. Avik Bhattacharya, IIT, Bombay; Prof. Dr. Ghanshyam Singh from Malaviya National Institute of Technology (MNIT) Jaipur; Prof. Dr. Sathish Kumar from Manipal Institute of Technology (MIT), Manipal; and Prof. (Dr.) Dr. J. Prakash from, Bangalore Institute of Technology, Bangalore, Karnataka.

A total of 135 papers have been received; out of which 59 papers have been accepted in the conference. Participants are coming from different parts of the country as well as across different countries. All these efforts undertaken by the Organizing Committees lead to a high-quality technical conference program, which featured high-impact presentations from keynote speakers and from paper presenters. The significant technical gathering in the conference is really carrying the message within the state of Sikkim where people are being conveyed about the global progression toward smart home, smart campus, and smart cities.

Lastly on behalf of the ICCDN organizing committee, we would like to thank Springer and IEEE Kolkata section for the kind cooperation. We also would like to thank the Patrons, General Chairs, the members of the Technical Program Committees and Advisory Committees, and reviewers for their excellent and tireless work for this very successful conference.

Majitar, India	Dr. Rabindranath Bera
Majitar, India	Dr. Prashant Chandra Pradhan
Taipei, Taiwan	Dr. Chuan-Ming Liu
Majitar, India	Dr. Sourav Dhar
Majitar, India	Dr. Samarendra Nath Sur

Contents

About the Editors

Dr. Rabindranath Bera has been a Professor at the ECE Department, Sikkim Manipal Institute of Technology since 2004. His current research interests include millimeter wave wireless system development, radiometer, radar, 4g/5g, and signal processing. He received a Bose fellowship from URSI, Japan, and an IAS fellowship from Govt. of India. During 35 years of extensive research, he has completed 7 major projects for MIT, AICTE, DRDO, TISCO, and DST. He has published more than 175 articles in prestigious journals and at conferences.

Dr. Prashant Chandra Pradhan is a Professor at the ECE Department, Sikkim Manipal Institute of Technology, India. He has published more than 20 papers in various prestigious journals and at international conferences. His current research interests are single electronics devices, multi-gate MOSFET, and photovoltaic devices. He is a life member of ISTE, India.

Dr. Chuan-Ming Liu is a Professor at the Department of CSIE, National Taipei University of Technology, Taiwan, where he was the Department Chair from 2013 to 2017. Currently, he is the Head of the Extension Education Center at the same school. He has published more than 80 papers in various prestigious journals and at international conferences. His current research interests include big data management and processing, uncertain data management, data science, spatial data processing, data streams, ad hoc and sensor networks, and location-based services.

Dr. Sourav Dhar is currently a Professor and Head of the ECE Department, SMIT. His current research interests include IoT, WSN, remote sensing, and microwave filter design. He is a member of IEEE, the IEEE-GRSS society, and IEI, India. He has published more than 30 papers in SCI/Scopus indexed international journals and at conferences. He also serves as a reviewer for Wireless Personal Communication, IEEE Transactions on Vehicular Technology, and several other journals and conferences.

Dr. Samarendra Nath Sur is an Assistant Professor at the ECE Department, Sikkim Manipal Institute of Technology, India. His current research interests include broadband wireless communication, advanced digital signal processing, and remote sensing. He is a member of IEEE, IEEE-IoT, IEEE-SPS, IEI, India, and IAENG. He has published more than 40 papers in SCI/Scopus indexed international journals and at conferences. He also serves as a reviewer for the International Journal of Electronics, IET Communication, Ad Hoc Networks, and IEEE Transactions on Signal Processing.

Real Power Transmission Loss Minimization and Bus Voltage Improvement Using UPFC

Bhaskar Gaur, Ravi Ucheniya, and Amit Saraswat

Abstract A shunt–series type of flexible AC transmission system named as Unified Power Flow Controller (UPFC) has an ability to manage real as well as reactive power in the power system network in a simultaneous manner. In this paper, a simulation model of UPFC is tested for the IEEE 14 bus standard test system on the DIgSILENT power factory software. Moreover, an optimum reactive power dispatch problem in presence of UPFC has been solved to reduce the real power loss in transmission lines. The proposed UPFC-based ORPD problem has been solved using the interior point method. The proposed approach is simulated under different loading conditions of the network. A comparative analysis of the obtained simulation results for each loading condition shows the effectiveness of UPFC for real power losses reduction.

Keywords Optimal reactive power dispatch · UPFC · IEEE 14 bus system · Real power loss minimization · DIgSILENT power factory

1 Introduction

Some of the challenges in a modern power system are being overcome by flexible AC transmission system (FACTS) [1]. These challenges are becoming more complex with increasing power demand. Proper management of reactive power in the network is one of them. Due to improper management of reactive power in the network, voltage instability, and real power losses are increasing day by day. Similar to traditional reactive power sources [2], FACTS devices are also capable of compensating reactive power in networks. In this paper, a series–shunt type of FACTS device, namely Unified Power Flow Controller (UPFC) [3, 4], has been used to reduce the total active

B. Gaur · R. Ucheniya (✉) · A. Saraswat (✉)
Department of Electrical Engineering, Manipal University Jaipur, Jaipur, Rajasthan, India
e-mail: raviu85@gmail.com

A. Saraswat
e-mail: amit.saraswat@jaipur.manipal.edu

B. Gaur
e-mail: bhaskargaur13@gmail.com

© Springer Nature Singapore Pte Ltd. 2020
R. Bera et al. (eds.), *Advances in Communication, Devices and Networking*,
Lecture Notes in Electrical Engineering 662,
https://doi.org/10.1007/978-981-15-4932-8_1

1

power loss in transmission lines. Optimal reactive power dispatch (ORPD) is a traditional and well-known optimization framework, by which the active transmission power loss is minimized, and hence the profile of all bus voltages is also improved [2, 5]. The structure of UPFC is an extraordinary combination of shunt and series elements. Due to which, it can regulate various transmission parameters such as line impedance, node bus voltage as well as angle. It can be able control both real as well as reactive power flow in the transmission line and improve the performance of the grid [1, 6]. In this paper, the performance of an UPFC connected in IEEE 14 bus test system [7] is investigated on the DIgSILENT power factory software simulation platform [8].

2 Mathematical Modeling of UPFC

UPFC has dual-voltage sources such as series and shunt voltage sources; therefore, it is capable of adjusting the flow of complex power in a transmission network. In UPFC, the series voltage source plays a vital role in controlling the complex transmission line power flow. The connection diagram of an UPFC to the given transmission network is presented in Fig. 1. Furthermore, according to the requirement of the series voltage source, the required amount of active power is supplied by the shunt source of voltage in the power network [9]. The static model of UPFC connected between the two buses is presented in Fig. 2.

The expression for injected voltage of UPFC is presented as follow:

$$V_{inj} = V_{sh} + V_{se} \tag{1}$$

$$V_{inj} = [V_{sh}(\cos \delta_{sh} + j \sin \delta_{sh})] + [V_{se}(\cos \delta_{se} + j \sin \delta_{se})] \tag{2}$$

where, V_{se} and δ_{se} are series injected controllable voltage magnitude and phase angle, respectively. V_{sh} and δ_{sh} are shunt injected an adjustable voltage magnitude and its phase angle, respectively. Moreover, the reactance of UPFC series and shunt coupling transformer denotes by X_{se} and X_{sh} respectively.

Fig. 1 Connection diagram of UPFC to the network

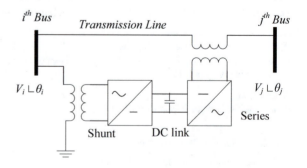

Fig. 2 A static model of UPFC

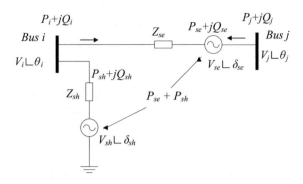

The conductance and susceptance for shunt and series coupling transformer for UPFC may be denoted by G_{sh}, B_{sh}, G_{se} and B_{se}, respectively. For a static model of UPFC, the power injections at i^{th} buses are mathematically expressed as follow [9, 10]:

$$P_i = G_{ii} V_i^2 + V_j V_i [G_{ij} \cos(\theta_j - \theta_i) - B_{ij} \sin(\theta_j - \theta_i)]$$
$$+ V_i V_{se} [G_{ij} \cos(\theta_i - \delta_{se}) - B_{ij} \sin(\theta_{se} - \delta_i)]$$
$$+ V_i V_{sh} [G_{sh} \cos(\theta_i - \delta_{sh}) - B_{sh} \sin(\theta_{sh} - \delta_i)] \qquad (3)$$

$$Q_i = -B_{ii} V_i^2 + V_j V_i [G_{ij} \sin(\theta_i - \theta_j) - B_{ij} \cos(\theta_i - \theta_j)]$$
$$+ V_{se} V_i [G_{ij} \sin(\theta_i - \delta_{se}) - B_{ij} \cos(\theta_i - \delta_{se})]$$
$$+ V_{sh} V_i [G_{sh} \sin(\theta_i - \delta_{sh}) - B_{sh} \cos(\theta_i - \delta_{sh})] \qquad (4)$$

where as the power injections at j^{th} buses are mathematically expressed as follow:

$$P_j = G_{jj} V_j^2 + V_{se} V_j [-B_{jj} \sin(\delta_{se} - \theta_j) + G_{jj} \cos(\delta_{se} - \theta_j)]$$
$$+ V_j V_i [-B_{ij} \sin(\theta_j - \theta_i) + G_{ji} \cos(\theta_j - \theta_i)] \qquad (5)$$

$$Q_j = -B_{jj} V_j^2 - V_j V_{se} [B_{jj} \cos(\theta_j - \delta_{se}) - G_{jj} \sin(\theta_j - \delta_{se})]$$
$$- V_j V_i [B_{ji} \cos(\theta_j - \theta_i) - G_{ji} \sin(\theta_j - \theta_i)] \qquad (6)$$

Also, the real and reactive power injected by series and shunt converters are as follows:

$$P_{se} = G_{jj} V_{se}^2 + V_j V_{se} [G_{jj} \cos(\delta_{se} - \theta_j) + B_{jj} \sin(\delta_{se} - \theta_j)]$$
$$+ V_i V_{se} [B_{ij} \sin(\delta_{se} - \theta_i) + G_{ij} \cos(\delta_{se} - \theta_i)] \qquad (7)$$

$$Q_{se} = -B_{jj} V_{se}^2 - V_j V_{se} [-G_{jj} \sin(\delta_{se} - \theta_m) + B_{jj} \cos(\delta_{se} - \theta_j)]$$
$$- V_i V_{se} [B_{ij} \cos(\delta_{se} - \theta_i) - G_{ij} \sin(\delta_{se} - \theta_i)] \qquad (8)$$

$$P_{sh} = -G_{sh}V_{sh}^2 + V_{sh} V_i [G_{sh} \cos(\delta_{sh} - \theta_i) + B_{sh} \sin(\delta_{sh} - \theta_i)] \qquad (9)$$

$$Q_{sh} = G_{sh}V_{sh}^2 + V_{sh} V_i [G_{sh} \cos(\delta_{sh} - \theta_i) - B_{sh} \cos(\delta_{sh} - \theta_i)] \qquad (10)$$

For a lossless converter, a real power provided by the shunt element (P_{sh}) is equal to a real power consumed by the series element (P_{se}). Therefore,

$$P_{sh} + P_{se} = 0 \qquad (11)$$

UPFC can simultaneously inject real power as well as reactive power in a given network. Hence, the Jacobian matrix of power flow analysis has been modified. The modified Jacobian matrix is represented by (12). Here, ΔP_{kk} is the real power mismatch of both the series as well as shunt converters.

$$
\begin{bmatrix} \Delta P_i \\ \Delta P_j \\ \Delta Q_i \\ \Delta Q_j \\ \Delta P_{ji} \\ \Delta Q_{ji} \\ \Delta P_{kk} \end{bmatrix}
=
\begin{bmatrix}
\frac{\partial P_i}{\partial \theta_i} & \frac{\partial P_i}{\partial \theta_j} & \frac{\partial P_i}{\partial V_i} & \frac{\partial P_i}{\partial V_j} & \frac{\partial P_i}{\partial \delta_{se}} & \frac{\partial P_i}{\partial V_{se}} & \frac{\partial P_i}{\partial \delta_{sh}} \\
\frac{\partial P_j}{\partial \theta_i} & \frac{\partial P_j}{\partial \theta_j} & \frac{\partial P_j}{\partial V_i} & \frac{\partial P_j}{\partial V_j} & \frac{\partial P_j}{\partial \delta_{se}} & \frac{\partial P_j}{\partial V_{se}} & 0 \\
\frac{\partial Q_i}{\partial \theta_i} & \frac{\partial Q_i}{\partial \theta_j} & \frac{\partial Q_i}{\partial V_i} & \frac{\partial Q_i}{\partial V_j} & \frac{\partial Q_i}{\partial \delta_{se}} & \frac{\partial Q_i}{\partial V_{se}} & \frac{\partial Q_i}{\partial \delta_{sh}} \\
\frac{\partial Q_j}{\partial \theta_i} & \frac{\partial Q_j}{\partial \theta_j} & \frac{\partial Q_j}{\partial V_i} & \frac{\partial Q_j}{\partial V_j} & \frac{\partial Q_j}{\partial \delta_{se}} & \frac{\partial Q_j}{\partial V_{se}} & 0 \\
\frac{\partial P_{ji}}{\partial \theta_i} & \frac{\partial P_{ji}}{\partial \theta_j} & \frac{\partial P_{ji}}{\partial V_i} & \frac{\partial P_{ji}}{\partial V_j} & \frac{\partial P_{ji}}{\partial \delta_{se}} & \frac{\partial P_{ji}}{\partial V_{se}} & 0 \\
\frac{\partial Q_{ji}}{\partial \theta_i} & \frac{\partial Q_{ji}}{\partial \theta_j} & \frac{\partial Q_{ji}}{\partial V_i} & \frac{\partial Q_{ji}}{\partial V_j} & \frac{\partial Q_{ji}}{\partial \delta_{se}} & \frac{\partial Q_{ji}}{\partial V_{se}} & 0 \\
\frac{\partial P_{kk}}{\partial \theta_i} & \frac{\partial P_{kk}}{\partial \theta_j} & \frac{\partial P_{kk}}{\partial V_i} & \frac{\partial P_{kk}}{\partial V_j} & \frac{\partial P_{kk}}{\partial \delta_{se}} & \frac{\partial P_{kk}}{\partial V_{se}} & \frac{\partial P_{kk}}{\partial \delta_{sh}}
\end{bmatrix}
\begin{bmatrix} \Delta \theta_i \\ \Delta \theta_j \\ \Delta V_i \\ \Delta V_j \\ \Delta \delta_{se} \\ \Delta V_{se} \\ \Delta \delta_{sh} \end{bmatrix}
\qquad (12)
$$

3 Problem Formulation

In this paper, minimizing the total real power loss has been considered as an objective function. The mathematical expression of the objective function has adopted from [2] and is presented as follows:

$$f = \min(P_{loss}) \qquad (13)$$

where

$$P_{loss} = \sum_{k=1}^{N_{TL}} G_k (V_i^2 + V_j^2 - 2V_i V_j \cos\theta_{ij}) \qquad (14)$$

The above objective function is minimized while satisfying the following equality constraints (15)–(16) as well as inequality constraints (17)–(25):

$$P_{gen,i} - P_{load,i} = \sum_{j=1}^{N_B} |Y_{ij}| \, |V_j| \, |V_i| \, \cos \left(\delta_i - \delta_j - \theta_{ij} \right) \tag{15}$$

$$Q_{gen,i} - Q_{load,i} = \sum_{j=1}^{N_B} |Y_{ij}| \, |V_j| \, |V_i| \, \sin \left(\delta_i - \delta_j - \theta_{ij} \right) \tag{16}$$

$$P_{Gen,i}^{\min} \leq P_{Gen,i} \leq P_{Gen,i}^{\max} \; i \in N_{PV} \tag{17}$$

$$Q_{Gen,i}^{\min} \leq Q_{Gen,i} \leq Q_{Gen,i}^{\max} \; i \in N_{PV} \tag{18}$$

$$V_{Gen,i}^{\min} \leq V_{Gen,i} \leq V_{Gen,i}^{\max} \; i \in N_{PV} \tag{19}$$

$$T_j^{\min} \leq T_j \leq T_j^{\max} \; j \in N_T \tag{20}$$

$$q_{cap,i}^{\min} \leq q_{cap,i} \leq q_{cap,i}^{\max} \; i \in N_{cap} \tag{21}$$

$$S_{TL,i} \leq S_{TL,i}^{\max} \; i \in N_{TL} \tag{22}$$

$$V_L^{\min} \leq V_L \leq V_L^{\max} \; L \in N_{PQ} \tag{23}$$

$$V_{sh}^{\min} \leq V_{sh} \leq V_{sh}^{\max} \; ; \; \delta_{sh}^{\min} \leq \delta_{sh} \leq \delta_{sh}^{\max} \tag{24}$$

$$V_{se}^{\min} \leq V_{se} \leq V_{se}^{\max} \; ; \; \delta_{se}^{\min} \leq \delta_{se} \leq \delta_{se}^{\max} \tag{25}$$

4 Solution Methodology

In this paper, an IEEE 14 bus test system is considered to analyze the performance of UPFC. The single-line diagram of the UPFC-connected standard test system is presented in Fig. 3. A detailed description of standard test systems (such as branch data, bus data, and generator data) has been adopted from [7]. The permissible limits of all control and state variables are also adopted from [7]. In the present simulations, it is assumed that an UPFC has been connected between buses 9 and 14 [11] as shown in Fig. 3. It is also assumed that this is an optimum location for the UPFC which has been adopted as in line with [11]. The weakest line in the network for UPFC is identified based on (a) voltage collapse point indicators (VCPI) and (b) line stability indices such as a line index (LQP) [11]. The simulations includes the interior point solution method [9] for ORPD problem in the presence of an UPFC as presented in subsequent section.

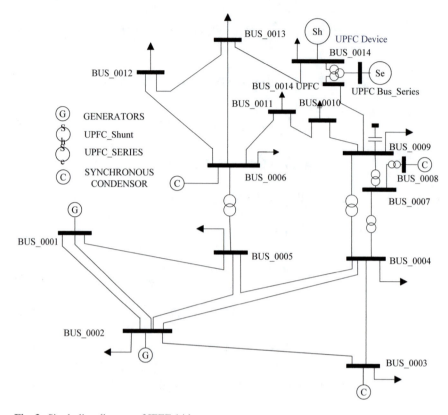

Fig. 3 Single-line diagram of IEEE 14 bus system

5 Simulation Results

The UPFC-based modified ORPD problem has been tested on a standard IEEE 14 bus test system and simulated the system under heavy loading conditions. For this, a total of ten loading conditions have been created and solved the modified ORPD problem using the interior point method. The entire simulation work is divided into two cases; the first is modified ORPD with UPFC and the second is without UPFC. Comparative analysis of simulation results has been performed. A comparative analysis based on real power losses for each loading condition is depicted in Fig. 4. It shows the effectiveness of the UPFC for a significant reduction in its real power transmission loss. Furthermore, the comparison of the voltage profile for two extreme situations such as base loading condition and extreme loading condition is presented in Fig. 5. The optimal setting of the control variables under each loading condition is presented in Table 1.

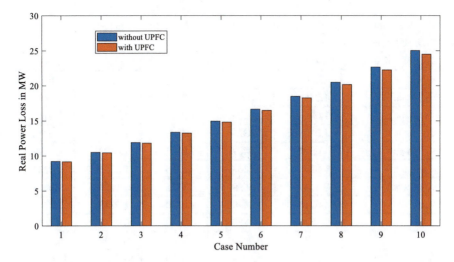

Fig. 4 Comparison of real power losses without and with UPFC device

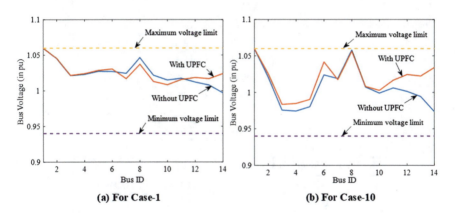

Fig. 5 The best voltage profiles obtained in case-1 and case-10

6 Conclusion

In this paper, the UPFC-based modified ORPD problem has been successfully solved by interior point method. A comparative analysis of the obtained simulation results gives a clear indication that the power loss in transmission network is effectively reduced by UPFC. Moreover, it is also observed that UPFC can provide controlled voltage support as well as it minimize the real power losses under the systems overloading conditions with no violation in any system constraint.

Table 1 Optimal settings of control variable considering UPFC for loss minimization

Optimal settings with UPFC					
Control variables	Case-1	Case-2	Case-3	Case-4	Case-5
V_1 (in pu)	1.0600	1.0600	1.0600	1.0600	1.0600
V_2 (in pu)	1.0453	1.0447	1.0441	1.0436	1.0428
V_3 (in pu)	1.0216	1.0196	1.0178	1.0160	1.0142
V_6 (in pu)	1.0305	1.0324	1.0333	1.0344	1.0354
V_8 (in pu)	1.0372	1.0407	1.0442	1.0470	1.0497
V_{sh} (in pu)	1.0240	1.0257	1.0265	1.0276	1.0286
V_{se} (in pu)	0.3537	0.3699	0.3915	0.4116	0.4364
T_{4-7}	1.0210	1.0170	1.0230	1.0240	1.0240
T_{-9}	0.9820	0.9780	0.9750	0.9710	0.9670
T_{5-6}	1.0020	0.9980	0.9940	0.9900	0.9850
Q_{c-9} (in MVar)	14.0000	15.0000	15.5000	16.5000	17.0000
P_{se}, (in MW)	0.3695	0.4033	0.4452	0.4874	0.5372
P_{loss} (in MW)	9.1595	10.4299	11.7950	13.2548	14.8147
Control variables	Case-6	Case-7	Case-8	Case-9	Case-10
V_1 (in pu)	1.0600	1.0600	1.0600	1.0600	1.0600
V_2 (in pu)	1.0416	1.0394	1.0359	1.0310	1.0253
V_3 (in pu)	1.0121	1.0089	1.0030	0.9941	0.9836
V_6 (in pu)	1.0358	1.0368	1.0381	1.0399	1.0416
V_8 (in pu)	1.0518	1.0535	1.0550	1.0559	1.0564
V_{sh} (in pu)	1.0291	1.0301	1.0312	1.0326	1.0341
V_{se} (in pu)	0.4642	0.4962	0.5326	0.5774	0.6273
T_{4-7}	1.0230	1.0200	1.0130	1.0030	0.9910
T_{4-9}	0.9640	0.9600	0.9560	0.9500	0.9430
T_{5-6}	0.9800	0.9730	0.9640	0.9520	0.9390
Q_{c-9} (in MVar)	17.5000	18.0000	18.5000	18.5000	18.5000
P_{se}, (in MW)	0.5938	0.6579	0.7306	0.8173	0.9150
P_{loss} (in MW)	16.4788	18.2552	20.1623	22.2354	24.4975

References

1. E.H. Watanabe et al., Flexible AC transmission systems, in *Power Electronics Handbook* (Elsevier, 2018), pp. 885–909
2. A. Saraswat, A. Saini, Multi-objective optimal reactive power dispatch considering voltage stability in power systems using HFMOEA. Eng. Appl. Artif. Intell. **26**, 390–404 (2013)
3. A. Nabavi-Niaki, M.R. Iravani, Steady-state and dynamic models of unified power flow controller (UPFC) for power system studies. IEEE Trans. Power Syst. **11**, 1937–1943 (1996)
4. M. Abdel-Akher et al., Developed generalised unified power flow controller model in the Newton-Raphson power-flow analysis using combined mismatches method. IET Gener. Transm. Distrib. **10**, 2177–2184 (2016)

5. M.S. Saddique et al., Solution to optimal reactive power dispatch in transmission system using meta-heuristic techniques status and technological review. Electr. Power Syst. Res. (2020)
6. B. Bhattacharyya, V.K. Gupta, S. Kumar, UPFC with series and shunt FACTS controllers for the economic operation of a power system. Ain Shams Eng. J. **5**, 775–787 (2014)
7. R.D. Zimmerman, C.E. Murillo-Sanchez, R.J. Thomas, MATPOWER: steady-state operations, planning, and analysis tools for power systems research and education. IEEE Trans. Power Syst. **26**, 12–19 (2011)
8. DIgSILENT GmbH: DIgSILENT power factory version 17. User's Manual, Gomaringen, Germany, March 2017
9. S.S. Shrawane Kapse et al., Improvement of ORPD algorithm for transmission loss minimization and voltage control using UPFC by HGAPSO approach. J. Inst. Eng. Ser. B. **99**, 575–585 (2018)
10. D. Prasad, V. Mukherjee, Solution of optimal reactive power dispatch by symbiotic organism search algorithm incorporating FACTS devices. IETE J Res. **64**, 49–160 (2018)
11. S. Ahmad et al., A placement method of fuzzy based unified power flow controller to enhance voltage stability margin, in *16th European Conference on Power Electronics and Applications* (IEEE, 2014), pp. 1–10

Heterogeneous Networks
in LTE-Advanced: A Review

**H. Srikanth Kamath, Samarendra Nath Sur, Hmdard Singh,
and Aayush Khanna**

Abstract Currently, the networks deployed in the LTE architecture are homogeneous, i.e. all mobile base stations are macro cells. This poses some limitations on user's connectivity, quality of service (QoS) and data speed which brings out the immense need of Heterogeneous Networks (HetNet). The advantages that are provided by HetNet is wider spectrum, higher data rates, lower latency, eliminates undesirable bandwidth segmentation, flexible and low-cost deployments. HetNet consists of heterogeneous cells of distinct transmit powers which causes interference between neighbouring cells. Therefore, various interference management techniques have been adopted to mitigate the effect of interference.

Keywords QoS · HetNet · LTE · LTE-A

1 Introduction

Long-Term Evolution Advanced (LTE-A) was first introduced by 3GPP in release 10 paper. The concept of LTE Advanced-based Heterogeneous Networks (HetNet) is about improving spectral efficiency per unit area. As the spectral efficiency per link is approaching its theoretical limits, thus there is a need to increase the deployment density of nodes [1]. When macro base stations are deployed, they are carefully planned such that their spectrum does not cause interference with the neighbouring base station's spectrum. If any interference occurs then the sectored antenna's beamwidth would be electronically adjusted or an appropriate mechanical tilt would be made to overcome the interference. Hence, the location of macro base station is properly chosen with careful network planning and proper radio frequency planning.

H. Srikanth Kamath (✉) · H. Singh · A. Khanna
Electronics and Communication Engineering Department, Manipal Institute of Technology, Manipal, India
e-mail: srikanth.kamath@manipal.edu

S. N. Sur
Electronics and Communication Engineering Department, Sikkim Manipal Institute of Technology, Sikkim Manipal University, Majitar 73736, Sikkim, India

© Springer Nature Singapore Pte Ltd. 2020   11
R. Bera et al. (eds.), *Advances in Communication, Devices and Networking*,
Lecture Notes in Electrical Engineering 662,
https://doi.org/10.1007/978-981-15-4932-8_2

The problem faced with macro networks is that in an urban dense area where network is very congested, the signal is not reachable within the private premises of building. Adding another base station can cause inter-cell interference or cell splitting in an already dense network is also not feasible [2]. So, there is a need for an alternative approach where the cells are flexible and can also be placed within the network with less frequency planning and should also be economically feasible. Such challenges are overcome by deploying low-power base stations within the coverage area of high-power base stations. Such networks with varying transmit power base stations are called HetNet.

2   HetNet Overview

HetNet consist of distinct nodes of varying transmission power composed of larger and smaller cells in the decreasing order of their transmission power. The former cells are macrocells and usually have transmission power between 5 and 20 W, whereas the latter are pico, femto, relay nodes which have a 10 mW–2 W transmission power. These smaller cells are deployed in the coverage of macro cells to overcome the outage holes caused by the macro base stations. For instance, in private premises like basements, or offices where signal level is quite low so to enhance the connectivity of LTE users, femto or pico cells can be installed to increase the signal strength. Therefore, the overlaying of these smaller cells within the coverage of larger cells ensures higher spectral reuse (Fig. 1).

The following discusses the details regarding each of the smaller cells

Pico nodes: Unlike macro cells which are used in conventional LTE network with high transmit power, in LTE-A, pico cells can be deployed for both indoor and outdoor purposes where the transmission power varies according to the environment in which the cell is deployed [1].

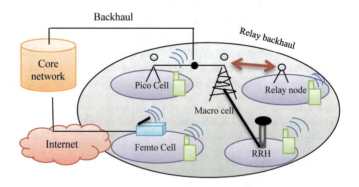

Fig. 1 Heterogeneous architecture composed of macro, femto, pico, relay nodes

Femto nodes: These nodes are smaller base stations that can be deployed in residential, office or any other private premises for consumer purposes in order to enhance the connectivity of users unlike pico cells which can be deployed for both indoor and outdoor purposes. These nodes are also called as Home eNodeB [HeNB] [3]. They provide cellular services from service providers via DSL or home cable to the users directly. This is an advantage of femto cells over pico cells. They are broadly classified as open and closed femto cells. In the former case, the user can get connected to femto base station without any access and is free of charge, whereas the latter case provides access only to the closed subscriber groups (CSG). Femto cells have omnidirectional antennas with transmission power of not more than 100 mW [4].

Relay nodes (RN): These nodes communicate with the Donor eNodeB (DeNB) via the relay backhaul link. These relays extend the coverage of macro base stations and acts as an aid for transmitting signals between the macro base station and user equipment [5]. The relay backhaul link is established between relay base station and macro base station through which exchange of information takes place via the air interface. Relay base stations offer additional flexibility where wireline backhaul is unavailable or economically non pragmatic [4].

Remote Radio Head (RRH): RRHs are connected via a dedicated backhaul, for instance, optical fibre to macro base stations. These stations have minimum autonomous intelligence and act as extensions of base station antenna ports.

3 Features of HetNet

Various features have been adopted in the 3GPP release 10 paper. Some features have been proposed in order to overcome interference from neighbouring cells. Inter-cell interference coordination (ICIC) is a devised technique that can be used to overcome interference due to neighbouring cells. User Equipment (UE) at the edge of the cell's network experience interference due to the neighbouring base stations. In order to accommodate cell edge users, cell range expansion which is another feature is adopted in HetNet where the smaller cells expand their range to accommodate the current users in its own coverage. These features are discussed in detail in the following sections.

3.1 Cell Range Expansion

As mentioned earlier that smaller cells are deployed in the coverage area of macro cells. But there can be interference between them, especially if co-channel deployment is used. As the downlink signal strength of macro cells is more than that of femto/pico cells, they will draw maximum number of user equipment towards them. But this is not the case for the uplink signal strength which is received equally by

Fig. 2 Limited footprints of
pico cells due to strong
macro signal

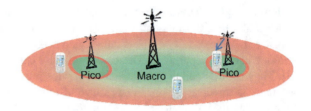

all stations. Therefore, there will be discrepancies between the uplink and downlink handover boundaries and there will be difficulties in server selection for the
UE [6]. If selecting a server was predominantly based on the choice of downlink
signal strength, then this will lead to underutilization of femto/pico cells and it will
certainly increase the load on the macro cells which will in turn cause frequent handovers which is undesirable. To overcome the above problem, cell range expansion
is used. Cell range expansion is enabled through cell biasing [4].

Figure 3 shows that as users approach the boundary of the pico cell's network,
it has to expand its network's range to ensure that there is equitable distribution of
resources amongst the neighbouring cells. In this case, the range expansion was done
on the basis of path loss and equitable distribution of resources [7]. Path loss is a
constraint which is imposed instead of downlink signal strength in order to ensure
that pico cells are effectively utilized and the load is equally distributed between the
macro and pico cells (Fig. 2).

A. *Inter-Cell Interference Coordination (ICIC)*

For pico cells to enable range expansion, they have to ensure that the data and control
transmissions do not interfere with the macro base station transmissions which in
general have more downlink signal strength. In heterogeneous networks, the resource
coordination amongst cells is a key feature in heterogeneous networks. The following
shows scenarios in which ICIC techniques are immensely required

- If the UE is served by pico cell and the downlink signal strength of macro base
  station is high then the macro base station causes interference between the pico
  base station and UE.
- If the UE is within the coverage of closed femto cell but barred from accessing
  it, then the UE will be served by macro base station without any further thought.
  Even though, the downlink signal strength of femto base station is less compared

Fig. 3 Increased footprint of
pico cells with range
expansion [7]

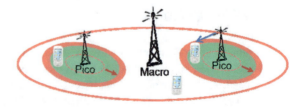

to macro base station, it still causes interference since the UE is within the femto cell's coverage.

As discussed above, in the former case, the interferer is macro base station and the victim is the pico base station. In the latter case, the interferer is femto base station and the victim is macro base station.

In order to overcome these problems, resource partitioning has to be implemented carefully through cell biasing. Resource partitioning can be done in time domain, frequency domain or spatial domain [8]. As discussed in [8], resource partitioning in time domain is most favourable for user distribution and spatial domain and ideally suits well to the spectrum constrained markets.

In time domain partitioning, macro base station uses 10-ms radio frames for communication with pico base stations. The radio frame is divided into ten subframes in which some of the subframes are used for control channel and reserved for pico base station. These subframes are called Almost Bank Subframes (ABS). During the transmission of ABS, the eNodeB (eNB) cannot transmit data but only control channel and the common reference signal (CRS), which is used for measurements and demodulation of both; the downlink (DL) control and the traffic channel [4]. Figure 4 shows that 50% of DL subframes are reserved for data and the other half for control channel in both pico and macro base stations.

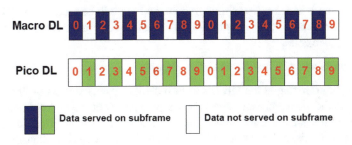

Fig. 4 DL radio frame structure for both macro and pico base stations [4]

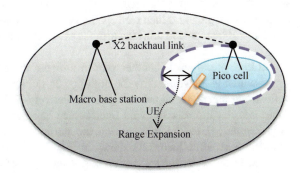

Fig. 5 A Macro base station interfering with a pico cell UE during its range expansion

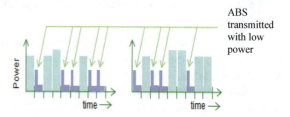

Fig. 6 The graph shows that data channel subframes are transmitted with high power and ABS is transmitted with low power [9]

The subframe on which data is not being served is the ABS and will be used for CRS and control channel purposes. When the ABS is transmitted, it is unicasted with low power [9] as shown in Fig. 6. Figure 5 shows a scenario by considering macro base station, pico base station and UE. In order to manage the interference between these two base stations, the macro base station sends radio frame to pico base station via the X2 backhaul link. This radio frame will consist of a combination of ABS and data subframes as discussed earlier. During the ABS, the macro base station informs pico base station about the interference pattern and at the same time it expands its range to serve cell edge UE. The ABS is transmitted with low power to pico cell via X2 backhaul link [10] as shown in Fig. 5. It is transmitted with low power because it contains information which is not necessary for UE to intercept and data subframe is transmitted with high power which will be received by UE. So in this way, interference management takes place during cell range expansion between macro and pico cells.

B. *Advanced Interference Cancellation Receiver*

Reducing interference at the UE ensures efficient range expansion of the cell. The UE detects the week cell and measures the parameters of the channel and sends a feedback to the core of the network which ensures proper handovers. In heterogeneous networks, the interference on UE is mainly due to acquisition channel and CRS interference. The interference affects the amount of data flow over the network and handover. If there is any interference from neighbouring cells, the interfering base station is detected. First, the received signal will be decoded, followed by estimation of the channel gain and then cancels the interfering signal as shown in Fig. 7. This process will be repeated until the acquisition channel of the serving cell is detected [4].

Furthermore, there is another way to mitigate the effects of interference. CRS signal which is transmitted by the high-power and low-power nodes is transmitted with strong and weak signal strength, respectively. This leads to interference with data tone. So, in order to avoid interference between them, CRS tone shifting is implemented [4]. In CRS tone shifting, the two tones are sent on different frequencies in order to avoid collision. As shown in Fig. 7, the strong CRS signalling tone is detected followed by the estimation of channel gain of the interfering channel and then CRS signal is cancelled out. This entire process is repeated until all the CRS signals are cancelled out.

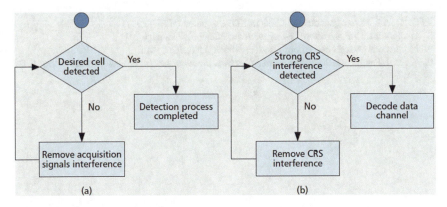

Fig. 7 Interference cancellation of acquisition channels and CRS at UE [4]

4 Conclusion

HetNet are a necessity and their deployment will significantly improve the quality of service, user connectivity and enhance the data rate. The UEs that are close by or within the coverage will be served by low-power nodes and those that are far away will be served by high-power nodes. Although HetNet bring many advantages into mobile communication networks, the interference due to it is also increased. In order to tackle the interference caused by HetNet, various techniques like cell range expansion are adopted in which the low-power nodes expand their range to accommodate cell edge users in order to reduce the load from high-power base stations. Since the DL signal strength of high-power node is more, it causes interference to the UE during the range expansion of the pico cell. In order to mitigate the effects of interference, ICIC is adopted which is done by resource partitioning and use of ABS. Furthermore, advanced interference cancellation systems are used in UE which further mitigate the effects of interference due to high power-base stations. These three features which have been discussed are primary elements of HetNet which will form the basis of the 5G network in future.

References

1. *GPP, 3GPP Rel-9 and Beyond* (Feb 2010)
2. J. Wannstrom, masterltefaster.com and K. Mallinson, Wise Harbor, for 3GPP, HetNet/Small *Cells* (August 18, 2014)
3. M. Nohrborg, for *3GPP*, LTE (December 2008)
4. A. Damnjanovic, J. Montojo, Y. Wei, T. Ji, T. Juo, M. Vajapeyam, Yoo, O. Song, D. Malladi, Qualcomm inc, A survey on 3GPP heterogeneous networks (IEEE 2011)
5. J. Wannstrom, for *3GPP*, LTE-advance (June 2013)
6. A. Khandekar, N. Bhushan, J. Tingfang, V. Vanghi, LTE-advanced heterogeneous networks (IEEE, 2010)

7. S. Brueck, Heterogeneous networks in LTE-advanced (IEEE, 2011)
8. Qualcomm, *LTE Advanced: Heterogeneous Networks* (January 2011)
9. http://www.sharetechnote.com/html/Handbook_LTE_ICIC.html
10. T. Hu, J. Pang, H.-J. Su, *LTE-advanced heterogeneous networks: release 10 and beyond* (IEEE, 2012)

A Novel Ultra-Wideband CMOS Low-Noise Amplifier for Wireless Communication

Sankhayan Chakrabarty

Abstract The paper presents the investigation of a novel ultra-wideband CMOS low-noise amplifier for receiver system used for wireless communication. The design is based on a multistage cascade and cascode amplifier cell with a capacitive gain-peaking technique offering gain-enhanced noise cancellation that gives a band of 16.2 GHz, maximum forward gain of nearly 8 dB, NF of 3 dB and impedance matching of nearly 55 Ω is obtained. The proposed architecture is a wideband amplifier in conjunction with the previously proposed cell giving much wider band of 21 GHz. The NF of 1.6 dB is achieved with a maximum gain of approximately 19 dB, while the impedance matching is nearly 50 Ω. The LNA is implemented using design kit in ADS.v.2 2013 platform and 180-nm design technology.

Keywords Wideband amplifier · Cascade amplifier · Cascode amplifier · Capacitive gain peaking

1 Introduction

The LNA plays a vital role in deciding the performance of a receiver system. In the field of RF communication, the superior quality receivers are required to be designed which must offer numerous features such as wide bandwidth, high noise cancelling approach, high gain with good impedance matching. One receiver system cannot offer all the desired parameters; hence, a trade-off is maintained in many of the important characteristics. Starting from the noise cancelling technique, it is an efficient method of cancelling noise but does not offer good gain flatness as it uses inductive gain peaking technique [1]. A low-noise amplifier produces a forward gain of 22 dB with a noise figure (NF) of 3.5 dB but offers narrow band [2]. In other CMOS wideband amplifiers using dual-feedback loops in the Kukeilka configuration, good bandwidth and return loss are obtained with compromise in gain [3–5]. Similarly, a

S. Chakrabarty (✉)
Department of Electronics and Communication Engineering, JIMS Engineering Management
Technical Campus, Greater Noida, UP, India
e-mail: chakrabartysankhayan@gmail.com

© Springer Nature Singapore Pte Ltd. 2020
R. Bera et al. (eds.), *Advances in Communication, Devices and Networking*,
Lecture Notes in Electrical Engineering 662,
https://doi.org/10.1007/978-981-15-4932-8_3

19

linearized UWB LNA with low power is presented over a 3–11-GHz bandwidth while maximum forward gain is not very high with its maximum value of 10 dB with NF 2.9 dB [6]. The use of passive components gives rise to non linearity and undesired parasitic effects. Hence, inductor less LNA is preferred that offers gain-enhanced noise cancelling architecture [7]. The other approach of active inductor (AIND) and negative capacitors (NCAP) can also be followed where CMOS devices are used in a way that they work as inductors. It not only improves gain and bandwidth but also gives higher power efficiency [8–14]. This approach also offers a limitation that since the transistors that are used to form active inductors are frequency dependent; they offer change in impedance and admittance in a non-linear manner with a slight drift in frequency. Even a compact wide band CMOS LNA for gain flatness enhancement uses active inductor [15]. This paper yields a new approach to design a new LNA where a cascaded cascode amplifier is used in conjunction with a wideband amplifier that offers gain flattening and noise cancellation. The initial stage called LNA segment here offers the band of 16.2 GHz, while the combination of the segment and wideband LNA gives a wider range of 21 GHz.

2 Design Consideration of LNA

The motive of this architecture is to incorporate the multiple parameters of LNA in an efficient manner. The first architecture proposed here is a four-stage cascade amplifier where, at the final stage, a cascode amplifier is used. The benefits of these amplifiers are evident that the cascade and cascode stages offer higher gain with good noise cancellation.

2.1 Proposed Segment Amplifier

Figure 1 represents the proposed schematic of segment of LNA. In the architecture shown, the overall gain is very high because of the cascading of four stages. The middle two stages use common source amplifiers with capacitive peaking which has been derived from the source degeneration technique. Figure 2 shows the schematic of capacitive gain peaking stage. The gain of such stage can be given as:

$$A_v = R_L \left(1/g_m + R_P \| C_P\right)^{-1}$$

The concept comes from resistive source degeneration method where the gain is compromised against the improved linearity. Here, the use of C_p offers the short circuit path to the RF current by bypassing it from the source resistance resulting in improved gain at the output. It is the ultimate approach to get the flatness in gain with a sacrifice with absolute gain. The capacitive peaking technique not only decides the

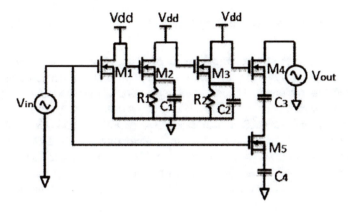

Fig. 1 Proposed Schematic of LNA segment

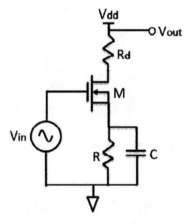

Fig. 2 Schematic of capacitive gain peaking stage [7]

gain and its flatness but plays an important role to decide the impedance matching and noise cancelling features [7]. The first transistor affects maximally on the noise and impedance matching parameters, so, here the capacitive degeneration technique is not used. M2 and M3 stages in cascade not only enhances the gain but also improves noise and matching performances with the help of source degeneration capacitive peaking technique. Finally, the cascode stage is used as noise cancelling stage that offers wide range of band which is found to be lying from 8.8 GHz to 25 GHz with a band of 16.2 GHz, whereas the maximum gain of 17 dB with its theoretical value of approximately 21 dB. The impedance matching is also measured in terms of $Z_{in.}$ The real input impedance is ranging from 71 to 44 Ω with imaginary part as zero. The minimum values of voltage standing wave ratios are 2.6 and 1.2 for $VSWR_{out}$ and $VSWR_{in,}$ respectively. The stability factor is also nearly one.

2.2 Proposed Low-Noise Amplifier

Figure 3 represents the proposed schematic of LNA. Here, a lot of techniques are being used to get a wideband low-noise amplifier with all the desired features. In conjunction with the previously discussed segment of low-noise amplifier, the complete amplifier consists of ten CMOS devices. Apart from the lower five CMOS, M6 and M7 are forming a capacitive peak cascode stage with M2 and M3, respectively. The M6 stage is used to find the gain increment factor directly. The use of R9 divides the current in the branches with the supply of 1.5 V. The tail resistor R8 is connected here to control the DC current of transistors M3 and M7. The upper transistors M7 and M8 are used as input matching stage. The shunt feedback resistance R12 assists

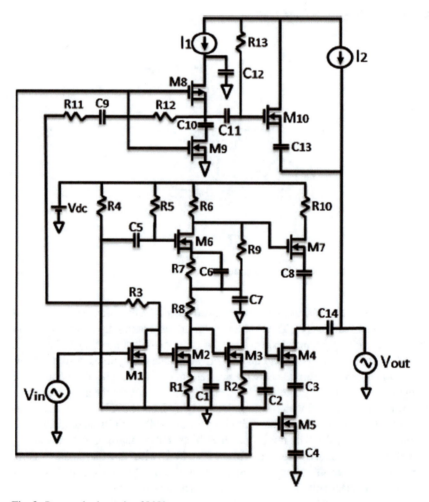

Fig. 3 Proposed schematic of LNA

to get the flat wideband. At the final stage offering noise cancellation along with wideband, an active inductor can be connected which is used to push the frequency by splitting the pole so that a wideband is achieved [7, 15]. In this proposed LNA, this active inductor has been replaced with a set of components having active and passive one to offer much wider band by using various other band extension techniques such as capacitive peaking technique with shunt resistance feedback and using cascading of the amplifiers with the combination of cascoding. The noise cancelling amplifier stage gets extra bias current by the use of current bleeding technique from the source amplifier. The most important parameter for getting the wideband with superb noise cancellation is the width and length of the CMOS transistors. It not only enhances these factors, but also supports other factors such as impedance matching, stability factor and voltage standing wave ratio. The simulation results are shown below with the comparison of previously suggested segment of LNA.

Comparing the noise performance in terms of NF, it can be concluded from the graph in Fig. 4, that the maximum NF for segment of LNA is 3.1 dB, while, for full LNA, it is only 1.6 dB. The maximum forward gain is coming out to be 19 dB shown in Fig. 5. In terms of bandwidth, the full LNA proposed here gives superior band as for segment it is only 16.2 GHz ranging from 8.8 to 25 GHz, while, for that of complete schematic, it is observed as 21.05 GHz ranging from 2.95 to 24 GHz as evident from the graph shown in Fig. 6. The maximum return loss (S_{11}) is nearly 13 dB, while S_{22} is almost the same depicted in Fig. 7. The stability factor of the amplifiers is supposed to be nearly equal to one which is almost same for both types of the proposed LNA which can be observed from the graph shown in Fig. 8 that the two curves almost superimpose on each other. The voltage standing wave ratios are also depicted in the graph shown in Fig. 8 where it should be noted that the

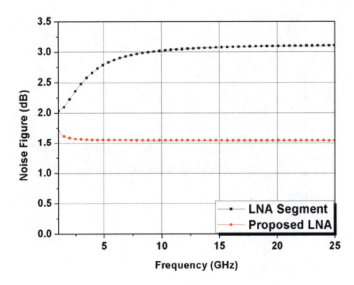

Fig. 4 Comparison of NF

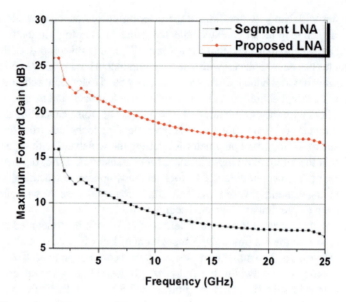

Fig. 5 Comparison of S_{21}

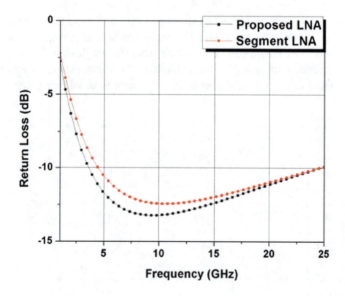

Fig. 6 Comparison of S_{11}

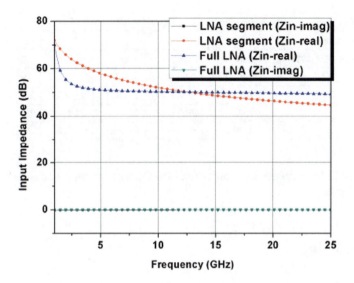

Fig. 7 Comparison of input impedance

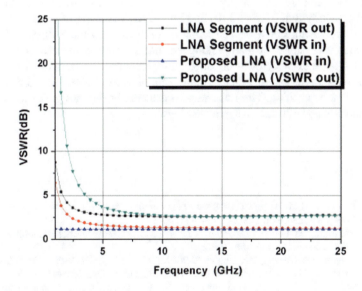

Fig. 8 Comparison of stability factor

$VSWR_{in}$ for segment varies between 1.2 dB and 1.3 dB for the middle of the band, while $VSWR_{out}$ is in the range of 1.1–1.2 dB for the complete schematic of LNA for about mid range of band. Finally, the impedance matching parameter is also taken into account where Z_{in} (real) ranges from 44 to 71 Ω for the segment LNA, while,

Table 1 Comparison of LNA with other reported papers

Design parameters	[1]	[3]	[7]	[15]	Present work
Technology (CMOS) (nm)	250	180	130	180	180
Frequency (GHz)	0.02–1.6	0.04–0.9	2–9.6	0.05–9	2.95–24
NF (dB)	1.9–2.4	3.8–5.0	3.6–4.8	3.8–5.0	1.6
S_{21} (dB)	13.7	20.3	11	20.3	19
S_{11} (dB)	<−8	<−7.4	<−8.3	<−7.4	<−10

for the full LNA, it ranges from 50 to 70 Ω with imaginary part as zero for both the circuits (Table 1).

3 Conclusion

This paper has demonstrated a novel ultra wideband CMOS low-noise amplifier for receiver system used in wireless communication. The design was based on a multi-stage cascade and cascode amplifier cell that gave a band of 16.2 GHz. Maximum forward of nearly 8 dB, NF of 3 dB and impedance matching of nearly 55 Ω. The proposed architecture of wideband amplifier used the previously discussed cell in conjunction with capacitive gain peaking technique to elevate the design parameters. Hence, the band extension was obtained with maximum spectrum of 21 GHz. The NF of 1.6 dB was achieved with maximum gain of approximately 19 dB along with the impedance matching. The LNA was implemented using design kit in ADS v.2 2013 platform and 180-nm design technology.

References

1. F. Bruccolen et al., Wide band CMOS low noise amplifier exploiting thermal noise cancelling. IEEE J. Solid State Circuits 39(2), 275–282 (2004)
2. D.K. Shaeffer, T.H. Lee, D.K. Shaeffer, A 1.5 V, 1.5 GHz CMOS low noise amplifier. IEEE J. Solid State Circuits 32, 745–759 (1997)
3. D.R. Huang, H.R. Chuang, Y.K. Chu, A 40-900 MHz broadband CMOS differential LNA with gain control for DTV RF tuner. IEEE Asian Solid State Conf. Dig. 465–468 (2005)
4. F.T. Chien, Y.J. Chan, Bandwidth enhancement of transimpedance amplifier by capacitive peaking design. IEEE J. Solid State Circuits 34, 1167–1170 (1999)
5. Y.-C. Chen, S.-S. Lu, Analysis and design of CMOS broadband amplifier with dual feedback loops, in *Proceedings of IEEE Asia-Pacific Conference on Advanced System Integrated Circuits*, pp. 245–248 (2002)
6. H. Zhang, X. Fan, E.S. Sinencio, A low-power, linearized, ultra-wideband LNA design technique. IEEE J. Solid State Circuits 43, 320–329 (2009)
7. Q. Li, Y.P. Zhang, A 1.5 V 2-9.6 GHz inductorless low noise amplifier in 0.13-micro metre CMOS. IEEE Trans. Microw. Theory Tech. 55, 2015–2023 (2007)

8. S. Kolev, B. Delacressonniere, J.-L. Gautier, Using a negative capacitance to increase the tuning range of a varactor diode in MMI technology. IEEE Trans. Microw. Theory Techn. **49**(12), 2425–2430 (2001)
9. J.F. Carpentier, C. Tilhac, G. Caruyer, F. Dumont, G. Parat, P. Ancey, A tunable bandpass BAW-filter architecture and its application to WCDMA filter, in *IEEE MTT-S International Microwave Symposium Digest* (2005), pp. 221–224
10. Y. Wu, X. Ding, M. Ismail, H. Olsson, RF bandpass filter design based on CMOS active inductors. IEEE Trans. Circuits Syst. II, Analog Digit. Signal Process. **50**(12), 942–949 (2003)
11. P. Vincent et al., A 1 V 220 MHz-tuning-range 2.2 GHz VCO using a BAW resonator, in *IEEE International Solid-State Circuits Conference on (ISSCC) Digital Technical Papers* (2008), pp. 478–479
12. Y. Song, S. Lee, E. Cho, J. Lee, S. Nam, A CMOS class-E power amplifier with voltage stress relief and enhanced efficiency. IEEE Trans. Microw. Theory Techn. **58**(2), 310–317 (2010)
13. A. Kaya, E.Y. Yüksel, Investigation of a compensated rectangular microstrip antenna with negative capacitor and negative inductor for bandwidth enhancement. IEEE Trans. Antennas Propag. **55**(5), 1275–1282 (2007)
14. B. Georgescu, H. Pekau, J. Haslett, J. McRory, Tunable coupled inductor Q-enhancement for parallel resonant LC tanks. IEEE Trans. Circuits Syst. II, Analog Digit. Signal Process. **50**(10), 705–713 (2003)
15. Y.-H. Yu, Y.-S. Yang, Y.-J. Emery Chen, A compact wideband CMOS low noise amplifier with gain flatness enhancement. IEEE J. Solid-State Circuits **45**(3), 1808–1814 (2010)
16. A. Ismail, A.A. Abidi, A 3–10-GHz low-noise amplifier with wideband LC-ladder matching network. IEEE J. Solid-State Circuits **39**(12), 2259–2268 (2004)

Analysis and Evaluation of Power Plants: A Case Study

Sudeep Pradhan, Dipanjan Ghose, and Shabbiruddin

Abstract Key factors to evaluate parameters of any power plant include, low cost for investing, minimum operation and maintenance cost, and maximum efficiency. This study is about determining and agreeing to the power plants set up in Tamil Nadu, India using observational data followed by the application of Multi-Criteria Decision-Making (MCDM) analysis to rank the existing power plants according to their optimal usage and generation. Various power plants based on different sources of energy installed in the state are taken into consideration, Complex Proportional Assessment (COPRAS) method is used to specify which among them the most suitable one for optimized usage is. The acquired solution will also help in deciding the future prospects about the power plants. The methodology used for the study is a basic approach toward handling such a decision problem and can be further applied to any study of similar nature.

Keywords Energy systems · MCDM analysis · COPRAS · DEMATEL · Sustainable development

1 Introduction

The diverse aspect of environmental problems including overutilization of fossil fuels leads to a need for the development of alternate energy sources for efficient economic development and future survival [1, 2]. Advancement in the approach of renewable energy sources (RES) besides being prominently accepted all around has

S. Pradhan · D. Ghose (✉) · Shabbiruddin
Department of Electrical and Electronics Engineering, Sikkim Manipal Institute of Technology,
Sikkim Manipal University, Majitar, Rangpo, Sikkim, India
e-mail: ghosedipanjan1998@gmail.com

S. Pradhan
e-mail: sudeeppradhan76@gmail.com

Shabbiruddin
e-mail: shabbiruddin85@yahoo.com

© Springer Nature Singapore Pte Ltd. 2020
R. Bera et al. (eds.), *Advances in Communication, Devices and Networking*,
Lecture Notes in Electrical Engineering 662,
https://doi.org/10.1007/978-981-15-4932-8_4

29

also attracted research interest, bringing up more efficient power plants, reducing the environmental impact through improved energy efficiency by the deployment of low-carbon technologies [3]. The consequence of extensive research with various approaches has been written in relevant literature focusing on the most effective method for the generation of energy [4].

Power plants are generally considered the most prominent way of electricity generation. The need for supplying the energy and its increasing demand leads to its direct proportionality with the number and efficiency of power plants being set up. Hence, they are undoubtedly one of the most important ingredients in shaping and developing a nation as a whole [5].

Employment of MCDM methods has brought a solution for multiple problems with numerous conflicting criteria [6, 7]. There have been various works conducted using MCDM analogies in similar genres and a few of them can be elaborated as follows: in 2012, a study was conducted to evaluate power plants in Turkey using Analytic Network Process (ANP) and MCDM procedure. Criteria such as technological, economic, socioeconomic impacts and life quality were well thought out and the results provided with the information that nuclear power plant is the best energy-producing plant in Turkey [8]. In another research, it was investigated about the prioritization of renewable energy sources by employing MCDM technology. It was concluded that hydropower plant was most suitable as an alternative energy source for Turkey [9].

Evidently, it is not uncommon to find usage of MCDM techniques in analyzing power plants in a particular location among its other implementations [10–13]. In this work, four well-functioning power plants in the Indian state of Tamil Nadu are taken up and evaluated with the COPRAS method, in order to obtain the most favorable and efficient plant through weightage of considered criteria. The criteria selected consisted mostly of technological and economic impacts. Criteria considered were assigned weights through the DEMATEL Method.

2 Methodology

2.1 Criteria Selection

Criteria selection requires distinct parameters that should be reliable, comparable, and should have a restriction in measurement. Here, four generalized criteria are considered for the evaluation of the plants. The favored criteria are primarily divided into three groups: technical (power, capacity factor), social (emission of pollutants, social acceptance), and economic (capital cost, operation and maintenance cost, and fuel cost). The capacity factor of a power plant is defined as the ratio of energy produced by the plant in a particular time with respect to the actual amount of energy to be produced by the plant in that given amount of time.

Power is the defined number registered for classifying the power output of a power station and is expressed in megawatts (MW) [14]. All the expenses associated with the purchase, technical installation, unit construction, buildings, road connection to the local grid network, and payment for engineering, drilling and other casual constructions come under the *"capital cost"*. Operation and maintenance cost (O&M) of a power plant is defined as the cost required for operating and maintaining that particular power plant. O&M cost plays a vital role in determining which generating facilities are cheapest to run and maintain in power technology [14].

2.2 COPRAS Method

In this method, first, the alternatives are calculated in accordance with the separate criteria, the positions of which can be obtained by comparing them to the ideal alternative, here the alternatives being different kinds of power plants. The formed decision matrix after comparison of the alternatives is then normalized to obtain the normalized matrix [15]. From the normalized matrix, by multiplying the weights of the criteria obtained from the DEMATEL's method to each element, the weighted normalized matrix is obtained. The sum S_{i+} and S_{i-} of weighted normalized values are computed for both beneficial and unbeneficial criteria, respectively. For beneficial criteria, higher value is better and for the non-beneficial criteria, lower values are better for the attainment of the goal [15]. Then the relative importance or priorities of candidate alternative Q_i is determined by using the following Eq. (1):

$$Q_i = S_{i+} + \frac{\sum_{i=1}^{m} S_{i-}}{S_{i-} \sum_{i=1}^{m} 1/S_{i-}} \qquad (1)$$

Then, the performance index of each alternative is calculated using

$$P_i = \left[\frac{Q_i}{Q_{\max}} \right] \times 100\% \qquad (2)$$

2.3 DEMATEL Weighting Method

This method is basically relied on the acceptance that a greater dispersal in the values of criteria would make the precedent more compelling [11]. Since it formalizes all the criteria and alternatives, the weighting method has a very meaningful impact on the overall performance of the optimum power plants. Hence, the following procedure is applied to calculate the criteria weight, considering A, B, C, D, E, F, G as the criteria given that A is Power, B is Capital Cost, C is Capacity factor, D is Operation and maintenance cost, E is the Emission of Pollutants, F is the Social Acceptance of the plant, and G is the Fuel Cost. The DEMATEL matrix works on the procedure of relating the values with each other. The diagonal matrix is always zero because it would turn out to be the relation with each other [11]. While other spaces in the matrix are marked with the values with the intensity of relationship with each other and the values are assigned between the degrees from 0 to 5.

3 Calculations

For the DEMATEL's method, at first, the direct influence matrix [11] was formed as shown in Table 1.

After the application of the DEMATEL's method and following the procedure [11] as described in the above section, the weights for the considered four criteria were obtained as shown in Table 2.

On having obtained the weights for each of the considered criteria, the COPRAS method was adopted to further carry forward the calculation processes. COPRAS is a very befitting technique in multi-criteria decision-making, in particular when the one who is making the decision is not very clear to identify or prioritize the criteria at the initial stage of formulating the energy system. The compromise solution could be the substructure for negotiating and bettering the decision-maker's importance based on the calculation of weights criteria.

For application of the COPRAS method, the initial decision matrix was first formed for each of the considered power plants as discussed in Sect. 4 using the four considered criteria. Using this decision matrix as shown in Table 2, the normalized and weighted normalized matrices were obtained to carry forward the COPRAS analogy and get the resultant priority vectors of each power plants.

Table 1 Direct influence matrix for DEMATEL's method

	A	B	C	D	E	F	G
A	0	6	5	8	8	5	3
B	3	0	5	7	7	6	4
C	4	8	0	9	8	3	5
D	2	7	4	0	8	3	3
E	5	4	2	4	0	5	5
F	3	3	3	2	3	0	4
G	3	4	3	2	4	6	0

Table 2 Weights of criteria after DEMATEL's method

Criteria	A	B	C	D	E	F	G
Weight	0.142403	0.15975	0.150603	0.149047	0.15747	0.12088	0.119849

4 Power Plants Considered

4.1 Tuticorin Thermal Power Station

Tuticorin thermal power plant is a coal-based thermal power plant, located near the Newport of Thoothukudi on the seashore of Bay of Bengal in Tamil Nadu. This power station has a capacity of 3000 MW [16]. Supply of coal is brought via rail from the coalfields of Odisha, Bengal, and Bihar to the ports of Paradip, Haldia, and Vizag. From these ports, coal is transported to Tuticorin with the help of a ship [17].

4.2 Kundankulam Nuclear Power Plant

Considered to be the single largest nuclear power plant in India, the Kundankulam nuclear power plant is located in the Tirunelveli district of Tamil Nadu. The power plant has four units which are expected to deliver the capacity of 6800 MW, out of four, two units are fully functional and the remaining two are under construction, the first two deliver the capacity of 2000 MW [18].

4.3 Kamuthi Solar Power Plant

The main aim of the Kamuthi solar power project was to produce clean electricity using renewable solar energy sources. The project is installed over an area of 2,500 acres in Kamuthi in the Ramanathapuram district of Tamil Nadu. It has a generating capacity of 648 MW [19]. Generation of the energy takes place without any serious harm to the environment, in a sustainable way.

4.4 Muppandal Wind Power Plant

Wind power plants work on the basic principle, where the energy of blowing wind rotates three propeller-like blades of a wind tower around a rotor. The rotor is connected to the main shaft, which spins a generator for the generation of electricity [20]. Muppandal wind farm is situated in Kanyakumari district, Tamil Nadu, India. It is the biggest operational onshore wind farm in India and has an installed capacity of 1,500 MW [20].

Table 3 shows data for the power plants considered based on the criteria of the COPRAS method. References for the available data are mentioned within the table. For the analytical data (Emission of Pollutants and Social Acceptance), opinion of

Table 3 Data for each considered power plant based on the criteria of calculation

Plant name	Type	Power (MW)	Capital cost	Capacity factor	O&M cost (in crore ₹ for 2018–19)	Emission of pollutants	Social acceptance	Fuel cost (crore ₹) [21]
Tuticorin Thermal Power Station	Thermal	3000 [16]	₹4910 crore [16]	78% [16]	204	9	3	0.43
Kundankulam Nuclear Power Plant	Nuclear	2000 [18]	₹57017 crore [18]	31.25% [22]	90 [23]	9	1	0.143
Kamuthi Solar Power Plant	Solar	648	₹4550 crore [19]	24% [19]	802	1	8	0
Muppandal Wind Power Plant	Wind	1500 [24]	₹6750 crore [25]	33% [25]	515	1	7	0

an expert from the Ministry of Power, India was taken and they were ranked on a scale from 0 to 9, 0 being the least favorable and 9 the most.

5 Results and Discussions

Figure 1 shows the arrangement of the considered power plants of Tamil Nadu according to their preferential hierarchy owing to the priority vectors obtained after the COPRAS analogy. The calculation is directed at acquiring the alternatives rating based on the MCDM method. The fundamental objective of this analysis is to use the procedure based on factual numerical information for various drawn-out criteria. The analysis here is based on a few chosen criteria, which was observed to be important and accordingly the rank was set up. Considering all the criteria in their optimized presence, Muppandal wind power plant has attained the first position, followed by

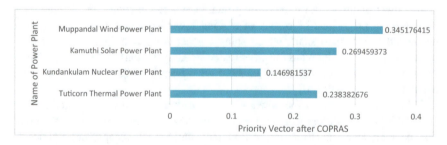

Fig. 1 Resultant plot of the power plants obtained after COPRAS method

the Kamuthi, Tuticorin, and Kundankulam power plants, respectively. Tamil Nadu has significant wind power potential, particularly in that region of the state which has an adequate amount of wind blowing throughout the year. Companies of the state and the nation have enough experience and knowledge in building wind power mills and the state with the country as a whole aim to maximize the amount of energy it produces from wind power. This study, however, due to constrictions in its scope, presents a generalized methodology to handle a decision-making problem as presented. Inclusion of a larger number of technical, environmental, and social criteria governing the working of a power plant will further extend the scope of this study, adding up to the betterment of the results obtained. Further, the number of plants considered can also be extensively increased to improvise on the resultant graph.

6 Conclusion

With never-ending demand for energy owing to increasing population and indus-trialization, the approach of increasing the energy planning system is of absolute interest. Multiple features considered in sustainable energy decision-making receive recognition. MCDM methods have been used in large aspects to sustainable energy decision-making considering multiple criteria. In this study, the evaluation of four different types of power plants, currently operating in Tamil Nadu, was done with respect to the specified key criteria such as power, operation and maintenance cost, capacity factor, cost of installation, fuel cost, social acceptance, and emission of pollutants. The evaluation was performed by using the DEMATEL weighting method and the COPRAS method. On the basis of available observational data, a conclusion is drawn that the Muppandal wind power plant is the most appropriate alternative for Tamil Nadu. The showcased scenario is only an alternative for the method to be used in multi-criteria evaluation of the power plants. Further, the following points were inferred:

- COPRAS method is one of the most reliable methods of decision-making for the provided multiple criteria of the various alternatives of power plants as it focuses on the ranking and selection from a given set of options. It helps us to find a negotiable answer to a conflicting problem, which lets the decision-maker reach a proper verdict.
- The legitimate numerical data information, which is used as an input criterion, has demonstrated that the result obtained is practical for Tamil Nadu.
- Here, it was found that the generalization of COPRAS methodology is tested in calculating the power plants and designing an optimized energy system.

At present, a wind power plant is considered to be one of the largest sources of alternate sources of energy generation. On the basis of modern tools and real appropriate data, it can be said that the future plan for the supply of energy in Tamil Nadu must be thought according to renewable energy sources, especially wind power plants. Apart from reducing the cost, it would also help in reducing the negative social and environmental impacts.

References

1. A. Alper, O. Oguz, The role of renewable energy consumption in economic growth: evidence from asymmetric causality. Renew. Sustain. Energy Rev. **60**, 953–959 (2016)
2. M. Gambini, M. Vellini, Hydrogen use in an urban district: environmental impacts a possible scenario based on coal. Int. J. Energy Environ. Econ. **2**(4), 13–31 (2011)
3. Z. Salameh, *Renewable Energy System Design* (Academic Press, 2014)
4. M.H. Rashid (Ed.), *Alternative Energy in Power Electronics* (Butterworth-Heinemann, 2014)
5. H. Louie, K. Anderson, *Analysis of Power Generation Forecast Utilization by Merchant Wind Plants* (2008), pp. 1–7
6. J. Wang, Y. Jing, C. Zhang, J. Zhao, Review on multi-criteria decision analysis aid in sustainable energy. Renew. Sustain. Energy Rev. **13**, 2263–2278 (2009)
7. S.D.A. Pohekar, M. Ramachandran, Application of multi-criteria decision making to sustainable energy planning d a review. Renew. Sustain. Energy Rev. **8**, 365–381 (2004)
8. E. Atmaca, H. Burak, Evaluation of power plants in Turkey using analytic network process (ANP). Energy **44**(1), 555–563 (2012)
9. M. Kabak, M. Dag, Prioritization of renewable energy sources for Turkey by using a hybrid MCDM methodology. Energy Convers. Manag. **79**, 25–33 (2014)
10. Z. Jiang, H. Zhang, J.W. Sutherland, Development of multi-criteria decision making model for remanufacturing technology portfolio selection. J. Clean. Prod. **19**(17–18), 1939–1945 (2011)
11. H.H. Lin, J.H. Cheng, Design process by integrating DEMATEL, and ANP methods, in *International Conference on Applied System Invention*, IEEE, Japan (2018)
12. D. Ghose, S. Naskar, Shabbiruddin, A.K. Roy, An open source software—Q-GIS based analysis for solar potential of Sikkim (India). Int. J. Open Source Softw. Process. (IGI Global) **10**(1), 49–68 (2019)
13. D. Ghose, S. Pradhan, Shabbiruddin, A fuzzy-COPRAS model for analysis of renewable energy sources in West Bengal, India, in *IEEE International Conference on Energy, Systems and Information Processing,* IIITD&M, Kancheepuram, Tamil Nadu, India, Proceedings to be published in IEEE (2019)
14. Power plant O&M: how does the industry stack up on cost? www.power-technology.com/features/featurepower-plant-om-how-does-the-industry-stack-up-on-cost-4417756/
15. R. Garg, R. Kumar, S. Garg, MADM-based parametric selection and ranking of E-learning websites using fuzzy COPRAS. IEEE Trans. Educ. (2018)
16. Tuticorn Thermal Power Station, https://economictimes.indiatimes.com/industry/energy/power/neyveli-lignite-corporations-rs-4910-crore-tuticorin-power-project-nearing-completion/articleshow/44223184.cms?from=mdr
17. Tuticorn Thermal Power Station, http://www.tangedco.gov.in/linkpdf/tttps.pdf
18. Kundankulam Nuclear Power Plant, https://en.wikipedia.org/wiki/Kudankulam_Nuclear_Power_Plant
19. Kamuthi Solar Power Project, https://en.wikipedia.org/wiki/Kamuthi_Solar_Power_Project
20. Muppandal Wind Farm, https://en.wikipedia.org/wiki/Muppandal_Wind_Farm
21. Fuel Cost, www.controlglobal.com/assets/Media/0809/CG0809_LLtable1.pdf
22. Nuclear Power Plants are cheaper to operate, https://marketrealist.com/2015/01/nuclear-power-plants-cheaper-operate
23. Report: Kundakulam Nuclear Power Project Units I and II, Union Government, Department of Atomic Energy, (2017), https://cag.gov.in/sites/default/files/audit_report_files/Report_No. 38_of_2017_-_Performance_Audit_on_Kudankulam_Nuclear_Power_Project%2C_Units_I_and_II_Department_of_Atomic_Energy.pdf

24. Generation cost calculation for 660 MW Thermal Power Plants, http://www.ijiset.com/v1s10/IJISET_V1_I10_94.pdf
25. V. Natarajan, J.C. Kanmony, Sustainability in India through wind: a case study of Muppandal wind farm in India. Int. J. Green Econ. **8**(1), 19 (2014)

Online Implementation of Cascade Predictive PI Control for Nonlinear Processes

Eadala Sarath Yadav, Indiran Thirunavukkarasu,
and Selvanathan Shanmuga Priya

Abstract This paper is concerned with design of predictive PI control algorithm for nonlinear cascade process. Conical tank level process and flow process are considered as inner and outer loops of cascade blocks, respectively. Control algorithm is extended version of smith predictor which reflects actuator action, whereas cascade approach deals with disturbance rejection and system performance. The experimental setup is interconnected in such a way that flow process uses the inflow from the conical tank's outflow. The methodology is implemented to minimize the energy consumption of final control element. Results are effective enough to validate the methodology and neglect input constraints for limited operating region.

Keywords Cascade control · Predictive PI · FOPDT · Nonlinear processes · Windup action and Skogestad's tuning

1 Introduction

Cascade control consists of two loops in which outer loop's controller acts as setpoint to the inner loop and the response of inner loop is fed as input to the outer loop's process. It is essential tool used for controlling chemical processes like heat exchanger, chemical reactor, distillation columns, etc. The approach is implemented

E. Sarath Yadav
Department of Electronics and Instrumentation Engineering, Vignan Institute of Technology and Science, Deshmukhi, Hyderabad 508284, Telangana, India
e-mail: sarath.eadala@gmail.com

I. Thirunavukkarasu (✉)
Department of Instrumentation and Control Engineering, Manipal Institute of Technology, Manipal Academy of Higher Education, Manipal 576104, Karnataka, India
e-mail: it.arasu@manipal.edu

S. Shanmuga Priya
Department of Chemical Engineering, Manipal Institute of Technology, Manipal Academy of Higher Education, Manipal 576104, Karnataka, India
e-mail: shan.priya@manipal.edu

© Springer Nature Singapore Pte Ltd. 2020
R. Bera et al. (eds.), *Advances in Communication, Devices and Networking*,
Lecture Notes in Electrical Engineering 662,
https://doi.org/10.1007/978-981-15-4932-8_5

39

for the processes with large overshoot and significant lags and this mechanism is used to improve disturbance rejection and enhance system performance. Santosh and Chidambaram [1] proposed a simple method to tune series cascade processes. Padhan and Majhi [2] implemented parallel cascade control strategy for stable, unstable, and integrating process. Modified smith predictor is used to improve inner loop performance and time delay term is considered using smith predictor approach. Sadasivarao and Chidambaram [3] had implemented genetic algorithm for tuning PID parameters for cascade process and Ziegler–Nichols tuning is used and proposed for selecting search region.

In this paper based on modified smith predictor, predictive PI has been implemented for inner loop to make the inner loop of the system more optimal. As the process variable of outer loop is flow, the final control element (actuator) is pneumatic control valve. The inflow through the pneumatic valve is given by the out flow of conical tank. Therefore, maximum pressure is required to allow the inflow through pneumatic control valve. It is attained only if there exist maximum level inside the conical tank. As the tank level reduces, the inflow through control valve of primary loop also reduces. Hence, it is difficult to control on higher setpoint values of the flow (primary loop) with current physical arrangement. But the approach is efficient enough for the lower setpoint values which are validated by comparing with single loop control design using 2DOF PID anti-reset windup approach by Satheesh Babu et al. [4]. Another aspect of predictive control scheme by formulating predictive gain into PI controller has been given by Eadala et al. [5]

2 Architecture and Model Identification

The block diagram of cascade control architecture embedded with prediction filter is shown in Fig. 1. Inner loop is conical level process (G_{p1}) and outer loop is flow control process (G_{p2}) which uses inductor motor as final control element for level control and pneumatic control valve flow process, respectively. Model identification is carried out by using two-point method for both processes [6].

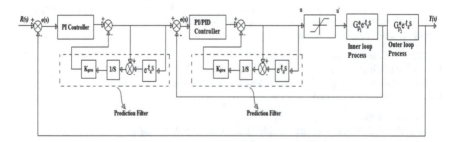

Fig. 1 Block diagram of cascade control architecture with prediction filter

Fig. 2 Experimental setup for cascaded level and flow processes

Figure 2 shows the physical arrangement of conical level and flow process station operates as cascaded process. Out flow from the conical tank is fed to inflow of the pneumatic control valve.

$$G_{p1}(s) = \frac{K}{\tau s + 1} e^{-t_d s} = \frac{0.925}{25.05s + 1} e^{-1.09s} \tag{1}$$

$$G_{p2}(s) = \frac{K}{Ts + 1} e^{-t_d s} = \frac{0.186}{1.35s + 1} e^{-4.25s} \tag{2}$$

3 Control Design

It is important to design a controller for cascade process in a systematic way. Primary objective of cascade control is to design inner loop controller first, then once the inner process becomes sustainable outer loop controller should be considered. Predictive PI control algorithm is implemented to both loops. Prediction filter $f(s)$ has two design parameters, prediction gain K_{pre} and dead time $e^{-t_d s}$, which are obtained by block reduction approach. The predictive filter design has been considered from Airikka [7]

Table 1 Controller parameters of inner loop

	Inner loop	Outer loop
K_p	2.0165	0.66
T_i	8.72	34
K_{pre}	0.114	0.029

$$R(s) = K_{PI}(s)f(s)e(s) \tag{3}$$

$$K_{PI}(s) = K_p(1 + \frac{1}{T_i s}) \tag{4}$$

$$f(s) = \frac{1}{1 + \frac{1}{s}K_{pre}(1 - e^{-t_d s})} \tag{5}$$

$$K_{pre} = \frac{\lambda}{4T_{cl}} \tag{6}$$

T_{cl} is factor of dead time estimate which is recommended as $2*t_d$ and assigned $\lambda = 1$. Cascade control is implemented to improve disturbance rejection and enhances process system performance. Table 1 shows controller parameters for inner (conical level control) and outer loop processes.

$$K_p = \frac{1}{(KT_{cl})} \tag{7}$$

$$T_i = 4T_{cl} \tag{8}$$

$$K_{pre} = \frac{\lambda}{4T_{cl}} \tag{9}$$

4 Result Analysis

Figure 3 shows cascade control response for both inner and outer loops, where inner loop response (level process) acts as controller to the flow process. Therefore, proper control of inner loop (level) reflects the efficient control of outer loop (flow). To test the effectiveness of the controller with respect to anti-reset windup, saturation limits from the algorithm have been eliminated and response has been observed. The controller is effective enough to operate within the boundary limits of saturation. As the flow process operates at lower region, the controller (inner loop response) will also operate at lower region.

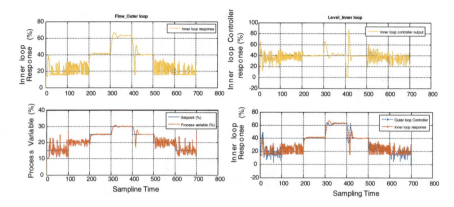

Fig. 3 Cascade control response for both inner and outer loops

Figure 4 depicts the smoothness of process as well as controller response for operating region of 25–35%. Cascade process response has been compared with 2DOF PID anti-integral windup control mechanism.

Figures 5 and 6 represent controller output and flow process response, respectively. It is observed that the implementation of predictive PI algorithm for cascade process eliminates the effect of windup action for actuator and makes it to operate smooth within the saturation bounds. Different performance measures have been calculated for the setpoint between 25 and 35 % of flow and comparison of performance indices for cascade predictive PI and 2DOF PID ARW is embodied in Tables 2, 3, respectively.

Fig. 4 Process variable (flow) response with inner loop output

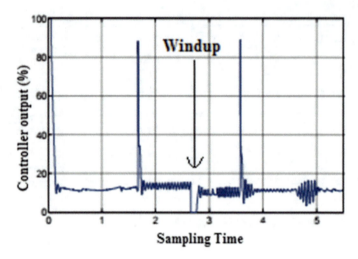

Fig. 5 Controller output

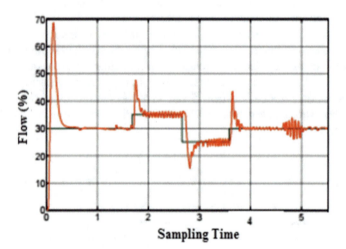

Fig. 6 Process response

Table 2 Performance measures of flow control

Setpoint (%)	Peak time (sampling time)	Rise time/fall time (sampling time)	Overshoot/Undershoot (%)	Settling time (sampling time)
25	16	5	1	9
30	18	7	2.3	12
25	16	9	32	30

Table 3 Comparison of performance indices for cascade predictive PI and 2DOF PID ARW

Setpoint (%)	Cascade predictive PI		2DOF PID anti-reset windup	
	IAE	ISE	IAE	ISE
25	122.29	961.1	1.3e+7	1.3e+10
30	171.19	1071.1	1.8e+7	3.8e+10
25	270	1444.1	3.4e+7	1.3e+11

5 Conclusion

Windup action for fast acting control loops is the critical issue in the real-time process because of integral action. Optimal tuning is essential to keep the controller in the operating region and eliminate the effect of windup. Cascade control is implemented to improve the system performance and escalates the efficiency of the process. Smith predictor-based predictive PI control algorithm has been implemented and response is compared with 2DOF PID anti-reset windup. The implemented algorithm gave better response in different aspects like controller action, system performance via performance indices, and time-domain performance analysis. It is observed that energy utilization of actuator using cascade control is less compared with 2DOF PID anti-reset windup method. Since the setup is single-input single-output (SISO) process, as a future scope the implemented algorithm should be revised for MIMO process. As the model is linearized at particular operating region, implementation of algorithm is limited to that operating region alone.

Acknowledgements The authors would like to thank DST, Government of India for the sanction of this project under the wide circular #SB/FTP/ETA-0308/2012 dated May 25, 2013. The authors also would like to thank MIT, Manipal University for providing the infrastructure facility for carrying out this project.

References

1. S. Santosh, M. Chidambaram, A simple method of tuning series cascade controllers for unstable systems. J. Control Theory Appl. **11**(4), 661–667 (2013)
2. D.G. Padhan, S. Majhi, An improved parallel cascade control structure for processes with time delay. J. Process Control **22**, 884–898 (2012)
3. M.V. Sadasivarao, M. Chidambaram, PID controller tuning of cascade control systems using genetic algorithm. J. Indian Inst. Sci. **86**, 343–354 (2006)
4. R. Satheesh Babu, C. Shreesha, I. Thirunavukkarasu, Comparative realization of 2DOF PID controllers for flow process with measurement delay. Int. J. Adv. Eng. Res. **7**(1), 36–46 (2014)
5. E.S. Yadav, T. Indiran, S.S. Priya, G. Fedele, Parameter estimation and an extended predictive-based tuning method for a lab-scale distillation column. ACS Omega **4**(25), 21230-21241 (2019)
6. E.S. Yadav, T. Indiran, Servo mechanism technique based anti-reset windup PI controller for pressure process station. Indian J. Sci. Technol. **9**(8) (2016)
7. P. Airikka, Predictive PID controller for integrating processes with long dead times, in *Mediterranean Conference on Control and Automation* (2014), pp. 475–480

Optimized Design of 60° Bend in Optical Waveguide for Efficient Power Transfer

Kamanashis Goswami, Haraprasad Mondal, and Mrinal Sen

Abstract Optimized design of 60° bend optical waveguide has been proposed in this work. Two-dimensional (2D) triangular lattice Photonic Crystal (PhC) with holes in slab structure is used to design the waveguide having two 60° bends. A single hole in each of these bending regions is optimized to make the proposed design simple and easily fabricable. Plane Wave Expansion (PWE) algorithm has been used for evaluating the band diagram of the PhC. Finite Difference Time Domain (FDTD) method has been used to measure and analyze the wave propagation profile of the waveguide. Optical power transfer efficiency of the waveguide has been found as approximately 98% for two standard optical operating wavelengths.

Keywords Photonic crystal · Optical waveguide · Photonic integrated circuits · Finite difference time domain · Plane wave expansion

1 Introduction

PhC [1] is an alternative arrangement of two dielectric materials having different refractive indexes. Due to this alternative arrangement, it exhibits complete Photonic Bandgap (PBG) [2] which does not allow a range of optical frequencies to propagate through the structure. However, light can be propagated under certain conditions through a Photonic Crystal Waveguide (PCW) [3]. Simple way to create a PCW is by removing/replacing rods or holes from/to the photonic crystal. Now, for designing all-optical logic devices/switches [4–6] within a Photonic Integrated Circuit (PIC), an optical waveguide is often required to be bent in different angle. However, multiple

K. Goswami · H. Mondal (✉) · M. Sen
Electronics Engineering Department, I.I.T. (I.S.M), Dhanbad, Dhanbad, India
e-mail: mandal.haraprasad@gmail.com

K. Goswami
e-mail: kamanashis.goswami@gmail.com

M. Sen
e-mail: mrinal.sen.ahm@gmail.com

© Springer Nature Singapore Pte Ltd. 2020   47
R. Bera et al. (eds.), *Advances in Communication, Devices and Networking*,
Lecture Notes in Electrical Engineering 662,
https://doi.org/10.1007/978-981-15-4932-8_6

guided modes [7] are usually generated when a bend is created in a PhC waveguide. Hence, at bending regions, abrupt deviations of the light path from the straight waveguide may cause mode-mixing problems. This might result in back reflections and poor transmissions through the bends toward the output. This problem can be overcome by changing the geometrical conformation of the PhC in the bending region, which is very much challenging task for the researchers. In recent past, many research proposals have been reported to overcome this problem, some of which are briefed as follows.

Moghaddam et al. have reported two separate designs of 60° bends in [8]. In their first design, three small holes have been incorporated in the middle of the waveguide near the bending vicinity. In the second design, five small holes have been incorporated in the middle of waveguide in the bending region. Furthermore, in the same region, the size of few more holes at the border of waveguide has been optimized. However, the transmission efficiency is nearly 90% in both cases, which is not sufficient for a PIC. Moreover, the designs are pretty challenging from the fabrication point of view. On the other hand, a similar structure has been reported by Frandsen et al. [9] where the outer edge of each bend has been smoothened by applying a soft curvature, and one hole has been removed from inner side. But their structure does not support substantially high-power transmission which is ~77% and the design is little complex. Chen et al. [10] have reported a waveguide structure with two capsule-shaped defects at two bend regions and achieved a transmission ~95%. Although the efficiency in this report is quite high, fabrication of such capsules in proper orientation and shape is a bit complex.

Therefore, to overcome the abovementioned shortfalls, the authors have been motivated toward this work. In this work, a very simple design of a PhC-based optical waveguide having two 60° bends has been proposed by optimizing only one hole in each bending region to increase the transmission efficiency to a great extent. The design simplicity enhances the feasibility of realization of such structures.

2 Design and Analysis

Design of proposed optical waveguide is shown in Fig. 1. 2D-PhC (holes in slab) arranged in a triangular lattice has been considered for the design. The refractive index of silicon slab is considered as 3.46. The lattice constant (a) and the radius of the air holes (r_{nd}) of the PhC are chosen as 368 nm and 118 nm, respectively. The proposed waveguide with two 60° bends has been created by removing three rows of holes in Γ-M direction of the PhC which is shown in Fig. 1. To obtain maximum transmission of power, the radius of one hole (r_d) at the outer side of each bend has been optimized and their radii increased by 1.22 times ($r_d = 1.22 * r_{nd}$) of non-defect holes. 2D Plane Wave Expansion (PWE) algorithm [11] has been applied to analyze the photonic bandgap as well as projected bandgap of the PhC which has been shown in Fig. 2. The complete PBG of the non-defect structure is shown in Fig. 2a and projected band diagram of the defect structure is shown in Fig. 2b.

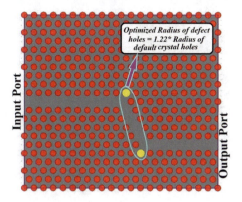

Fig. 1 Schematic of two 60° bend waveguides

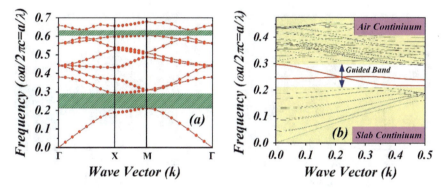

Fig. 2 Band diagram of **a** non-defect PhC structure, **b** defect PhC structure

The PBG has been obtained in Transverse Electric (TE) mode in the normalized frequency range (a/λ) 0.215 to 0.295 (i.e., 1247 nm to 1711 nm wavelength).

The 2D Finite Difference Time Domain (FDTD) [12] method has been applied to measure and analyze the performance of the proposed structure. Output power transmission profile for different values of diameter of corner holes has been shown in Fig. 3. From this figure, it has been observed that power transmission is substantially low (below 5%) when the diameter of the said hole is kept unchanged. Transmission starts increasing when the diameter of the hole is being increased and transmission reaches to its maximum value (~98%) when the diameter of corner holes is increased to 1.22 times of its normal value. Further increase in the diameter of corner holes decreases the output power transmission and it becomes zero at certain values of diameter of corner holes.

The output power transmission profile of the proposed structure for both wavelengths (1310 and 1550 nm) has been analyzed and depicted in Fig. 4. The lattice constant and radius of holes of the PhC are chosen as 368 nm and 118 nm, respectively, which offer maximum power transmission at 1550 nm wavelength through

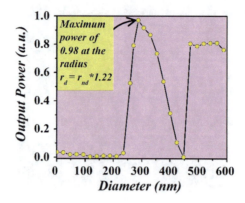

Fig. 3 Output power transmission profile with respect to various diameters of optimized holes

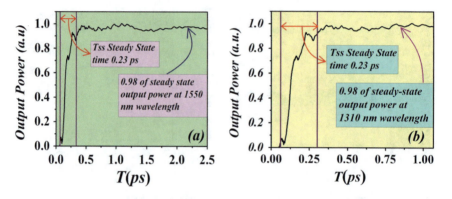

Fig. 4 Output power transmission profile for **a** 1550 nm wavelength, **b** 1310 nm wavelength

the waveguide. Figure 4a shows output power transmission profile, where 1550 nm wavelength signal has been launched at the input of the waveguide and at steady-state condition ~98% of the input power has been received at the output of the waveguide. Similarly, all the parameters (i.e., lattice constant and radius of holes) of the PhC have been changed and the PhC has been designed, where maximum power transmission has been achieved through the waveguide in 1310 nm wavelength. Figure 4b shows the output power transmission profile, where 1310 nm signal has been applied at the input and ~98% of input power has been received at the output of the waveguide.

Transmittance [13, 14] characteristics of the proposed waveguide with respect to various wavelengths have been calculated which are shown in Fig. 5. An optical Gaussian pulse has been launched at the input of the waveguide and frequency response of that pulse has been observed at the output port of the waveguide. Frequency response of that pulse has been obtained based on Fast Fourier Transform (FFT) method. From the transmittance characteristics of the device, it has been observed that a maximum transmittance of 0.98 has been obtained at the output of the waveguide.

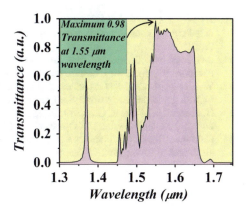

Fig. 5 Transmittance characteristics of the structure with respect to various wavelengths

3 Conclusion

In this work, a structure of waveguide having two 60° bends has been proposed. The work is done on 2D PhC where holes are arranged in a triangular lattice. The beauty of this design is that it is perhaps the simplest structure so far according to author's knowledge. Only one single hole at each bending region has been optimized by increasing 1.22 times of its normal (non-defect) size. It is offering a throughput of 99% at each bending for both 1310 nm and 1550 nm wavelength optical signals. Therefore, the simplicity of the structure makes it appropriate for designing waveguide where such bending is required in integrated circuit.

References

1. M. Sen, M.K. Das, High-speed all-optical logic inverter based on stimulated Raman scattering in silicon nanocrystal. Appl. Opt. **54**, 9136–9142 (2015)
2. J.H. Yuan, Y.Y. Lu, Photonic bandgap calculations with Dirichletto-Neumann maps. J. Opt. Soc. Am. **23**, 3217–3222 (2006)
3. A. Mekis, J.C. Chen, I. Kurland, S. Fan, R. Villeneuve, J.D. Joannopoulos, High transmission through sharp bends in photonic crystal waveguides. Phys. Rev. Lett. **77**, 3787–3790 (1996)
4. H. Mondal, S. Chanda, M. Sen, T. Datta, All optical AND gate based on silicon photonic crystal, in *Proceedings of the ICMAP*, Dhanbad, India (2015), pp. 1–2
5. H. Mondal, K. Goswami, C. Prakash, M. Sen, An all-optical ultra-compact 4-channel wavelength de-multiplexer, in *Proceedings of the ICMAP*, Dhanbad, India (2018), pp. 1–2
6. D. Gogoi, K. Das, H. Mondal, P. Talukdar, K. Hussain, Design of ultra-compact 3-channel wavelength demultiplexer based on photonic crystal, in *Proceedings of the ICADOT*, Pune, India (2016), pp. 590–593
7. P. Bettini, S. Boscolo, R. Specogna, M. Midrio, Design optimization of waveguide bends in photonic crystals. IEEE Trans. Magn. **45**, 1630–1633 (2009)
8. M.K. Moghaddam, M.M. Mirsalehi, A.R. Tari, A 60° photonic crystal waveguide bend with improved transmission characteristics. Optica Applicata **39**, 307–317 (2009)
9. L.H. Frandsen, A. Harpoth, P.I. Borel, M. Kristensen, Broadband photonic crystal waveguide 60° bend obtained utilizing topology optimization. Opt. Express **12**, 5916–5921 (2004)

10. J. Chen, Y. Huang, Y. Yang, M. Lu, J. Shieh, Design, fabrication, and characterization of Si-based arrow photonic crystal bend waveguides and power splitters. Appl. Optics **51**, 5876–5884 (2012)
11. S.G. Johnson, J.D. Joannopoulos, Block-iterative frequency domain methods for Maxwell's equations in a plane wave basis. Opt. Express **8**, 173–190 (2001)
12. W. Huang, S. Chu, S. Chaudhuri, A semi-vectorial finite difference time-domain method (optical guided structure simulation). IEEE Photonic Technol. Lett. **3**, 803–806 (1991)
13. H. Mondal, S. Mrinal K. Goswami, Design and analysis of all-optical 1 to 2 line decoder based on linear photonic crystal. IET Optoelectron. **13**, 191–195 (2019)
14. H. Mondal, M. Sen, C. Prakash, K. Goswam, C. Sarma, Impedance matching theory to design an all optical AND gate. IET Optoelectron. **12**, 244–248 (2018)

Rational Approximation of Fractional-Order Multivariable System

Jaydeep Swarnakar

Abstract The cognizance of the fractional-order system (FOS) is flourishing among the researchers of science and engineering. Infinite dimension of the FOS is no more a problem nowadays as far as their practical implementations are concerned due to the advent of many rational approximation methods. This paper revisits one such recognized approximation method, namely, the Oustaloup approximation method, to acquire the approximation of a multivariable FOS signified by a transfer function matrix (TFM). Essential simulation results are deliberated.

Keywords Fractional-order system · Multivariable system · Oustaloup approximation

1 Introduction

Fractional calculus theory has become popular at present, although the topic is virtually 300 years old [1]. Many physical systems are modeled using the transfer functions or the state spaces possessing fractional orders. The applications of the FOS are prevalent in control theory [2], signal processing [3], biomedical engineering [4], and many more to mention. To cast the infinite-dimensional FOS for handling the problem like controller design and realization, synthesis, etc., the prerequisite is to find its rational approximation. This imposes to express the FOS as an equivalent integer-order system (IOS) within a specified frequency range so that the IOS replicates the frequency-domain characteristics of the original FOS. In literature, various approaches have been suggested for satisfying this very purpose [5]. Among all of them, Oustaloup approximation is the one which has been broadly used for approximating the FOS [6–10]. This paper focusses on approximating a multivariable FOS. To satisfy this objective, the generalized Oustaloup

J. Swarnakar (✉)
Department of Electronics & Communication Engineering, School of Technology, North-Eastern Hill University, Shillong 793022, Meghalaya, India
e-mail: jaydeepswarnakar@gmail.com

© Springer Nature Singapore Pte Ltd. 2020   53
R. Bera et al. (eds.), *Advances in Communication, Devices and Networking*,
Lecture Notes in Electrical Engineering 662,
https://doi.org/10.1007/978-981-15-4932-8_7

method [3] has been revisited to approximate the individual fractional-order single-input-single-output (SISO) transfer functions prevailing in the original multivariable system. Finally, the ideal multivariable system and its approximated counterpart have been compared to examine the precision of the approximation based on frequency responses attained. The paper is outlined in four sections. Introduction is given in Sect. 1. Section 2 describes the technique to approximate the fractional-order multivariable system from its transfer function matrix (TFM). Section 3 discusses some pertinent simulation results. The concluding comments are briefed in Sect. 4.

2 Approximation of Fractional-Order TFM

The commensurate multivariable FOS is modeled using state space as given below [10]:

$$D^\alpha x = Fx + Gu \tag{1}$$

$$y = Hx + Ku \tag{2}$$

where u, x, and α indicate the vectors implying the input, state, and fractional order, respectively. On the other hand, F, G, H, and K symbolize the matrices, namely, state, input, output, and the direct transmission correspondingly. The TFM of the multivariable system is given below:

$$M(s) = H(s^\alpha I - F)^{-1}G + K \tag{3}$$

Let us assume that the multivariable FOS is having the TFM as follows:

$$M(s) = \begin{bmatrix} \frac{A_1}{B_1 s^{\alpha_1}+1} & \frac{A_2}{B_2 s^{\alpha_2}+1} \\ \frac{A_3}{B_3 s^{\alpha_3}+1} & \frac{A_4}{B_4 s^{\alpha_4}+1} \end{bmatrix} \tag{4}$$

Equation (4) consists of four fractional-order transfer functions of SISO form, where each of them is having a fractional power term s^{α_k} $(k = 1, 2, 3, 4)$. The entire fractional-order TFM is approximated to its integer order equivalent by applying Oustaloup approximation individually on s^{α_k} $(k = 1, 2, 3, 4)$. The mathematical formula of the Oustaloup approximation [3] is used to find the Lth-order approximant of s^{α_k} within a band (ω_m, ω_n) as given below:

$$s^{\alpha_k} \approx R \prod_{c=1}^{L} \frac{s + \omega_p'}{s + \omega_p} \tag{5}$$

where ω_m is the low frequency, ω_n is the high frequency, $R = \omega_n^\alpha$, $\omega_j = \sqrt{\omega_n/\omega_m}$, $\omega_p' = \omega_m \omega_j^{(2c-1-\alpha)/L}$, and $\omega_p = \omega_m \omega_j^{(2c-1+\alpha)/L}$. So, the overall approximant of $M(s)$ will result in the following form:

$$M(s) \approx \begin{bmatrix} M_{11}(s) & M_{12}(s) \\ M_{21}(s) & M_{22}(s) \end{bmatrix} \tag{6}$$

$M_{bc}(s)$ ($b = 1, 2$ and $c = 1, 2$) is the equivalent rational approximant of $A_j/(B_j s^{\alpha_j} + 1)$, where $j = 1, 2, 3,$ and 4, respectively. The frequency responses of all the individual fractional-order transfer functions pertaining to the TFM are to be verified with the frequency responses of their corresponding rational approximants to investigate the usefulness of the approximation.

3 Simulation and Results

The TFM of the multivariable FOS is considered as given below [11]:

$$M(s) = \begin{bmatrix} \frac{1.2}{2s^{0.5}+1} & \frac{0.6}{3s^{0.7}+1} \\ \frac{0.5}{s^{0.8}+1} & \frac{1.5}{3s^{0.6}+1} \end{bmatrix} \tag{7}$$

Taking $L = 5$, $\omega_m = 0.01$ rad/s, and $\omega_n = 100$ rad/s, the approximations of $s^{0.5}$, $s^{0.7}$, $s^{0.8}$, and $s^{0.6}$ are calculated first using Eq. (5) as noticeable in Table 1. Employing Table 1, the final approximation of the original multivariable fractional-order system $M(s)$ has produced four rational transfer functions as given in Table 2. The frequency responses of these approximants have been compared with their corresponding original FOSs as shown from Figs. 1, 2, 3, and 4. From each and every figure, it is clearly visible that the magnitude approximations are comparatively better than the phase approximations as the magnitude graphs of the approximants are almost analogous to their original counterpart. Phase approximations are reasonable in the low- and mid-frequency ranges. However, the phase graphs show significant deviations in

Table 1 Fifth-order approximations of $s^{0.5}$, $s^{0.7}$, $s^{0.8}$, and $s^{0.6}$ within 0.01–100 rad/s using Oustaloup method

Fractional power terms	Equivalent integer-order approximation
$s^{0.5}$	$\dfrac{20s^5+597s^4+2436s^3+1537s^2+149.94s+2}{2s^5+149.94s^4+1537s^3+2436s^2+597s+20}$
$s^{0.7}$	$\dfrac{25.12s^5+623.6s^4+2117s^3+1111s^2+90.14s+1}{s^5+90.14s^4+1111s^3+2117s^2+623.6s+25.12}$
$s^{0.8}$	$\dfrac{39.81s^5+901.4s^4+2790s^3+1336s^2+98.83s+1}{s^5+98.83s^4+1336s^3+2790s^2+901.4s+39.81}$
$s^{0.6}$	$\dfrac{15.85s^5+431.4s^4+1606s^3+924s^2+82.2s+1}{s^5+82.2s^4+924s^3+1606s^2+431.4s+15.85}$

Table 2 Approximations for each fractional-order transfer function existing in TFM

Fractional-order transfer function	Equivalent integer-order approximation
$\frac{1.2}{2s^{0.5}+1}$	$M_{11}(s) = \frac{1.2s^5+89.97s^4+922.3s^3+1462s^2+358.2s+12}{21s^5+671.9s^4+3205s^3+2755s^2+448.4s+12}$
$\frac{0.6}{3s^{0.7}+1}$	$M_{12}(s) = \frac{0.6s^5+54.08s^4+666.5s^3+1270s^2+374.2s+15.07}{76.36s^5+1961s^4+7461s^3+5449s^2+894s+28.12}$
$\frac{0.5}{s^{0.8}+1}$	$M_{21}(s) = \frac{0.5s^5+49.42s^4+667.8s^3+1395s^2+450.7s+19.91}{40.81s^5+1000s^4+4126s^3+4126s^2+1000s+40.81}$
$\frac{1.5}{3s^{0.6}+1}$	$M_{22}(s) = \frac{1.5s^5+123.3s^4+1386s^3+2409s^2+647.1s+23.77}{48.55s^5+1376s^4+5741s^3+4378s^2+678s+18.85}$

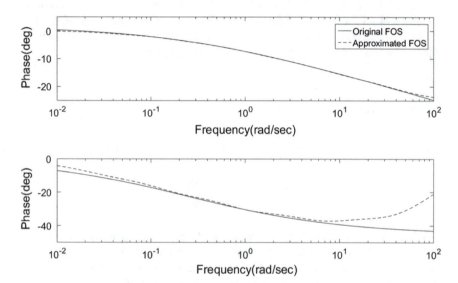

Fig. 1 Frequency responses of $1.2 / (2s^{0.5} + 1)$ and $M_{11}(s)$

the high-frequency range. Overall approximation of the multivariable system $M(s)$ seems to be quite acceptable employing the Oustaloup approximation method.

4   Conclusions

In this paper, a multivariable FOS is approximated using the Oustaloup approximation method. The TFM of the FOS comprises of four transfer functions which are approximated separately and then the resulting frequency responses have been compared with the corresponding underlying fractional-order systems. It has been shown that the magnitude approximation is substantial enough in the entire frequency range, whereas the phase approximation results error in the high-frequency range while maintaining adequate consistency in the lower and middle half of the overall frequency range.

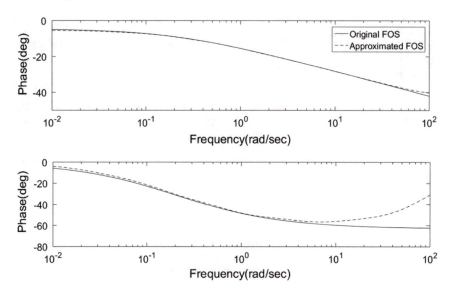

Fig. 2 Frequency responses of $0.6/\left(3s^{0.7}+1\right)$ and $M_{12}(s)$

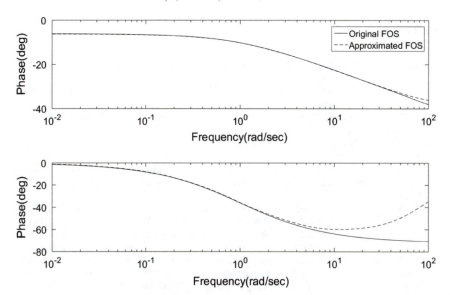

Fig. 3 Frequency responses of $0.5/\left(s^{0.8}+1\right)$ and $M_{21}(s)$

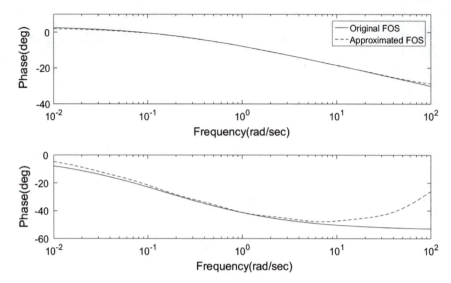

Fig. 4 Frequency responses of $1.5 / \left(3s^{0.6} + 1\right)$ and $M_{22}(s)$

References

1. Y.Q. Chen, I. Petras, D. Xue, Fractional-order control—a tutorial, in *Proceeding of the American Control Conference (ACC)*, St. Louis, USA (2009), pp. 1397–1411
2. I. Podlubny, Fractional-order systems and $PI^\lambda D^\lambda$ controllers. IEEE Trans. Autom. Control **44**(1), 208–214 (1999)
3. H. Sheng, Y.Q. Chen, T.S. Qiu, *Fractional Processes and Fractional-Order Signal Processing* (Springer, London, 2011)
4. B.T. Krishna, Studies on fractional-order differentiators and integrators: a survey. Sig. Process. **91**(3), 386–426 (2011)
5. B.M. Vinagre, I. Podlubny, A. Hernande, V. Feliu, Some approximations of fractional-order operators used in control theory and applications. Fract. Calculus Appl. Anal. **3**(3), 231–248 (2000)
6. A. Oustaloup, *CRONE Control: Robust Control of Non-integer Order* (Hermes, Paris, 1991)
7. J. Baranowski, W. Bauer, M. Zagórowska, T. Dziwiński, P. Piątek, Time-domain oustaloup approximation, in *Proceedings of the IEEE 20th International Conference on Methods and Models in Automation and Robotics (MMAR)*, Miedzyzdroje, Poland (2015), pp. 116–120
8. J. Swarnakar, P. Sarkar, L.J. Singh, Realization of fractional-order operator in complex domains—a comparative study, in *Lecture Notes in Electrical Engineering*, vol. 462, ed. by R. Bera, S. Sarkar, S. Chakraborty (Springer, Singapore, 2018), pp. 711–718
9. J. Swarnakar, P. Sarkar, L.J. Singh, Rational approximation methods for a class of fractional-order SISO system in delta domain, in *Lecture Notes in Electrical Engineering*, vol. 537, ed. by R. Bera, S.K. Sarkar, O.P. Singh, H. Saikia (Springer, Singapore, 2019), pp. 395–402
10. M. Rachid, B. Mammbar, D. Said, Comparison between two approximation methods of state space fractional systems. Sig. Process. **91**(3), 461–469 (2011)
11. Z. Li, Fractional-order modeling and control of multi-input-multi-output processes. Doctoral dissertation, UC Merced (2015)

FO 2-(DOF) Control of Non-commensurate Fractional Order (FO) Plants

Reetam Mondal and Jayati Dey

Abstract Servo control of prevailing industrial control applications demands the improvement of the feedback control action while maintaining the loop robustness goals as well as to shape and forge the output response of the system according to the predefined requirements. On this basis, fractional order (FO) 2-DOF (degree of freedom) control methods adopted which are more general than the classical conventional order counterparts. Numerical example and simulation results are explained to demonstrate the method put forward.

Keywords Servo control · Regulatory control · Robustness · Non-commensurate · FO 2-DOF (Degree of Freedom)

1 Introduction

An exemplary matter of difficulty in control engineering is that to devise a controller which can modulate the system without being affected by uncertainty in the plant or controller parameters [1]. The 1-DOF control in this case may expediently endanger standards lower than that desirable and hence 2-DOF control [2]. In this view, recently, FO 2-DOF control has scooped much importance in the control fraternity. The 2-DOF control techniques reported in available literatures mostly employed by the conventional PID and FO $PI^\lambda D^\mu$ controllers [3]. Analytical 2-DOF control scheme for Integral plus time delay plants using PI and PID controllers are proposed in [4] and PD controller as a feed-forward compensator in [5]. It is revealed in [6] that fractional order $PI^\lambda D^\mu$ (FOPID) controllers associated with the 2-DOF control scheme has added advantage over classical integer-order PID controllers for better reference tracking, noise suppression, and disturbance rejection. Meta-heuristic

R. Mondal (✉) · J. Dey
National Institute of Technology (NIT), Durgapur, West Bengal, India
e-mail: reetammondal2008@gmail.com

J. Dey
e-mail: deybiswasjayati@gmail.com

© Springer Nature Singapore Pte Ltd. 2020
R. Bera et al. (eds.), *Advances in Communication, Devices and Networking*,
Lecture Notes in Electrical Engineering 662,
https://doi.org/10.1007/978-981-15-4932-8_8

nature inspired algorithms has been developed in [7, 8] to focus on the application and design of these FO 2-DOF controllers on AGC of multi-area power system.

An elementary design method of FO 2-(DOF) control is recommended here for non-commensurate linear time-invariant (LTI) systems extending the formulation plan of fractional order (FO) compensators in [9]. Large overshoot and settling time is obtained using 1-DOF controller. As a remedy another pre-filter in the feed-forward loop is blended with the 1-DOF feedback controller to execute smooth setpoint tracking, thus implementing the 2-DOF control strategy.

2   FO 2-DOF Control of Non-Commensurate Order Plants

A 2-DOF controller comprises of a serial FO pre-filter and a feedback FO compensator in the closed-loop structure as in Fig. 1.

A tradeoff scheme may be necessary to achieve and maintain robust loop goal performance in addition to acceptable output regulation which calls for the FO 2-DOF structure. For, an SISO FO-LTI plant defined as,

$$G(s) = \frac{N(s)}{D(s)} \tag{1}$$

The FO compensator transfer function taken as [9],

$$C_f(s) = \frac{H_f(s)}{F_f(s)} = \frac{1 + \alpha T s^\beta}{1 + T s^\beta} \tag{2}$$

with the additional degree of freedom, $\beta \in (0, 2)$ [9]. At a selected new gain crossover frequency (ω'_{cg}) the desired stability margins of,

$$L_f(s) = C_f(s)G(s) = \frac{H_f(s).B(s)}{F_f(s).A(s)} \tag{3}$$

is fulfilled ensuring sensitivity peak magnitudes ≤ 2 in the mid-frequency region. The characteristic equation is given by,

Fig. 1 2-DOF control schematic diagram

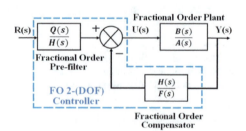

$$\Delta_f(s) = D(s)F_f(s) + H_f(s)N(s) \tag{4}$$

The characteristic equation above in general form can be written as [10],

$$a_n s^{\alpha_n} + a_{n-1} s^{\alpha_{n-1}} + \cdots + a_1 s^{\alpha_1} + a_0 s^{\alpha_0} = 0 \tag{5}$$

which may possess an infinite number of roots. Now, the above polynomial in s-domain can be viewed as,

$$\sum_{i=0}^{n} a_i s^{\alpha_i} \tag{6}$$

where, $\alpha_i = \frac{u_i}{v_i}$ with v as the number of sheets in the Riemann surface which can be mapped into a complex $w-$ plane by $w = s^{1/m}$ with m is the least common multiple (LCM) of v, in which the arguments, $\varphi_w = |\arg(w)|$ of the closed-loop poles represented by the sector of the principle Riemann sheet as [10],

$$-\frac{\pi}{m} < \varphi_w < \frac{\pi}{m} \tag{7}$$

The segment $|\varphi_s| > \frac{\pi}{2}$ of the complex s-plane which is stable, transforms to $|\varphi_w| > \frac{\pi}{2m}$ in the $w-$ plane. Hence, the system will be stable if the roots of the principle sheet reside in the region where [10],

$$|\varphi_w| > \frac{\pi}{2m} \tag{8}$$

Now, for a Type 0 plant, a FO PI^λ controller is entailed to be administered with the controller in Eq. (2) to eliminate the steady-state error as,

$$C_f(s) = \frac{H_f(s)}{F_f(s)} = \frac{1 + \alpha T s^\beta}{1 + T s^\beta} \cdot \frac{K_P s^\lambda + K_I}{s^\lambda} \tag{9}$$

in which K_I/K_P is placed far below, ω'_{cg}, the phase lag will have negligible effect on the phase of the system to be compensated near ω'_{cg}. In case of any Type 1 plant the compensator in Eq. (2) can be implemented alone. The closed-loop characteristic equation according to Eq. (4) will have some roots near the principal sheet defined by the segment (7) which are actually responsible for the different dynamics and oscillations in the system. To restrain from this, the FO pre-filter for this controller in Eq. (9) is set as,

$$PF(s) = \frac{Q(s)}{F(s)} = \frac{\chi(q_1 s^\beta + 1)(q_0 s^\lambda + 1)}{(T s^\beta + 1)(s^\lambda)} \tag{10}$$

in which the numerical parameters q_1 or q_0 of the polynomial $Q(s)$ properly tuned generates improved transient response in FO 2-DOF framework. If the plant $G(s)$ is Type 1 then the expression of the algebraic terms of $Q(s)$ of the FO pre-filter will be reduce to, $Q(s) = q_1 s^\beta + q_0$ where q_0 is set to establish reference tracking. Robust stability with respect to plant parameter variation can be examined by ensuring $\max_\omega |S(j\omega)| \leq 2, \forall \omega$ for the sensitivity function,

$$S(s) = \frac{A(s).F(s)}{D(s)} \tag{11}$$

It is notable here that the $(1 - S(s))$ is a transfer function from noise to the output.

3 Numerical Example and Simulation Results

A Non-Commensurate FO-LTI system is considered here as [10],

$$G(s) = \frac{5s^{0.6} + 2}{s^{3.3} + 3.1s^{2.6} + 2.89s^{1.9} + 2.5s^{1.4} + 1.2} \tag{12}$$

An FO PI^λ controller in cascade with FO compensator in Eq. (9) is applied here. Now, with the choice of $\omega = \omega_c = \frac{K_I}{K_P} = 2$ rad/s and $\varphi = -5^o$ to -6° at ω_c the explication the numerical term of λ is obtained as 1.3 which makes the FO proportional integral controller as,

$$PI(s^\lambda) = \frac{s^{1.3} + 2}{s^{1.3}} = \frac{2(0.5s^{1.3} + 1)}{s^{1.3}} \tag{13}$$

To relocate the frequency at which the magnitude plot crossed 0 dB at $10 rad/s$ with the Phase Margin (PM) intended to be at 45° the non-integer-order lead compensator is designed as [9],

$$C(s) = \frac{13.326s^{0.99834} + 1}{0.0007561s^{0.99834} + 1} \tag{14}$$

The variation of the magnitude and phase of $C(s).G(s)$ conforming to Eq. (3) is portrayed through Fig. 2a which suggests that the desired frequency domain features are well satisfied. The sensitivity characteristic following Eq. (11) is depicted in Fig. 2b. The closed-loop poles locations satisfying $|\varphi_w| > \pi/2m$, with $m = 100$ in the complex w-plane are visible from the zoomed plot in Fig. 2c which confirms stability [10].

The pre-filter transfer function below is adopted in accord to (10) is determined as,

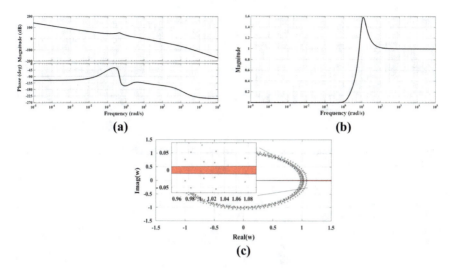

(a)   (b)

(c)

Fig. 2 **a** Variation of magnitude and phase with frequency of $C(s).G(s)$. **b** Sensitivity plot of $|S(j\omega)|$. **c** Position of roots in the w-plane

$$PF(s) = \frac{Q(s)}{F(s)} = \frac{2 \times (13.326s^{0.99834} + 1)(0.4s^{1.3} + 1)}{(0.0007561s^{0.99834} + 1)(s^{1.3})} \tag{15}$$

where, $q_0 \in R^+$ variation of which reduces the peak overshoot and steady-state error as shown in Fig. 3. The performance of the plant in (12) to a standard step input is presented in Fig. 4a using the FO compensator in Eq. (9) shows that the settling time reduced from 0.6 to 0.4 s and the % Overshoot reached is also reduced considerably from 25.6 to 2.4% applying the FO pre-filter $PF(s)$ in Eq. (10) in 2-DOF configuration.

It is observed that the FO pre-filter can cancel out the unwanted zeros and poles of closed-loop transfer function that can render improved system response. The maximum extent of control effort vindicated in Fig. 4b is considerably low for FO 2-DOF controller in variance to the FO 1-DOF scheme.

Fig. 3 System response employing FO 2-DOF controller for variation of q_0

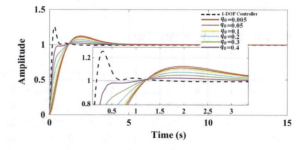

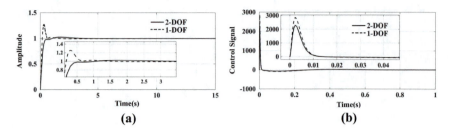

Fig. 4 **a** Performance reaction to step input and **b** Control effort by FO 2-DOF controller compared to FO 1-DOF controller

Fig. 5 Transient response with and without time delay utilizing the FO 2-DOF controller

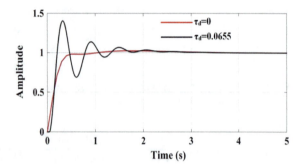

The delay margin with the compensator in Eq. (9) is calculated mathematically as, $\tau_d = (\pi \times PM)/(180 \times \omega_{cg}) = 0.0785$ [9] which predicts that it will be able to bear this uncertainty and preserve the stability of the plant with this time delay, beyond which it will become unstable. To substantiate this, a transport delay block $e^{-s\tau_d}$ in incorporated in the forward path of the closed loop in MATLAB/SIMULINK where, the value of the delay is increased from zero till the system becomes unstable.

This has led to determination of the delay margin through simulation to be 0.0655 s. The nominal step response and the response of the FO plant with the time delay employing the FO controller and the FO pre-filter are displayed through Fig. 5.

4   Concluding Remarks

A comprehensible design perspective of an FO 2-DOF controller for a non-commensurate plant is described. It is observed from the numerical analysis that the incorporation of the 2-DOF strategy together with the FO operators has enhanced the system robustness and the fitting reference tracking by imparting ancillary versatility in the synthesis of the numerical factors of the controller. An effort is thus made here to place on record a scheme satisfying the two-fold purpose of an efficient control with lower control input against plant parameter perturbations in the system.

References

1. M. Li, P. Zhou, Z. Zhao, J. Zhang, Two degree of freedom fractional order PID controller design for fractional order processes with dead time. Elsevier ISA Trans. **61**, 147–154 (2016)
2. J. Dey, T.K. Saha, Design and performance analysis of two degree of freedom (2-DOF) control of DC-DC boost converter, in *IEEE International Conference on Industrial Technology (ICIT)* (2013), pp. 493–498
3. A. Ates, C. Yeroglu, Online tuning of two degree of freedom fractional order control loops. Balk. J. Electr. Comput. Eng. **4**(1), 5–11 (2016)
4. B. Zhang, W. Zhang, Two degree of freedom control scheme for processes with large time. Wiley Asian J. Control **8**(1), 50–55 (2006)
5. H. Taguchi, M. Araki, Two degree of freedom PID controllers-their functions and optimal tuning. Elsevier IFAC Proc. Vol. **33**(4), 91–96 (2000)
6. N. Pachauri, V. Singh, A. Rani, Two degree of freedom fractional order proportional-integral-derivative based temperature control of fermentation process. J. Dyn. Syst., Meas. Control (ASME) **140**, 1–10 (2018)
7. S. Debbarma, L.C. Saikia, N. Nidul Sinha, Automatic generation control of multi-area system using two degree of freedom fractional order PID controller: a preliminary study, in *IEEE (PES) Asia Pacific Power and Energy Engineering Conference (APPEEC)* (2014), pp. 1–6
8. S. Debbarma, L.C. Saikia, N. Sinha, Automatic generation control using two degree of freedom fractional order PID controller. Elsevier Electr. Power Energy Syst. **58**, 120–129 (2014)
9. J. Dey, R. Mondal, S. Halder, Generalized phase compensator of continuous time plants. ISA Trans., Elsevier **81**, 141–154 (2018)
10. C.A. Monje, Y.Q. Chen, B.M. Vinagre, D. Xue, V. Feliu, *Fractional Order Systems and Controls-Fundamentals and Applications*, in Advances in Industrial Control (AIC) (Springer, Berlin, 2010)

Industry 4.0: Communication Technologies, Challenges and Research Perspective Towards 5G Systems

Lalit Chettri and Rabindranath Bera

Abstract The Fifth Generation (5G) is the most promising technology in designing and developing industrial IoT (IIoT). The industrial IoT developed using 5G technology is called as industry 4.0. The Internet plays a major role in industry 4.0 by connecting and communicating between various machines, sensors, devices and controllers. With the context of 5G enabling industry 4.0, new solutions for automation can be achieved by lean production, whenever important industrial resources like finance, labour and material are concerned. So, it is obvious to use lean production to save production and running cost. In this context, the article emphasizes on the key challenges, vision of industry 4.0 using 5G wireless communication technology and 5G framework for industry 4.0. Some of the key technologies like time-sensitive network (TSN), ultra-reliable low-latency communication (URLLC), industrial Ethernet, spectrum usage and low-power wide-area networks (LPWAN) for 5G communication used in industry 4.0 are presented in this article. Moreover, we also provide a research direction in 5G industry 4.0 and its impact on socio-economic and environmental impacts.

Keywords Industry 4.0 · 5G · IoT · Low-power wide-area networks · Smart factories · Automation · Mm wave · MIMO · Time-sensitive network · Industrial Ethernet

1 Introduction

5G is a prime enabling technology to facilitate the industrial transformation to industrial 4.0 by providing seamless communication within the factory via low-power wide-area networks [1]. The connectivity within industry 4.0 architecture is based on global standards. Before introducing 5G and industry 4.0, new concepts and

L. Chettri (✉) · R. Bera
5G IoT Lab, Sikkim Manipal Institute of Technology, Majhitar, Sikkim, India
e-mail: lalitelectricalengineer@gmail.com

R. Bera
e-mail: rbera50@gmail.com

© Springer Nature Singapore Pte Ltd. 2020
R. Bera et al. (eds.), *Advances in Communication, Devices and Networking*,
Lecture Notes in Electrical Engineering 662,
https://doi.org/10.1007/978-981-15-4932-8_9

driving technologies should be known first. The unprecedented reliability, robotic internet, extremely low latency and the comprehensive IIoT connectivity are the prime technological drivers in deploying industry 4.0 [2]. The industry 4.0 is conceptualized as smart factories where information and communication technology (ICT) makes a way for evolution in additive manufacturing in industries. Nowadays, the industrial development has reached the next level of digital transformation and the future factories have been pictured. More recently, the industry 4.0 concepts have become an important issue of discussion between the researchers, industrialist and academicians. Despite the continuous discussion over this topic, there are still some challenges about new manufacturing techniques, business framework, workload and wireless accessibility.

The concept of industry 4.0 allows the companies to make a change and to bring digital evolution in the industrial sector by developing suitable framework for efficient manufacturing processing and control [1]. The technology behind industry 4.0 is Internet of things, which allows every machine, sensor, controller and people are connected between them via low-power wide-area networks (LPWAN) based on 5G wireless accessibility. The goal of industry 4.0 is to maximize efficiency in production and create full transparency in process and control. The purpose of this article is to provide a comprehensive understanding of industry 4.0 based on 5G wireless communication technology. We also described the challenges and potential technological requirements in industry 4.0. We have also provided an overview on how 5G technology can be implemented in digital transformation of industry 4.0. Finally, we have described the use cases of 5G industry 4.0, and the future research directions are discussed and presented with better understanding to the readers.

1.1 Challenges and Benefits of 5G Enabling Industry 4.0

As we are moving into the era of the fourth industrial revolution, which is commonly known as industry 4.0, the industry 4.0 is the use of automation information using 5G internet of things (IoT) technology and exchange of data in the manufacturing process. Of course, industry 4.0 is a great opportunity and has been a platform in realizing 5.0, but every opportunity comes with a challenge. In 5G industry 4.0, IoT will connect every machine, sensors, robots and devices used in the factories and allow the seamless communication and data transmission between them. The industry 4.0 will open a door for a new business model, production management, monitoring and automation. Some of the key challenges in realizing 5G industry 4.0 are

- Interconnection between the devices, machines, robots and humans of all the departments operating the industry is the key challenge in Industry 4.0.
- IoT experts including big data and cloud computing are to be recruited by the industry for proper installation of digital transformation and maintenance.

- The understanding of the business model is a key issue where the organization needs to rethink and process to maximize new things and opportunities.
- High-speed Internet connectivity and ultra-reliable low-latency communication, better spectrum usage and robotic Internet are the key requirements in 5G industry 4.0, and it can only be achieved with the deployment of promising technologies such as 5G new radio (NR), low-power wide-area networks, mm wave communication and MIMO antenna.
- Cyber threat and security measures are serious concerns in industry 4.0, the proper cybersecurity measures and design pattern should be investigated and verified for secured communication between the partnered industry, workers and industrial unit.

Apart from challenges, there are several benefits of industry 4.0. With the help of digital transformation, the benefits of 5G IoT enabling industry 4.0 will ultimately help business plan and production process to monitor smartly and efficiently. Some of the benefits of 5G IoT industry are presented below:

- Enhancement of quality products through optimization and automation. The quality of products can be boosted by real-time monitoring through 5G enabled IoT.
- The continuity in business can be achieved through advance maintenance and monitoring accessibility.
- Faster exchange of real-time data and information between the devices, machines and robots through low power networks. The use of LPWAN can be beneficial in power consumption and energy efficiency.
- Better working condition and less human requirement; because of the use of robots it can be collaborated between robots, devices and humans.
- Due to the better efficiency and reliability of products, customer satisfaction can be justified.

1.2 Proposed Framework for 5G Enabled Industry 4.0

The industry 4.0 virtualizes every 5G technology as a guide towards fully smart factories. In order to achieve full success in digital transformation, it is necessary to prepare a technology framework in a most efficient and accurate manner [1]. The framework should address the technology requirement, its integration and the outcomes with some current situations, and pave the way for 5G industry 4.0. In this section, we propose a framework for 5G technology that enables industry 4.0. The proposed framework consists of three phase: technological phase also called as strategy phase, integration phase and sustainable outcomes as shown in Fig. 1 [1–4].

In the technological phase, the collective planning and procedure are executed, while preparing the framework for industry 4.0, complete analysis for next year has to be considered. Many industries are already establishing their industry with digitally

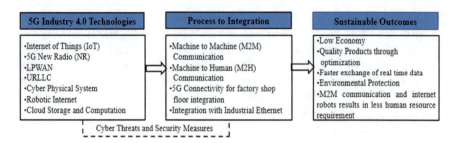

Fig. 1 Proposed framework of 5G enabling Industry 4.0

equipped smart factories. While designing a plan and proposal they must consider some parameters such as clients, suppliers and technology partners. In the technological phase, some of the recent 5G technologies that enable industry 4.0 have to be investigated. Some of the key 5G technologies for industry 4.0 are 5G NR, IoT, low-power wide-area networks (LPWAN) and ultra-reliable low-latency communication (URLLC). While considering these technologies, the industries should consider new application organization value chain.

In the process development phase, we have considered the integration of some key 5G innovation such as machine to machine (M2M), and machine to human (M2H) communication and integration to industrial Ethernet. In this phase, industrial automation can be achieved with the key enabler without human interruption. The Ethernet connection will enable transmission of data at a faster rate and communication between the industrial persons can be achieved. In sustainable outcome, we have considered some key benefits of industry 4.0 such as low economy, exchange of real-time data, environment protection, less requirement of human resources and delivery of quality products through optimization and operation.

1.3   5G Technology Drivers for Industry 4.0

5G wireless technology plays a major role in realizing industry 4.0. 5G enables the industrialist to overcome some of the existing challenges by making fully digitally equipped smart factories. Some of the key 5G technology drivers for industry 4.0 are discussed below.

1.3.1   Ultra-Reliable Low-Latency Communication (URLLC)

URLLC is one of the service requirements of 5G wireless communication technology with suitable requirements of latency availability and reliability. 5G NR is a major technology used in URLLC and it will pave the way for large-scale deployments of advance antenna that enables beamformation and MIMO technology. MIMO and mm wave communication technology in 5G NR is considered as the powerful tool for increasing throughput, coverage and capacity. MIMO is important in URLLC because it improves reliability. The scalable numerology will be the key parameter in achieving low latency and large sub-carrier spacing reduces the time interval reduction in latency and reliability of MAC and physical layer. The heterogeneous networks (HetNets) are introduced in 3GPP release 15 and release 16. The purpose of 3GPP release 16 is to achieve 0.5–1 ms latency with 99.99% reliability which is the key consideration for industrial automation. The latency requirements for motion control for machines are <1 ms with the average of >100 devices connected covering 100 m*100* 30 m surface area. The time cycle requirement for industrial robots is 1 ms with <1 km^2 typical surface area.

1.3.2 Integration with Industrial Ethernet and 5G NR

Many industries are already substantially automated with the help of new technology to pursue better efficiency and greater production speed. 5G can be an enabler for new operating models in industry 4.0 [1]. In realizing 5G industry 4.0, firstly, 5G should separate the functionality of wired system to support controllers, sensors, switches and actuators. Secondly, 5G should be made as the integral part of the industrial automation and integration. There are many existing Ethernet system such as Profinet and EtherCAT, which is designed for industrial communication and real-time robotic equipment control. In this technology, the configuration of production is fixed and it is not possible to reconfigure. The solution for this problem is to replace such existing wired communication by 5G wireless networks called heterogeneous networks (HetNets). The deployment of 5G industry 4.0 is governed by 5G alliance for connected industries and automation (5G-ACIA). The industry 4.0 also uses time-sensitive networks (TSN) for developing industrial networks and the integration of 5G NR and TSN will greatly improve the development in industrial internet connectivity.

1.3.3 Spectrum for Industrial Users and Interference Control

The availability of spectrum resources usage is the key requirement in achieving capacity, bitrates and latency. The reliable network operating in 6 GHz can be operational in unlicensed frequency bands. To provide reliable service in 5G, the spectrum resource needs to be managed efficiently. The performance of 5G NR depends upon factors such as spectrum band, duplex mode, RAT, and coexistence scenario that are used in spectrum [5, 6]. Mid band spectrum (2–6 GHz) is very well suitable for indoor deployments, because it provides good coverage with limited set of transmission points. In 5G NR, mm wave frequencies are highly preferable because in this frequency band both licensed and unlicensed spectrum are available for usage. The unlicensed spectrums are sensed by cognitive radio technology by using MIMO antenna which is capable of beamformation. Some of the advance 5G technologies such as antenna array design, beamformation, beam tracking and tracing, and coordinated multipoint (CoMP) are responsible in overcoming the challenges of spectrum usage and system design with the opportunity to deploy 5G NR in industrial use cases.

1.3.4   Proposed 5G Architecture for the Factory Shop Floor

Different factories have adopted 5G networks for wireless automation. The factory shop floor with 5G connectivity allows communication between different sensors, devices, machines, robots and each terminal are connected to coordinate and share data. The devices used in 5G shop floor can be directly connected or connected through gateways. The shop floor can support massive machine type communication (mMTC) and can easily be integrated with the service provider through low-power wide-area networks (LPWAN) [7, 8]. The typical architecture of factory for the future with 5G connectivity shop floor is shown in Fig. 2 [8–13].

The optimal local connectivity solution requires a well-planned 5G network with a desired 5G radio system using licensed spectrum to enable ultra-low latency. The virtualization of core networks supports and controls the user plane. The core user network user plane is necessary to deploy in the factory to achieve high availability of spectrum resource security and privacy. The local management system is required to monitor and manage end to end connectivity between the devices and local network architecture. The management system also needs to integrate with other elements of the operating technology (OT) system and the IT industries.

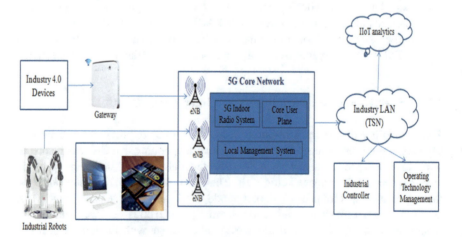

Fig. 2 5G connectivity solution for the future factory shop floor

1.3.5   Mapping Time-Sensitive Network (TSN) to 5G NR

TSN is similar to 5G NR, which uses time slot reservation like the numerology and frame structure used in 5G NR. Some of the technologies like transporting and Ethernet payload, time-sensitive process are not the part of 5G NR and 3GPP. Hence it is necessary to map time-sensitive network (TSN) to 5G new radio (NR). If 5G and TSN are mapped many industries will be benefitted from economies of scale, innovation and deployment of industry 4.0 with open standards. The TSN 5G integration is summarized in three key points.

Integration between TSN AND 5G core network: The TSN is used to generate information and control and it should be passed through 5G network. The critical TSN information ensuring quality of service (QoS) should be transmitted to the 5G system for proper verification and installation of networks in industry 4.0.

Integration between 5G device (UE) and Industrial equipment: 5G NR with Ethernet adopter must be integrated. For ensuring better industrial optimization, the new industrial equipment is integrated to the existing equipment or replaced with the new. Integration within the 5G system: In order to achieve QoS for TSN, Ethernet frame must be mapped to 5G frame. The 5G NR must also deliver a very precise time signal generated by master clock in TSN system.

2   Role of 5G in Industry 4.0

Robotics is a part of industry 4.0. The industrial markets including autonomous robots 5G wireless technologies are growing which is a faster rate for deploying industry 4.0. The robots used in industrial operation are automated machines with embedded intelligence and are used in effective and robust manufacturing process [14, 15]. To ensure more production at less time, the industries have shifted to use robots instead of human beings. The industrial robots are designed for certain allocated task and cannot be reconfigured for other operations. In addition, the robots are to automatically detect the failure and maintenance. The robots are designed so intelligently with the collaboration with humans and other robots which allow self adaptable on new production process and manufacturing. The production and industrial automation have become very efficient due to collaboration of robots to some promising technologies like cloud computing, big data analytics. In successful assemble of robots, some technological components are required such as actuators and sensors machines. Since smart sensors are used in data collection, processing and sharing of data that are stored in cloud and industry 4.0 will fully rely upon advance smart sensors.

Sensors are for visual perception, sensing and obstacle detection; sensors are used in robotic operation in industrial use. The other technological requirements in robots are artificial intelligence [15]. This technology enables robots to monitor, understand and optimize the production process. Thus the robots are developed with self awareness and maintenance ability. With the use of cloud computing and big data analytics, the industrial robots do not only assist humans, but it helps in monitoring production process with its ability to analyze the situation and make decisions. In Internet of robotics (IoR) all the devices, sensors, robots are connected to each other for sharing information. In cloud robotics, the robots will access the information from big data and cloud with further processing as per the industrial requirements. The data from the clouds are used to access the robots. The cloud robotics is used to connect the physical layer things such as sensors, robots, devices, over a cloud through a local server. The Internet of robots is classified into five layers and is discussed below.

Robotic Things Layer: It consists of hardware like robots, sensors, machines, devices, vehicles, etc.

Network Layer: Low-power wide-area networks (LPWAN) like Wi-Fi, ZigBee, Bluetooth, LoRa and NB-IoT are considered in this layer [16].

Internet Layer: This layer is responsible for overall communication within the networks.

Infrastructure Layer: Cloud computing and big data analytics are considered in this layer [9, 17, 18].

Application Layer: Application of robots such as monitoring, human assistance, controlling, automation and production are considered in this layer.

2.1 Research Direction and Future Scope

The demand of 5G industrial IIoT or Industry 4.0 is to provide fully digitally equipped smart factories. It is necessary to address the key challenges, its benefits and driving technologies that support 5G connectivity in industry 4.0. In this section, we present some key challenges and future research directions.

2.1.1 Securing 5G Networks Against Cyber Threats

Cybersecurity is a major challenge in 5G connectivity for securing data and information [1]. Cyber-attack could be devastating to the industries' administration. The cyber-attacks must be detected quickly and prevented with the security plan and measures. Firstly, the industries must have a complete framework of digital assets to develop an understanding of the networks. Secondly, the industries should have an authentication process to guard the physical and digital system. Thirdly, the industries should have strategies to detect and avoid the cyber threats. The effective way to overcome the challenges of cyber threat is to deploy a continuous monitoring system. The industries should have a contingency plan to quickly respond and recover from a potential cyber-attack. The other research measures in securing industry 4.0 from cyber threat are not to connect the machines, sensors, devices and control network directly. This allows attackers to find a gap and give physical damages to the control system. The precaution measures in securing industry 4.0 are to use firewall a software program, tray and reduce the number of routes as much as possible to restrict the traffic between the networks. The use of virtual private network (VPN) is another security measure to reduce the number of IP addresses. The strong password deployment could be other preventive measures in industry 4.0

2.1.2 Designing Low Power Device Networks

Low power consumption is the key requirement in industry 4.0. Not all technologies and network protocols stay fit in every scenario of industry 4.0. The industrialist should be aware of connectivity about wired or wireless communication while designing networks. We need to understand the technology and protocols that are suitable in MTC application in industry 4.0. Since, real-time communication is vital in industries, so we must adopt advance technology and service level arrangements. So, to overcome the challenges of connectivity, we must design a network with industrial Ethernet which uses more robust cabling and connectors. The low power networks can also be designed using low-power wide-area networks (LPWAN) devices such as ZigBee, LoRa, Bluetooth, Wi-Fi and NB-IoT. These devices are responsible to transmit the data in a timely and efficient way and some improvements should be made in the existing network architecture.

2.1.3 Data Analytics for Factory Automation

Through the data analytics of industry 4.0, we can understand the system operational state and system overall performance. We can also identify the pattern of behaviour of machines, sensors and devices under various conditions. This helps us to understand how to optimize machine uptime and efficiency to increase throughput. The critical problems are missed due to insufficient diagnostics resulting in system failure and costly repair. With the use of data analytics process, production line sensor, optimal data are analyzed perfectly [7]. The production issue can be identified before system failure is resolved quickly to avoid production interruption. The data analytics are categorized into four types. Firstly, the descriptive analysis in industry 4.0 is used for a good picture of what is happening in the factories. A useful data of visualization tool must be used to enhance the understanding. Secondly, the diagnostics analysis is used to determine the state of processed data. It is also used to identify the problems in the proper functioning of the industries. Thirdly, the predictive analytics uses the algorithm and past data which helps in predicting the future of factories. It is used to predict the failure in devices, machines and sensors for preventive maintenance. Lastly, the prescriptive analysis is used for a complete understanding of the factory operation. And it identifies the next set of action, this enables the analyst to recommend the necessary action to improve factory manufacturing operation. So, for a complete data analytics process, the designer should proceed with an algorithm that follows the above four analysis process.

2.1.4   5G Connectivity in Smart Factories

The role of 5G networks in industry 4.0 is vital in factory automation. While designing 5G systems for industries, we must have a complete understanding about the technology that drives 5G wireless communication system. 5G and IoT will be key in enhancing and enabling the technologies in processing and manufacturing. 5G networks help industrialists to build smart factories and take advantage of promising technologies like automation, artificial intelligence, augmented reality, internet of things (IoT) and internet of robots (IoR). Key aspects of 5G like low latency, spectrum usages, interference cancellation and data rates are key requirements to support critical application in factories. Thus, 5G technology will allow high flexibility, low-cost deployment and lesser time for factory floor production reconfiguration. In industry 4.0, 5G will enable the operator to address some technical challenges like industrial control, maintenance, planning and design. The selection of low power devices and collaboration of human to robots and robots to robots within 5G architecture enables factories for efficient production, maintenance and delivering of quality products. Therefore, it is necessary to have a complete understanding about the waveform selection, numerology and spectrum allocation for successful development of 5G core networks for industrial applications. Spectrum usages, low latency and higher data rates can be achieved by use of mm wave communication technology and MIMO antenna. Thus, it is recommended deploying 5G with smart technologies for better efficiency and reliability. The integration of 5G with new radio (NR) and time-sensitive network (TSN) must be adopted for realizing smart factories in industry 4.0.

3   Conclusion

Industry 4.0 is growing actively in all parts of the globe. In this section, we summarize the 5G technologies that govern industry 4.0. The industrial IoT is faced with many problems, such as low efficiency of resource use and maximum human requirement. Thus, to overcome these challenges, industry 4.0 can be realized using 5G wireless technology. Industry 4.0 with 5G has numerous advantages, such as low latency communication efficient energy consumption, smart monitoring and control, and delivering of quality products. The emerging issues about industry 4.0, its key driving technologies, framework and deployment of smart factories using 5G are described in this article. It is necessary to reduce latency and ensure massive connectivity of devices, sensors, machines and high data rates for successful deployment of efficient industrial automation. The benefits and challenge of industry 4.0 are outlined in this paper. The key technology, its deployment model and enabler of 5G system are highlighted in this article. Moreover, the areas of application in industry 4.0 are discussed. The technology roadmap and research direction are also provided in this paper and it helps the reader to realize 5G industry 4.0 where automation is a critical requirement.

References

1. A. Ustundag, E. Cevikcan, *Industry 4.0: Managing the Digital Transformation*. Springer Series in Advance Manufacturing (2018). ISBN 978-3-319-57869-9 ISBN 978-3-319-57870-5 (eBook) https://doi.org/10.1007/978-3-319-57870-5
2. A. Gilchrist, *Industry 4.0: The Industrial Internet of Things* (2016) ISBN-13 (pbk): 978-1-4842-2046-7 ISBN-13 (electronic): 978-1-4842-2047-4, https://doi.org/10.1007/978-1-4842-2047-4
3. H. Shariatmadari et al., Machine-type communications: current status and future perspectives toward 5G systems. IEEE Commun. Mag. **53**(9), 10–17 (2015)
4. M. Blanchet, T. Rinn, G. von Thaden, G. de Thieulloy, Industry 4.0: the new industrial revolution: how Europe will succeed, in Think Act, Roland Berger Strategy Consultants GmbH (2014)
5. L. Lyu, Student Member, IEEE, C. Chen, Member, IEEE, S. Zhu, Member, IEEE, X. Guan, Fellow, IEEE, 5G enabled codesign of energy-efficient transmission and estimation for industrial IoT systems. IEEE Trans. Ind. Inform. **14**(6) (2018)
6. B. Holfeld et al., Wireless communication for factory automation: an opportunity for LTE and 5G systems. IEEE Commun. Mag. **54**(6), 36–43 (2016)
7. S. Kurt, H.U. Yildiz, M. Yigit, B. Tavli, V.C. Gungor, Packet size optimization in wireless sensor networks for smart grid applications. IEEE Trans. Ind. Electron. **64**(3), 2392–2401 (2017)
8. S. Augustsson, J. Olsson, L.G. Christiernin, G. Bolmsjö, How to transfer information between collaborating human operators and industrial robots in an assembly. Nordichi **286–294**, 2014 (2014)
9. Gartner Newsroom, Press release: Gartner says 6.4 Billion Connected 'Things' will be in use in 2016, up 30% from 2015 (2015). http://www.gartner.com/newsroom/id/3165317
10. D. Mishra, A. Gunasekaran, T. Papadopoulos, S.J. Childe, Big data and supply chain management: a review and bibliometric analysis. Ann. Oper. Res. (2016), 1–24
11. GPP TS 22.368, *Service Requirements for Machine-Type Communications (MTC)*, vol. 13.1.0 (2014). http://www.3gpp.org. Accessed 30 Apr 2017
12. N. Al-Falahy, O.Y. Alani, Technologies for 5G networks: challenges and opportunities. IT Prof. **19**(1), 12–20 (2017)
13. P. Agyapong, M. Iwamura, D. Staehle, W. Kiess, A. Benjebbour, Design considerations for a 5G network architecture. IEEE Commun. Mag. **52**(11), 65–75 (2014)
14. E. Borgia, The internet of things vision: key features, applications and open issues. Comput. Commun. **54**, 1–31 (2014)
15. A. Bicchi, M.A. Peshkin, J.E. Colgate, Safety for physical human robot interaction, in *Springer Handbook of Robotics Heidelberg* (Germany, Springer, 2008), pp. 1335–1348
16. A. Cherubini, R. Passama, A. Meline, A. Crosnier, P. Fraisse, Multimodal control for human-robot cooperation, in *IEEE/RSJ International Conference on Intelligent Robots and Systems (IROS)* (2013), pp. 2202–2207
17. M.A.K. Bahrin, M.F. Othman, N.H.N. Azli, M.F. Talib, Industry 4.0: a review on industrial automation and robotic. J. Technol. 137–143 (2016) (198 B. Bayram, G. İnce)
18. J. Moyne, J. Iskandar, Big data analytics for smart manufacturing: case studies in semiconductor manufacturing. Processes **5**(3) (2017). http://dx.doi.org/10.3390/pr5030039

Performance Characterization of a Microstrip Patch Antenna on Multiple Substrate

Anukul Jindal, Khushal Kapoor, Tanweer Ali, Omprakash Kumar, and M. M. Manohara Pai

Abstract The paper deals with the design, analysis, and characterization of a rectangular microstrip patch antenna on different substrate for various parameters like S11, bandwidth, gain, and directivity. The various substrate used for designing the antenna are FR4 epoxy, RT/Duroid 5880, Taconic RF-60, teflon, polyimide with an operating frequency of 2.9 GHz (± 3%). The dimension of the radiating patch area is kept same, while the height is varied with the substrate. Analysis of the study helps with the selection of a particular substrate (considering parameters like gain, bandwidth, directivity, return loss) for the corresponding frequency range in order to meet necessary wireless applications requirements.

1 Introduction

The antenna is one of the critical components in any wireless communication system. Owing to its versatility and numerous advantages, microstrip antenna finds widespread usage in present and future wireless scenario [1, 2]. It consists of a radiating patch on one side of a dielectric substrate and has a ground plane on the other

A. Jindal · K. Kapoor · T. Ali (✉) · O. Kumar
Department of Electronics and Communication Engineering, Manipal Institute of Technology,
Manipal Academy of Higher Education, Manipal, India
e-mail: tanweer.ali@manipal.edu

A. Jindal
e-mail: anukuljindal001@gmail.com

K. Kapoor
e-mail: khushal29kapoor@gmail.com

O. Kumar
e-mail: omprakash.kumar@manipal.edu

M. M. Manohara Pai
Department of Information & Communication Technology, Manipal Institute of Technology,
Manipal Academy of Higher Education, Manipal, India
e-mail: mmm.pai@manipal.edu

© Springer Nature Singapore Pte Ltd. 2020
R. Bera et al. (eds.), *Advances in Communication, Devices and Networking*,
Lecture Notes in Electrical Engineering 662,
https://doi.org/10.1007/978-981-15-4932-8_10

side [3]. Microstrip patch antenna has wide application in the field of communication making them one of the most preferred antennas since they can operate on frequencies where other antennas are not feasible to design, they are easy to fabricate and can be easily etched on a PCB [4–6].

The first step in designing a microstrip antenna for any frequency band, is to choose an appropriate substrate. The substrate is generally required for the mechanical support of the antenna [6, 7]. To provide this support, the substrate should consist of a dielectric material, which may affect the electrical performance of the antenna [7–10]. With increase in dielectric constant both the resonant frequency as well as the bandwidth decreases. With increase in thickness of the substrate, the fringing increases which decreases the resonating frequency of the antenna. While parameters like gain, return loss, and bandwidth improves significantly [11].

Different substrates are used by the manufacturers to design and operate the patch antenna for a particular operating frequency range. In this research, we present a design of patch antenna for different substrates (FR4 epoxy, RT/Duroid 5880, Taconic RF-60, Teflon and Polyimide) with an operating frequency of 2.9 GHz ($\pm$ 3%). Slots are etched on the patch of the antenna in order to obtain a design model with desired results. The study provides valuable information to the antenna designer as one can select an appropriate design considering the various parameters.

2 Properties of the Substrates Used

2.1 Teflon

Polytetrafluoroethylene (PTFE) is one of the best materials to be used where low coefficient of friction is required [3]. It exhibits a very low loss tangent (0.001 at 10 GHz) and negligible water absorption characteristics making it the material of choice for military applications. With a very good resistance to processing chemicals, and high temperature resistance, PTFE substrates were able to meet all the technical requirements of RF/wireless design [7, 8].

2.2 Polyimide

Polyimide is a polymer of imide with its high heat resistance it is used as fuel cells, flexible displays. Polyamide can be prepared by reacting dianhydride and diamine or dianhydride and diisocyanate [3]. Due to their thermal stability and excellent chemical resistance, they make an ideal material to be used as substrate.

2.3　RT/Duroid 5880

RT/Duroid 5880 high-frequency laminates are PTFE composites reinforced with glass microfibers. The randomly oriented microfibers result in exceptional dielectric constant uniformity [3]. Due to its low water absorption characteristics, RT/Duroid 5880 laminates are ideal for applications in high-moisture environments. They are resistant to all solvents and reagents, hot or cold, normally used in etching printed circuits or in plating edges and holes [4–6]. RT/Duroid 5880 laminates have very low dielectric constant, and low dielectric loss making them well suited for high broadband applications where dispersion and losses need to be minimized.

2.4　FR-4 Epoxy

FR stands for flame retardant is a NEMA-grade designation for glass-reinforced epoxy laminate material. It is a high-pressure thermoset plastic laminate grade with good strength to weight ratio. It is also flame resistant due to its bromine content, a nonreactive halogen used in flame retardants. In prototyping where circuits in initial testing stages may be pushed to extreme temperature, it is advantageous to use FR-4 as a PCB material [2–4].

2.5　Taconic RF-60

RF-60As is an organic ceramic fiberglass-reinforced laminate. This product's unique composition results in low moisture absorption and uniform electrical properties. RF-60A exhibits very low moisture absorption, exceptional interlinear bonds, and solder resistance [8, 11] (Table 1).

3　Design of Patch Antenna Using Multiple Substrate

On simulating the antenna designed using microstrip design equations as given in [3], it is observed that the antenna does not resonate at the desired frequency. In order to shift it back to the desired frequency, slots are cut on the patch of the antenna. For the

Table 1 Characteristic properties of different substrates

Parameters	Teflon	Polyimide	RT/Duroid 5880	FR-4 Epoxy	Taconic RF-60
Dielectric constant	2.1	3.5	2.2	4.4	6.15
Loss Tangent	0.001	0.008	0.0009	0.02	0.0028

antenna to work properly, a resonating frequency of 2.9 GHz ($\pm 3\%$) is required with a return loss of more than -10 dB. RT/Duroid 5880 is taken as a reference substrate and the physical dimension of the patch is calculated using the design equations [3].

3.1 Teflon

The antenna designed using teflon substrate having dielectric constant -2.1, density -2.2gm/cm^3, and a loss tangent -0.001 is depicted in Fig. 1a. Initially, a rectangular patch antenna was taken which shows no operation (since S11 is not below -10 dB). In order to make antenna exhibit useful resonance, a slot is cut at the top-right section of the patch as depicted in Fig. 1b. The design provides better return loss than Antenna 1 but its S11 is still not below -10 dB. Finally, cutting one more slot on the top-right section of the patch as illustrated in Fig. 2, desired model is obtained with S11 equals to -12.9 dB (Fig. 3a).

Fig. 1 Microstrip patch antenna using Teflon. **a** Antenna 1, **b** Antenna 2

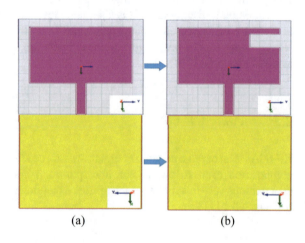

(a)   (b)

Fig. 2 Final design of microstrip patch antenna using Teflon substrate

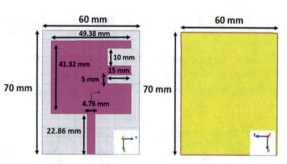

Fig. 3 S11 for different iterations

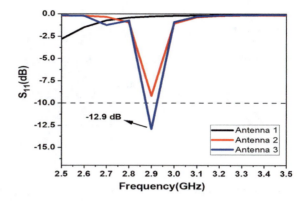

3.2 RT/Duroid 5880

The antenna designed using RT Duroid 5880 substrate having dielectric constant − 2.2, density −2.2 gm/cm³, and a loss tangent −0.0009 is depicted in Fig. 4a. Initially, a rectangular patch antenna was taken which shows no operations. In order to make antenna show useful resonance, a slot is cut at the bottom-left section of the patch as depicted in Fig. 4b. The design provides better return loss than Antenna 1 but its S11 is still not below −10 dB. On cutting one more slot in the mid-region of the patch (Fig. 4c), the desired result is still not obtained. Finally, on cutting one more slot near the top most region of the patch as illustrated in Fig. 5, desired model is obtained with S11 −17.5 dB (Fig. 6a).

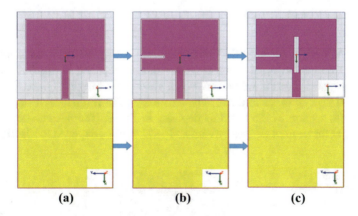

(a)   (b)   (c)

Fig. 4 Microstrip patch antenna using RT/Duroid 5880 with, **a** Antenna 1 (no slots cut), **b** antenna 2 (only one slot cut), **c** antenna 3 (two slots are cut)

Fig. 5 Final design of
microstrip patch antenna
using RT/Duroid 5880
substrate (Antenna 4)

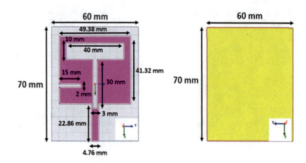

Fig. 6 S11 for different
iterations

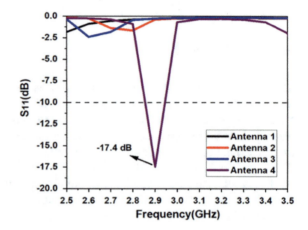

3.3 FR4 Epoxy

The antenna designed using FR4 Epoxy substrate having dielectric constant −4.4,
density −1.85 gm/cm^3, and a loss tangent −0.02 is depicted in Fig. 7. Initially, a
rectangular patch antenna was taken which shows no operation (since S11 is not
below −10 dB). In order to make antenna show useful resonance, a slot is cut at
the top-right section of the patch as depicted in Fig. 8. The design shows useful
resonance at the desired frequency with S11 −17.0 dB as depicted in Fig. 9a.

Fig. 7 Microstrip patch
antenna using Fr4 epoxy
with no slots cut (Antenna 1)

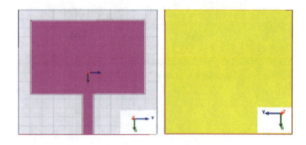

Fig. 8 Final design of microstrip patch antenna using Fr4 epoxy substrate (Antenna 2)

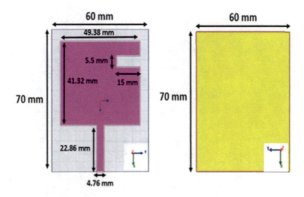

Fig. 9 S11 for different iterations

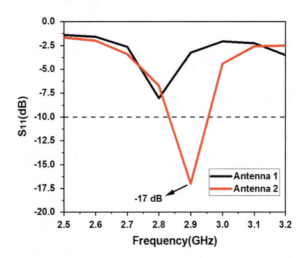

3.4 Taconic

The antenna designed using Taconic substrate having dielectric constant −6.15, density −2.8 gm/cm³, and a loss tangent −0.0028 is depicted in Fig. 10a. Initially, a rectangular patch antenna was taken which shows no operation (since S11 is not below −10 dB). In order to make antenna show useful resonance, a slot is cut at the top-left section of the patch as depicted in Fig. 10b. The design provides better return loss than Antenna 1 but its S11 is still not below −10 dB. On cutting one more slot at the top-right section of the patch (Fig. 10c), the desired result is still not obtained. Finally, on cutting one more slot near the bottom most region of the patch as illustrated in Fig. 11, desired model is obtained with S11 −15.8 dB (Fig. 12a).

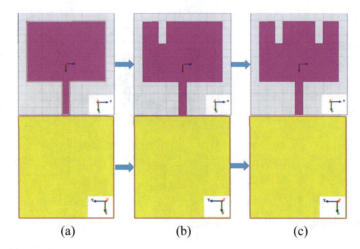

Fig. 10 Microstrip patch antenna using Taconic with, **a** Antenna 1 (No slots cut), **b** antenna 2 (Only one slot cut), **c** antenna 3 (Two slots are cut)

Fig. 11 Final design of microstrip patch antenna using Taconic substrate (Antenna 4)

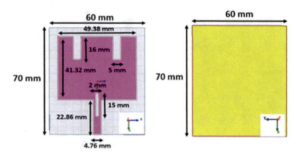

Fig. 12 S11 for different iterations

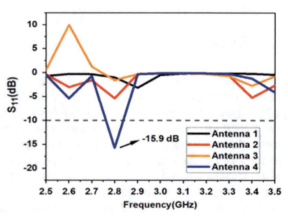

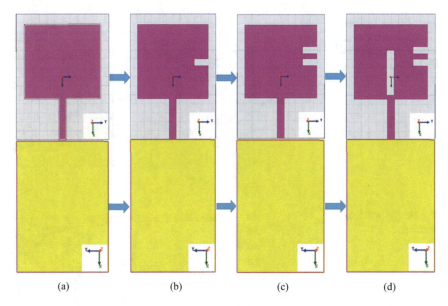

(a)   (b)   (c)   (d)

Fig. 13 Microstrip patch antenna using Polyimide with, **a** Antenna 1 (No slots cut), **b** antenna 2 (Only one slot cut), **c** antenna 3 (Two slots are cut), **d** antenna 4 (Three slots are cut)

3.5 Polyimide

The antenna designed using Taconic substrate having dielectric constant −3.5, density −1.42 gm/cm³, and a loss tangent −0.008 is depicted in Fig. 13a. Initially, a rectangular patch antenna was taken which shows no operation (since S11 is not below −10 dB). In order to make antenna show useful resonance, a slot is cut in the mid-region of the patch as depicted in Fig. 13b. The design provides better return loss than Antenna 1 but its S11 is still not below −10 dB. On cutting one more slot at the top-right section of the patch (Fig. 13c), the desired result is still not obtained. On further cutting one more slot on the patch as illustrated in Fig. 13d, no useful resonance is obtained. Finally, on cutting one more slot near the bottom most region of the patch as illustrated in Fig. 14, desired model is obtained with S11 −26.1 dB (Fig. 15a).

4 Results

The various microstrip antenna configurations using multiple substrate is designed using HFSS 13.0, using FEM method. The summarized results of each substrate are given below in Table 2.

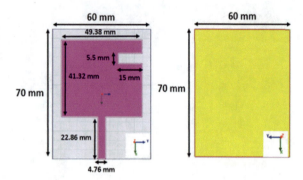

Fig. 14 Final design of microstrip patch antenna using Polyimide substrate (Antenna 5)

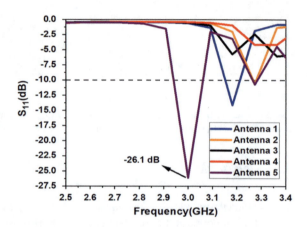

Fig. 15 S11 for different iterations

Table 2 Simulation results of microstrip antenna using different substrates

Parameters	Substrate dimension (mm³)	Operating frequency (GHz)	Return loss (dB)	Gain (dB)
RT/Duroid 5880	70 × 60 × 1.6	2.9	−12.9	1.26
FR-4 Epoxy	70 × 60 × 1.5	2.9	−17.0	−6.41
Taconic RF-60	70 × 60 × 0.64	2.8	−17.4	−5.69
Teflon	70 × 60 × 1	2.9	−13.0	−3.34
Polyimide	70 × 60 × 1	3.0	−26.1	−1.93

5   Conclusion

A rectangular microstrip patch antenna is designed using different substrates and then simulated. It is observed that antenna designed using polyimide substrate provides best S11 and directivity. Whereas antenna designed using FR4 epoxy substrate shows the highest value of bandwidth. Antenna designed using RT/Duroid 5880 substrate has highest gain (dB). The analysis of the optimize parameters can guide the user with the most appropriate design in order to meet the specific wireless requirements.

References

1. H. Kaschel, C. Ahumada, Design of a triband antenna microstrip for 2.4 GHz, 3.5 GHz and 5.7 GHz applied a WBAN, in IEEE Conference on Chilecon, (Oct 2017), pp. 1–7
2. M.Y. ElSalamouny, R. M. Shubair, Novel design of compact low-profile multi-band microstrip antennas for medical applications, in IEEE Antennas and Propagation Conference, (Nov 2015), pp. 1–4
3. C.A. Balanis, Antenna theory: analysis and design. Microstrip Antennas, 3rd edn. (Wiley, 2007)
4. A. Alhaddad, R. Abd-Alhameed, D. Zhou, C. See, I. Elfergani, P. Excell, Low profile dual-band-balanced handset antenna with dual arm structure for wlan application. Microwaves Antennas Propag. IET **5**(9), 1045–1053 (2011)
5. T. Ali, R.C. Biradar, A miniaturized decagonal Sierpinski UWB fractal antenna. Prog. Electromagnet. Res. **84**, 161–174 (2018)
6. T. Ali, M.S. AW, R. C. Biradar, A. Andújar, J. Anguera, A miniaturized slotted ground structure UWB antenna for multiband applications, Microwave Opt. Technol. Let. **60**(8), 2060–2068 (2018)
7. S.A. Mohammad, M.M. Khaleeq, T. Ali, R.C. Biradar, A miniaturized truncated ground plane concentric ring shaped UWB antenna for wireless applications, in 2017 2nd IEEE International Conference on Recent Trends in Electronics, Information and Communication Technology (RTEICT) (pp. 116–120), IEEE, 2017
8. T. Ali, S.A. Mohammad, R.C. Biradar, A novel metamaterial rectangular CSRR with pass band characteristics at 2.95 and 5.23 GHz. in 2017 2nd IEEE International Conference on Recent Trends in Electronics, Information and Communication Technology (RTEICT) (pp. 256–260), IEEE, 2017
9. M.K.A. Rahim, M.N.A. Karim, T. Masri, O. Ayop, Antenna array at 2.4 GHz for wireless LAN system using point to point communication, in 2007 APACE 2007, Asia-Pacific Conference on Applied Electromagnetics, pp. 1–4, IEEE, 2007
10. R. Nema, A. Khan, Analysis of five different dielectric substrates on microstrip patch antenna. **55**(14) (2012)
11. R. Masood, C. Person, R. Sauleau, A dual-mode, dual port pattern diversity antenna for 2.45-GHz WBAN. IEEE Antennas Wirel. Propag. Lett. **16**, 1064–1067 (2017)

Gender Differences in Patients' Assessment of Positive Outcomes of Online Direct-to-Consumer Promotion

Jaya Rani Pandey, Saibal Kumar Saha, Samrat Kumar Mukherjee, Vivek Pandey, and Ajeya Jha

Abstract Gender difference in help-seeking behavior has been an important research discussion and, generally, it has been discovered that men seek less help than women do. This is also true for seeking healthcare information where men generally have been found to be more reluctant than women in seeking health advice. This paper explores the gender difference that patients harbor vis-a-vis positive outcomes of online direct-to-consumer promotion (DTCP) of pharmaceutical products. The hypothesis is that *women find the facility provided through the arrival of internet more positively than their male counterparts*. Opinion of patients were taken with the help of interview. On the basis of this exploratory study, variables were developed. With the help of a 5-point Likert scale, these variables were incorporated in an interview schedule in order to determine the quantitative belief of patients. Only patients taking recourse to allopathy were taken as sample respondents for this research. The study concludes that women are more positive regarding online health search than men. The implications of the result are that promoters of DTCP are also advised to pay due weightage to the differing needs of genders and orient their promotion campaigns accordingly.

Keywords Key words · Gender difference · Positive outcomes · Direct-to-consumer promotion · Patients · Educative · Side effects

J. R. Pandey · S. K. Saha · S. K. Mukherjee · V. Pandey · A. Jha (✉)
Sikkim Manipal Institute of Technology, Sikkim Manipal University, Majitar, Sikkim, India
e-mail: ajeya611@yahoo.com

J. R. Pandey
e-mail: jayaranim@rediffmail.com

S. K. Saha
e-mail: saibal115@gmail.com

S. K. Mukherjee
e-mail: samrat.k@smit.smu.edu.in

V. Pandey
e-mail: vivekgorkhachettry@gmail.com

© Springer Nature Singapore Pte Ltd. 2020
R. Bera et al. (eds.), *Advances in Communication, Devices and Networking*,
Lecture Notes in Electrical Engineering 662,
https://doi.org/10.1007/978-981-15-4932-8_11

1   Introduction

Gender differences in seeking help have been studied by many researchers and who generally find that men seek less help than women do (Cadaret and Speight [1]; Wasylkiw and Clairo [2]; Brenner et al. [3]). This is particularly more prevalent in health-seeking behavior where men generally have been found to be more reluctant than women in seeking health advise. It may, therefore, be assumed that men and women will differ in their perceptions on the positive outcomes of online healthcare information that is generally being supported by pharmaceutical marketers. This paper explores the gender difference that patients harbour vis-a-vis positive outcomes of online DTCP of pharmaceutical products. The probable reasons could be that men are more reluctant healthcare interactions due to embarrassment and they are more inclined to endure symptoms much longer than women, particularly if the symptoms are relatively unrestrictive [4].

2   Literature Review

Gender differences in assessment of online healthcare information has concerned the attention of number of researchers across the world, particularly the developed nations (Joseph et al. [5]; Kontos et al. [6]; Mukherjee et al. [7]; Joseph et al. [8]; Bidmon and Terlutter [9]; Fox et al. [10]; Gauld and williams [11]; Yan [12]; Turget [13]; Mitchell and Boustani [14]; Nam et al. [15]; Vats[16]). Findings are mixed but generally difference is perceptible with women taking lead over men in online health search. This is not surprising because women have been found to seek healtcare information in general also (Mackenzie et al. [17]; Matheson et al. [18]; Verhaak et al. [19]; Carrière [20]).

It is being hypothesized that women find the facility provided through the arrival of the Internet more positively than their male counterparts. This study is based on Indian patients.

3   Methodology

This study is based on the opinion of patients on positive aspects of DTCP of drugs. Accordingly, the objective of this work is to measure the patients do not differentiate in gender while considering the positive aspects of online health-related information. Following is the null hypothesis for this study.

4 Ho1: Patients Do not Reflect Any Significant Gender-Based Difference in Considering DTCP to Be Positive

On the basis of exploratory study, patients were interviewed about their opinion of DTC and variables were developed for this study. As, other than impact of gender on DTCP, positive impacts also appeared, patients were asked to specify the reasons for considering it positive. With the help of a 5-point Likert scale, these variables were used in an interview schedule to determine the quantitative belief of patients.

Only allopathic patients were taken as sample respondents for this research. Out of 800 patients approached for screen survey, 44 responses were incomplete and 440 patients did not use the Internet. Hence, these responses were screened out and final sample size was 314. For measuring the reliability of data, Cronbach's alpha was calculated and was found to be 0.843. Cronbach's alpha values above 0.6 are acceptable for statistical analysis. Test statistics for the data followed normal distribution where kurtosis and skewness lie within the limits of 1 in most of the cases and within 2 in all. Hypothesis for this study was tested using 95% confidence level%. In order to determine the significance of variations between the beliefs of female and male patients, one sample t-Test was conducted. For achieving 95% confidence level for t-test, Z-value of the data should be more than ± 1.96 and its significance should be less than 0.05.

5 Result and Discussion

In this subsection of the study, gender differences regarding the perceived positive outcomes of the study have been focused upon. Results and discussion in this respect are as follows.

1. **Trust over the Internet information**: Table 1, shows that mean of females' response for this variable is 4.01 with Standard Deviation (SD) 0.782, while the mean of male is 4.05 with SD 0.77. Since the values are close, we may interpret that both females and males have same beliefs. As t-value is less than 1.96 (0.466) and the level of significance is more than 0.05 (0.642), we accept the null hypothesis. Not much difference exists across gender and both appear to be rather trustful of the DTCP information.

2. **Feeling of empowerment over information gathered from the Internet**: For this variable, t-value is less than 1.96 (1.026) and the level of significance that is lower than 0.05 (0.00). Hence, the null hypothesis is rejected. Women are more inclined to believe feeling empowered because of DTCP. This could be because health needs of women differ and perhaps they are more home-bound and have lesser access to health information. Women also consider themselves more responsible for health care of the children and the aged and hence information on fingertips certainly must be quite empowering for them.

Table 1 Patient advance (gender): positive aspect

S. n.	Variable	Female		Male		T-value	Significance
		Mean	SD	Mean	SD		
1	Trust over Internet information	4.01	0.782	4.05	0.77	0.47	0.642
2	Feeling of empowerment over information gathered from the Internet	3.83	1.026	3.25	0.98	5.17	0
3	Easy understandability of Internet information over technical language used by physician	4.23	0.701	4.13	0.78	1.16	0.247
4	Knowledge of side effects of drugs	4.18	0.756	3.97	0.8	2.4	0.016
5	Better diagnosis using information from the Internet	3.76	830	3.06	0.83	7.43	0
6	Online information helps to keep me healthy	4.05	0.781	4.03	0.79	0.22	0.788
7	Taking proactive steps based on online information	4.07	0.83	3.96	0.8	1.19	0.233
8	Online information is educative	3.75	0.91	3.08	1.04	6.06	0
9	Feeling of helplessness in the absence of right information online	4.06	0.78	3.99	0.79	0.2	0.843
10	Online health information is assuring	3.71	0.721	3.59	0.8	1.18	0.249
11	Online information helps in better understanding of the state of disease	3.83	0.83	3.09	0.83	7.43	0
12	Feeling of helplessness due to banning of health-related websites	3.91	0.849	3.75	0.97	1.54	0.125

3. **Easy understandability of Internet information over technical language used by physician**: For this variable, t-value is less than 1.96 (1.16) and the level of significance is more than 0.05 (0.247). Hence, the null hypothesis is accepted. Women consider DTCP information as more understandable than the language used by physicians, though this inclination has not found to be significant.

4. **Knowledge of side effects of drugs**: For this variable, t-value is more than 1.96 (2.4). Hence, the null hypothesis is rejected and the alternate hypothesis is accepted. The conclusion is confirmed as the significance value is less than 0.05 (0.016). Difference is marginal though significant. Women perhaps are more concerned with negative aspects of any issue. Higher awareness based of DTCP could be the reflection of this inclination.

5. **Better diagnosis using information from the Internet**: For this variable, t-value is more than 1.96 (7.428). Thus, the null hypothesis is rejected and the alternate hypothesis is accepted. The conclusion is confirmed as the significance value is less than 0.05 (0.0). Yet again, we find women are more inclined to believe that DTCP helps them in better diagnosis. We have already stated that women perhaps have very different needs in terms of health issues and which they find to find preferably in their privacy. Also online information provides them diagnosis opportunity for their family members too.

6. **Online information helps to keep me healthy**: For this variable, t-value is less than 1.96 (0.217). Thus, the null hypothesis is accepted. The conclusion is confirmed as the significance value is more than 0.05 (0.788). No significant gender differences can be noted over here and, hence, it is concluded that both men and women rather strongly agree to this statement.

7. **Taking proactive steps based on online information**: For this variable, t-value is less than 1.96 (1.194). Thus, the null hypothesis is accepted. The conclusion is confirmed as the significance value is more than 0.05 (0.233). No difference exists yet women are found to be slightly more prone to agree to the statement. Health issues are perhaps more a concern for women in their day-to-day life.

8. **Online information is educative**: For this variable, t-value is more than 1.96 (6.062). Thus, the null hypothesis is rejected and the alternate hypothesis is accepted. The conclusion is confirmed as the significance value is less than 0.05 (0.0). Women tend to agree more to the belief that DTCP is highly educative. Perhaps this stems from their greater need for health-based information.

9. **Feeling of helplessness in the absence of right information online**: For this variable, t-value is less than 1.96 (0.202). Thus, the null hypothesis is accepted. The conclusion is confirmed as the significance value is more than 0.05 (0.843). Hence, it can be concluded that hardly any gender difference is noticed for this variable.

10. **Online health information is assuring**: For this variable, the t-value is below 1.96 (1.18) so according to decision rule we accept the null hypothesis and reject the alternative. To confirm it, we see the level of significance which is 0.249 higher than 0.05. No significant differences in the expressed belief by men and women are visible in this instance.

11. **Online information helps in better understanding of the state of disease**: For this variable, the t-value is 7.428 much higher than 1.96. So, we reject the null hypothesis. It is confirmed by the significance level which is 0.00 lower than 0.05. Women, yet again, have been found to express significantly stronger affirmative response. Perhaps because otherwise less information is available to them in a traditional setup and also because they have a greater need for the same.

12. **Feeling of helplessness due to the banning of health-related websites**: For this variable, t-value is less than 1.96 (1.54). Thus, the null hypothesis is accepted. It is confirmed by the significance level which is more than 0.05 (0.125). Responses more or less mimic the results of previous statement. The opportunity provided by DTCP appears to be significant for both the genders in equal measures.

6 Discussion

The researcher found that women are more positive regarding online health search than men. In most of the statements, there is not much difference in the belief of men and women regarding positive aspect online health search. Few researchers have studied gender based differences. Bidman and Terlutter [9] found that women are more positive about use of internet for health information search. Mitchell and Boustani [14] reported gender being the most noteworthy demographic variable in this respect. Bidmon and Terlutter [9] reveals that women have more positive attitude toward internet than men. Nam et al. [15] found men less positive toward DTCP. Ghia et al. [21] found that the influence of DTCP is equal on both the genders.

We have differing views also. Vats [16] found no difference in the attitude of male and female regarding positive impact of DTCP [22].

What could be the implications for such conclusions? Indian policymakers are advised to make note of this gender difference and ensure that gender-specific in formations are made available in traditional setup also. Promoters of DTCP are also advised to pay due weightage to the differing needs of genders and orient their promotion campaigns accordingly. Such a positive belief about DTCP should be considered a hard won trust and must not be lost by casual DTCP approaches. Physicians perhaps are best positioned to appreciate the conclusions. This positive inclination towards DTCP particularly by women must not make them oblivious of the corresponding risks. Physicians may caution the patients particularly the women in this respect.

7 Conclusion

Out of the 12 variables taken in the study, it is found that significant differences exist between the expressed beliefs held by male and female patients in 5 variables. It is interesting to note that women appear to be more positive about DTCP than their male counterparts. It is important for regulatory authorities, physicians and companies to note that there is gender difference in search behavior of online health-related websites. Physicians may implement gender-based counseling of patients to ascertain encouragement of progressive aspects and such promotions.

References

1. M.C. Cadaret, S.L. Speight, An exploratory study of attitudes toward psychological help seeking among African American men. J. Black Psychol. **44**(4), 347–370 (2018)
2. L. Wasylkiw, J. Clairo, Help seeking in men: when masculinity and self-compassion collide. Psychol. Men Masculinity **19**(2), 234 (2018)
3. R.E. Brenner, K.E. Engel, D.L. Vogel, J.R. Tucker, N. Yamawaki, D.G. Lannin, Intersecting cultural identities and help-seeking attitudes: the role of religious commitment, gender, and self-stigma of seeking help. Ment. Health Relig. Cult. **21**(6), 578–587 (2018)
4. D.T. Doherty, Y. Kartalova-O'Doherty, Gender and self-reported mental health problems: predictors of help seeking from a general practitioner. Br. J. Health. Psychol. **15**(1), 213–228 (2010)
5. M. Joseph, G. Stone, J. Haper, E. Stockwell, K. Johnson, J. Huckaby, The effect of manufacturer-to-consumer prescription drug advertisements: an exploratory investigation. J. Med. Mark. **5**(3), 233–244 (2005)
6. E. Kontos, K.D. Blake, W.S. Chou, A. Prestin, Predictors of eHealth usage: insights on the digital divide from the health information national trends survey 2012. J. Med. Internet Res **16**(7), e172 (2014)
7. S.K. Mukherjee, J. Kumar, A.K. Jha, J.R. Rani, Role of social media promotion of prescription drugs on patient belief-system and behaviour. Int. J. E-Collab. (IJeC) **15**(2), 23–43 (2019)
8. M. Joseph, D.F. Spake, D.M. Godwin, Aging consumers and drug marketing: Senior citizens' views on DTC advertising, the medicare prescription drug programme and pharmaceutical retailing. J. Med. Mark. **8**(3), 221–228 (2008)
9. S. Bidmon, R. Terlutter, Gender differences in searching for health information on the internet and the virtual patient-physician relationship in Germany: exploratory results on how men and women differ and why. J. Med. Internet Res. **17**(6), e156 (2015)
10. S. Fox, L. Rainie, J. Horrigan, A. Lenhart, T. Spooner, M. Burke, O. Lewis, C. Carter (2000), *The online healthcare revolution: how the web helps Americans take better care of themselves.* Pew Internet and American Life Project, (Washington, D.C. 2000), (Online)
11. R. Gauld, S. Williams, Use of the Internet for health information: a study of Australians and New Zealanders. Inform. Health Soc. Care **34**(3), 149–158 (2009)
12. Y.Y. Yan, Online health information seeking behavior in Hong Kong: an exploratory study. J. Med. Syst. **34**, 147–153 (2010)
13. E. Turget, Online health information seeking habits of middle aged and older people: a case study (Master's thesis), 2010
14. V.W. Mitchell, P. Boustani, The effects of demographic variables on measuring perceived risk. in *Proceedings of the 1993 Academy of Marketing Science (AMS) Annual Conference*, (Springer International Publishing, 2015), pp. 663–669

15. S. Nam, P. Manchanda, P. Chintagunta, The effect of signal quality and contiguous word of mouth on consumer acquisition for a video-on-demand service. Mark. Sci. **29**(4), 690–700 (2010)
16. S. Vats, Impact of direct to consumer advertising through interactive internet media on working youth. Inte. J. Bus. Adm. Res. Rev. **1**(2), 88–99 (2014)
17. C.S. Mackenzie, W.L. Gekoski, V.J. Knox, Age, gender, and the underutilization of mental health services: the influence of help-seeking attitudes. Aging Ment. Health **10**(6), 574–582 (2006)
18. F.I. Matheson, K.L. Smith, G.S. Fazli, R. Moineddin, J.R. Dunn, R.H. Glazier, Physical health and gender as risk factors for usage of services for mental illness. J. Epidemiol. Community Health **68**(10), 971–978 (2014)
19. P.F. Verhaak, M.J. Heijmans, L. Peters, M. Rijken, Chronic disease and mental disorder. Soc. Sci. Med. **60**(4), 789–797 (2005)
20. G. Carrière, Consultations with doctors and nurses. Health Rep. **16**(4), 45 (2005)
21. C. Ghia, R. Jha, G. Rambhad, Impact of pharmaceutical advertisements. J. Young Pharmacists **6**(2), 58–62 (2014)
22. M.D. Shapiro, S. Fazio, From lipids to inflammation: new approaches to reducing atherosclerotic risk. Circ. Res. **118**(4), 732–749 (2016)

A Wearable Antenna Using Jeans Substrate Operating in Dual-Frequency Band

Shreema Manna, Tanushree Bose, Rabindranath Bera, and Sanjib Sil

Abstract In this paper, the research work has been carried out on designing wearable microstrip patch antenna which resonates at dual-frequency bands, which are 2.6 GHz and 5.8 GHz, respectively. Both the bands are permissible bands for wearable devices as recommended by Federal Communication Commission (FCC). Easily available textile material like Jeans is used as the substrate material and copper is used for the conducting part. The simulation result is satisfactory with respect to gain, bandwidth, return loss, directivity VSWR, and other antenna parameters.

Keywords FCC · Wearable antenna · Textile material · Return loss · Gain

1 Introduction

Nowadays wearable devices are gaining much importance mainly in remote monitoring of environmental conditions and different parameters of human body. With the use of these devices, it becomes easy to monitor the health conditions, especially for elderly people for sudden degradation of their health condition. Due to the rapid advent of the technology called the Internet of Things; the wearable devices are becoming an integral component of each and individual things. Antenna is the inherent integral part of all devices used for remote transfer of data. Wearable

S. Manna (✉) · T. Bose · R. Bera
Sikkim Manipal Institute of Technology, Majitar, Sikkim, India
e-mail: shreema123@gmail.com

T. Bose
e-mail: tanushree.contact@gmail.com

R. Bera
e-mail: rbera50@gmail.com

S. Sil
CIEM, Kolkata, India
e-mail: sanjib_sil@hotmail.com

© Springer Nature Singapore Pte Ltd. 2020
R. Bera et al. (eds.), *Advances in Communication, Devices and Networking*,
Lecture Notes in Electrical Engineering 662,
https://doi.org/10.1007/978-981-15-4932-8_12

antennas find much significant application in the area which includes rescue team, defense (where the wearable device can be a part of their clothing), persons who are working in the mines or far remote places where regular vigilance is impossible. One of the important requirements of antenna used in wearable devices which are attached to human body for health monitoring is that they must be flexible, easy to wear and compatible to human body. Jeans is one of such materials which satisfy these requirements. This material is very easily available and can be easily attached to clothing. These kinds of textile materials are also having the advantages like low cost, flexible and easily wearable. The textile materials are having the inherent advantage of low dielectric constant, which helps in improving the impedance bandwidth by reducing the surface wave loss. Antenna bending effect can also be measured and can be taken into account for antennas made up with textile materials.

Wearable antennas can be attached as a part of clothes and all other wearable things like belt, shoe, helmet, bag, tie, cap, etc. The availability of such a variety of textile materials made the rapid growth of wearable devices possible.

2 Antenna Design

The antenna design and simulation is carried out in HFSS13.0. In the first stage of antenna design, simple rectangular structure of patch is used. Copper is used as the conductor. Jeans is used as a substrate having 2 mm thickness. The dielectric constant used here is 1.76 and loss tangent 0.02. Overall dimension of the ground surface is 50 mm × 50 mm and the dimension of the patch is 34 mm × 34 mm. The simulation result shows that the antenna resonates at 2.6 and 5.8 GHz. Further to improve the structure in terms of gain, band width, directivity, and other parameters, a wide slot of 10 mm × 10 mm is inserted at the ground. The simulation result shows a return loss of −15 and −20 dB at respective resonating frequency bands of 2.6 and 5.8 GHz. Figures 1 and 2 show the corresponding antenna structure and simulation result.

3 Result and Discussion

The proposed antenna has been designed and simulated using HFSS software. The antenna designed is planner patch antenna with microstrip feed. The following Fig. 2 shows the S11 display of the designed antenna. The diagram clearly shows that the antenna resonates at 2.6 GHz and 5.8 GHz. The return loss achieved is around −15 dB and −20 dB, respectively.

Fig. 1 Antenna structure

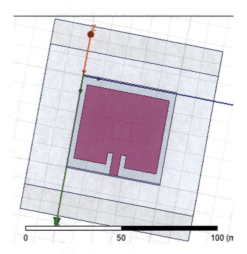

The antenna is resonating at 2.6 GHz and a very high gain of around 7.8 dB is achieved at this frequency. The frequency 5.8 GHz is the first harmonic of the resonating frequency. The return loss at this frequency is around −20 dB and gain is around −10 dB. The VSWR achieved is 0.84 which satisfies the acceptable limit for a good antenna performance. In this design, inset feed is used improve the impedance matching (Fig. 3).

The result of simulation of the designed antenna shows that it satisfies the circumstances of the antennas used in wearable devices. Jeans is used as substrate here which is easily available and can be easily attached or stitched to clothing.

4 Conclusion

This paper is the outcome of an effort given in the design of wearable antenna using textile material like jeans. This can be easily attached to clothing, washable, and also fashionable. The proposed antenna is resonating at dual-frequency bands of 2.6 and 5.8 GHz. Both the frequency bands are acceptable for wearable devices. This work can be further improved to get better impedance matching, better efficiency.

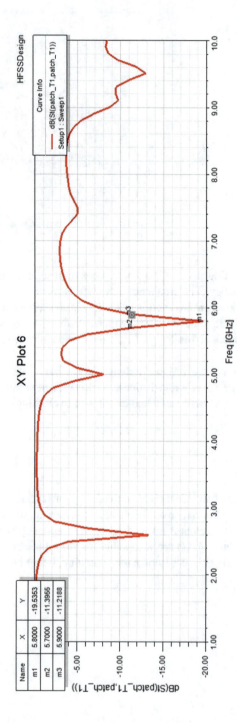

Fig. 2 S11 display of simulated antenna

Fig. 3 Radiation Pattern
display of simulated antenna

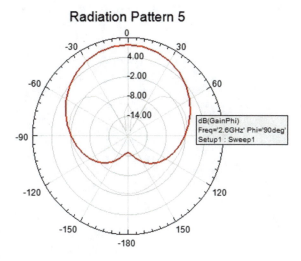

References

1. S.H. Li, J.S. Li, Smart patch wearable antenna on Jeans textile for body wireless communication, in *2018 12th International Symposium on Antennas, Propagation and EM Theory (ISAPE)* (IEEE, 2018), pp. 1–4
2. K.H. Wang, J.S. Li, Jeans textile antenna for smart wearable antenna, in *2018 12th International Symposium on Antennas, Propagation and EM Theory (ISAPE)* (IEEE, 2018), pp. 1–3
3. I. Agbor, D.K. Biswas, I. Mahbub, A comprehensive analysis of various electro-textile materials for wearable antenna applications, in *2018 Texas Symposium on Wireless and Microwave Circuits and Systems (WMCS)* (IEEE, 2018), pp. 1–4
4. M.I. Ahmed, M.F. Ahmed, A.E.H. Shaalan, Novel electro-textile patch antenna on jeans substrate for wearable applications. Prog. Electromagnet. Res. **83**, 255–265 (2018)
5. A. Tsolis, W.G. Whittow, A.A. Alexandridis, J.C. Vardaxoglou, Evaluation of a human body phantom for wearable antenna measurements at the 5.8 GHz band, in *2013 Loughborough Antennas & Propagation Conference (LAPC)* (IEEE, 2013), pp. 414–419
6. C.K. Nanda, S. Ballav, A. Chatterjee, S.K. Parui, A body wearable antenna based on jeans substrate with wide-band response, in *2018 5th International Conference on Signal Processing and Integrated Networks (SPIN)* (IEEE, 2018), pp. 474–477

InGaN/GaN Multiple Quantum Wells Solar Cell as an Efficient Power Source for Space Mission

Shingmila Hungyo, Khomdram Jolson Singh, Dickson Warepam, and Rudra Sankar Dhar

Abstract A changeable band gap and radiation hard ternary indium gallium nitride (InGaN) with multiple quantum well (MQW) structure-based solar cell is numerically modelled and analysed for a better power source in space mission using modern TCAD tool. Since InGaN has direct band gap varying from 0.7 to 3.4 eV covering nearly the complete solar spectrum, it can be one of the best candidates for making efficient tandem solar cells. It is well known that the combination of InGaN/GaN multiple quantum well (MQW) structures in GaN-based devices drastically decreases surface recombination thereby improving cell performance. Here, a numerical simulation study of MQW InGaN/GaN solar cell with an active region formed by a number of InGaN quantum wells (QWs) separated by GaN quantum barriers (QBs) is investigated. It is found that there is no significant variation in cell parameters under the influence of indium content up to 25%. But, with the increase in the numbers of QW periods, the photovoltaic parameters, especially conversion efficiency, increases significantly. Under space AM0 solar illumination, the cell efficiency increases up to 8.2% for 20 MQW with 20% indium content of InGaN/GaN structure. And also, it gives better external quantum efficiency (EQE) up to 60% at 380 nm wavelength range near UV region. This clearly demonstrates that InGaN with GaN top cell and silicon bottom cell can definitely give broader and higher quantum efficiency.

S. Hungyo · D. Warepam · R. S. Dhar
Department of Electronics and Communication Engineering, National Institute of Technology, Aizawl 796012, Mizoram, India
e-mail: shingmilahungyo@gmail.com

D. Warepam
e-mail: dicksonwarepam@gmail.com

R. S. Dhar
e-mail: rdhar@uwaterloo.ca

K. J. Singh (✉)
Department of Electronics & Communication Engineering, Manipur Institute of Technology, Canchipur, Imphal-03, Manipur, India
e-mail: jolly4u2@rediffmail.com

© Springer Nature Singapore Pte Ltd. 2020
R. Bera et al. (eds.), *Advances in Communication, Devices and Networking*,
Lecture Notes in Electrical Engineering 662,
https://doi.org/10.1007/978-981-15-4932-8_13

Keywords InGaN · Multiple quantum well (MQW) · Modelling · TCAD
(Technology computer-aided Design) · Shockley–Read–Hall (SRH) · Space solar
cell

1 Introduction

The Group III nitrides compound semiconductors (GaN, InN, AlN) and their alloys
have direct gaps which can cover the entire solar spectrum range. InN has optical gap
of 0.7 eV (1771 nm), GaN a gap of 3.4 eV (366 nm) and specifically the alloy InGaN
based on the indium composition, can cover a wide band gap range from 0.70 eV
up to 3.4 eV which would be ideal for high-efficiency solar cell. Theoretically, a
multijunction cell formed by the stacking of three $In_xGa_{1-x}N$ cells can yield up to 70%
of conversion efficiency [1]. On the other hand, single-junction p–i–n solar cells with
multiple quantum wells(MQW) InGaN/GaN has the same potential of multijunction
(MJ) cells but with a less complicated structure and, also, higher open-circuit voltage
[2] and less surface recombination velocities.

Recent technologies utilize the combination of different material systems, such
as the Groups III–V and Group IV, into MJ cells. The main hurdle is that the physical
properties of different material families can be radically different, causing serious
issues in device fabrication, performance and lifetime. Therefore, in this work, we
will look at photovoltaic properties of single-junction p–i–n InGaN/GaN MQW solar
cells, especially its band structure, I–V characteristics and quantum efficiency. And
to achieve further higher efficiencies, multiple layers with optimally chosen band
gap materials can be further used so that the device can effectively utilize the larger
fraction of the solar spectrum.

For space solar cell and array application, specific power (W/kg), power stowed
volume (W/m^3), lifetime and radiation hard material are of critical importance. In
addition, the nitride materials can offer exceptional radiation tolerance that is well
beyond what can be achieved with conventional solar cell materials currently flown
into space. However, the Group III nitride materials will not likely have as high
efficiency as their much more matured technology Groups III–V counterparts. At
the same time, however, the advantage is that they will degrade far less over the
entire life of the space mission. The main objective of this work is to determine the
feasibility of the InGaN material system with less complicated structure MQW solar
cells for effective space power source application.

2 Radiation-Hard Semiconductor Material

Indium gallium nitride (InGaN) is able to withstand a greater amount of radiation
when compared to other gallium arsenide (GaAs) and indium gallium phosphide
(GaInP). For any type of space applications, radiation-hard materials are very much

essential. InGaN "retains its optoelectronic properties at radiation damage doses at least two orders of magnitude higher than the damage thresholds of the materials (GaAs and GaInP) currently used in high-efficiency MJ cells" according to [3]. In the literature [3], the amount of radiation levels were 1 MeV electron, 2 MeV proton and 2 MeV alpha particle irradiation. This demonstrates that InGaN is not only a potentially high-efficient photovoltaic material, but it is also able to withstand the harsh space environment.

3   Indium Gallium Nitride

For space applications, high-efficient solar cell models are of great interest. By increasing the conversion efficiency of each photovoltaic cell, the total number of solar panels can be drastically decreased. Therefore, the overall weight that needs to be launched into space is reduced. A cost reduction can be achieved over the life of the space power system. A previous InGaN simulation has been performed [4] using fundamental semiconductor physics formulas.

$$n_i^{\text{InN}} = \frac{6.2 \times 10^{22}}{\text{cm}^3}$$

$$n_i^{\text{GaN}} = \frac{8.9 \times 10^{22}}{\text{cm}^3}$$

Critical data for InN and GaN are obtained from [5]. A phonon is required in addition to a photon in order to excite an electron from the valence band to the conduction band in an indirect band gap semiconductor. Therefore, a direct band gap semiconductor is generally better for optoelectronics application. GaAs and wurtzite InGaN are examples of direct band gap semiconductors.

The band gap of the semiconductor material determines the wavelength of light that meets the requirements to generate electrical energy. The conversion formula between band gap and wavelength is:

$$\lambda(\mu\text{m}) = \frac{hc}{Eg(\text{eV})} = \frac{1.24}{Eg(\text{eV})}$$

where λ is the wavelength in micrometres, h is Planck's constant, c is the speed of light in vacuum and Eg is the band gap in eV. The location of the solar cell affects the input solar radiation spectrum. We know that a solar cell on Mars receives a different (smaller) spectrum than a solar cell on a satellite that orbits Earth.

The energy received outside Earth's atmosphere is approximately 1365 W/m^2 and this particular spectrum is called Air Mass Zero or AM0. The data are obtained from the National Renewable Energy Laboratory (NREL) [6]. The InN fundamental band gap was approximately 0.77 eV at room temperature. Since GaN has a band gap of approximately 3.4 eV at room temperature, then InGaN can have a band gap ranging from 0.77 to 3.4 eV by changing the percentage composition of indium and gallium within InGaN. Figure 2 confirms that InGaN follows the pattern of ranging from 0.77 to 3.4 eV. According to [7], the following formula provides an approximation of InGaN band gap:

$$Eg(x) = 3.42x + 0.77(1 - x) - 1.43x(1 - x)$$

where Eg is the InGaN band gap, 3.42 eV is the GaN band gap, 0.77 eV is the InN band gap, 1.43 eV is the bowing parameter b, x is the Ga concentration and $(1 - x)$ is the In concentration.

3.1 Indium Gallium Nitride Challenges

InN has the highest electron affinity among all semiconductors [8]. Hence, InN has a tendency to be n-type. It is much more difficult to build p-type material with InN. However, evidence of p-type doping in InN and InGaN has been reported in [9] using magnesium. Further progress in this area dictates the creation of InGaN PN junctions.

3.2 Simulation Environment

The quantum simulator in SILVACO ATLAS environment provides a realistic platform for numerical simulation of all quantum effect underlying in the quantum well region of the device. There are various quantum models in SILVACO's quantum simulator, in this work parabolic quantum well model has been used to predict bound state energies in a region for subsequent use in modelling optoelectronic gain, radiative recombination and absorption as it solves Schrodinger equation to calculate the bound state energies and wave functions. The bound state energies are used to predict the gain and spontaneous recombination rates and the wave functions can be used to predict the overlap integral. The capture–escape model is used together with Quantum Well Model to predict the position of Fermi levels of quantum wells. The multiband Kronig–Penney (KP) models that take into account this inter-mixing of states to provide a much more accurate band structure and optical response over a wider region of the Brillouin zone. However, since the parabolic model implements the summation over states in momentum space analytically, it is much faster than the more general KP model. The OPTR determines the possibility that a photon is generated when an electron and hole recombine. Green has also shown that the OTPR model increases the accuracy of the photovoltaic cell simulation. The primary importance to the simulation of a PV cell is the accurate modelling of electron–hole pair generation [10]. LUMINOUS, the optoelectronic simulation module in

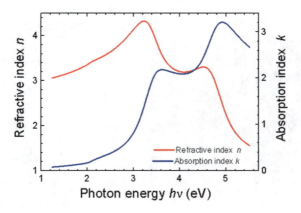

Fig. 1 N and K values of InGaN

ATLAS determines the photogeneration at each mesh point in an ATLAS structure by performing two simultaneous calculations. The refractive index n is used by LUMINOUS to perform an optical ray trace in the device. The extinction coefficient k is used to determine the rate of absorption and photogeneration (electron–hole pair generation) for the calculated optical intensity at each mesh point. Together, these simulations provide for wavelength-dependent photogeneration throughout a cell. These n and k values of our designs are based on calculation [11], interpolation and recent literatures shown in Fig. 1.

4 Modelling of InGaN MQW Solar Cell

As shown in Table 1, the proposed device cell structure consists of a 0.4-micron p-type GaN top layer with a doping level of 10e17 cm^{-3} and also 0.4-micron n-type GaN bottom layer with a doping level of 10e17 cm^{-3}. In between top and bottom layers exists an undoped multi-quantum wells (MQW) region, where In$_{0.20}$Ga$_{0.80}$N quantum wells (QWs) of 5 nm thick are separated by 5 nm thick GaN quantum barriers (QBs). We noted that the thickness of QWs and QBs are fixed in all subsequent simulations. The ATLAS developed single-junction InGaN MQW cell model as shown in Fig. 2a.

Table 1 Our MQW cell configuration

Cell type	Material	Eg (eV)	Doping Conc. (cm^{-3})	Thickness (μm)
Top layer	p-GaN	3.4	P+ 10e17	0.4
MQW (10)	In$_{0.20}$Ga$_{0.80}$N(QWs) GaN (QBs)	2.66 3.4	i-layer (undoped)	5 nm 5 nm
Bottom layer	n-GaN	3.4	n+ 10e17	0.4

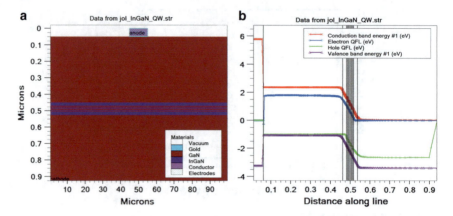

Fig. 2 **a** Simulated model of 10-well InGaN/GaN MQW solar cell, **b** band structure of 10 period In$_{0.2}$Ga$_{0.8}$N/GaN MQW solar cell

Optical data in the form of dielectric constants were found in [11] for In$_{0.20}$Ga$_{0.80}$N. This corresponds to a calculated band gap of 2.66 eV. The band diagram of the ten-well InGaN/GaN MQW solar cell is shown in Fig. 2a.

5 Result and Discussion

Figure 3a is the detailed photogeneration rate obtained under space AM0 solar spectrum illumination. The legend within the figure defines the InGaN cell's photogeneration rates. These are expressed using the log of the electron–hole pair generation rates that correspond to the colour-coded display. For example, the highest numerical value (e.g., 22.1) corresponds to the colour-coded horizontal layer that is generating

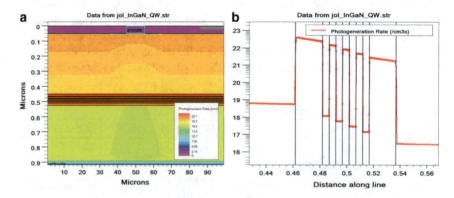

Fig. 3 **a** Photogeneration rate of whole InGaN/GaN MQW solar cell structure, **b** photogeneration rate increased inside the well of InGaN/GaN MQW solar cell

$10^{22.1}$ electron–hole pairs per cm^3. The dielectric constants were then converted to index of refraction (n) and extinction coefficient (k). The band gap and the optical data were entered into the input deck. It is further shown that photogeneration rate is relatively high in MQW region as expected shown in Fig. 3b. The combination of InGaN/GaN MQW structures in GaN-based devices drastically decreases surface recombination but improves recombination rate at critical MQW region as indicated in Fig. 4a, b. This will definitely improve the cell performance.

The simulation ran in Silvaco Atlas. From the log file, the current and voltage data were extracted. Figure 5a shows the extracted IV curves of 10 period In$_{0.2}$Ga$_{0.8}$N/GaN MQW solar cell. With the increasing numbers of QW/WB periods enhances cell photovoltaic parameters as given in Table 2. This work used the band gap formula from [7]. A common finding in [7] and this work is that both simulations show that

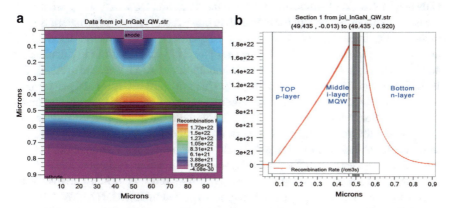

Fig. 4 **a** High recombination rate at i-layer MQW region predicted inside the cell structure, **b** high recombination rate at i-layer MQW region

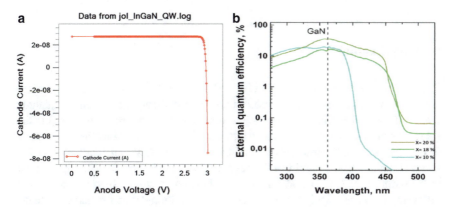

Fig. 5 **a** I–V characteristics of 10 period In$_{0.2}$Ga$_{0.8}$N/GaN MQW solar cell, **b** calculated quantum efficiency of 10 period InGaN/GaN MQW solar cell with different Indium content $x = 10$, 18 and 20% (Referencing with that of Valdueza [12])

Table 2 Performance parameters of MQW cell with different number of well

Number of well	1	5	10	15	20
Jcc (mA)	0.12	0.13	0.27	0.28	0.34
Voc (V)	2.72	2.78	2.95	3.1	3.6
FF (%)	88	90	94	94	95
Conversion efficiency (%)	2.93	3.64	5.59	7.7	8.2

InGaN MJ or MQW solar cells can provide a significant improvement in solar cell efficiency. The calculations involved simulating the response of an InGaN p–i–n device over an AM0 spectral range of 0–4 eV. In simulations, many of the material parameters of InGaN were estimated based on interpolation between parameters for GaN and InN, most of which are present in the literature. However, some parameters are not yet known, especially for InN, and, in those cases, values from materials were assumed. For interpolation, either a linear or quadratic approximation based on the alloy composition was used. Modelling in [4] used physics equations to calculate Isc and Voc. The input spectrum used in that simulation was AM1.5 instead of AM0. The InGaN band gap formula used was from [7]. This work thus used the band gap formula from [7]. It is found that there is no significant variation in cell parameters under the influence of indium content up to 25%. But, with the increase in the numbers of QW periods, the photovoltaic parameters, especially conversion efficiency increases significantly as indicated in Table 2. Under space AM0 solar illumination, the cell efficiency increases up to 8.2% for 20 MQW with 20% indium content of InGaN/GaN structure. And also, it shows better external quantum efficiency (EQE) up to 60% up to 380 nm wavelength range near UV region 10 period InGaN/GaN MQW solar cell with 20% indium content as shown in Fig. 5b (Referencing with that of Valdueza [12]). This clearly demonstrates that InGaN with GaN top cell and Silicon bottom cell can definitely give broader and higher quantum efficiency.

6   Conclusions

The results of this work conclude that Group III nitrides InGaN is potentially an excellent semiconductor photovoltaic material for space application and multiple quantum wells MQW InGaN/GaN have the same potential of its counterpart multijunction (MJ) cells but with a less complicated structure. It is also found that the combination of InGaN/GaN MQW structures in GaN-based devices drastically decreases surface recombination thereby significantly improving cell performance. With the increase in the numbers of QW periods, the photovoltaic parameters, especially conversion efficiency increases significantly. An essential development of a reliable simulated InGaN MQW model with parameter analysis using modern TCAD Silvaco ATLAS before actual fabrication is demonstrated with significant efficiencies for possible use as a power source for space mission.

7 Future Research

The model used default settings for other parameters, such as permittivity, affinity, radiative recombination rate, electron and hole lifetimes, electron and hole density of states and lattice constants. One area of future research is to obtain accurate measured data for the above parameters. This subject is critical in improving the model. In spite of that, quite low values of conversion efficiency for InGaN-based solar cells were published in the literature. This low value of performance is related to a poor absorption of solar spectrum by InGaN layers with low In-content and high dislocation density in In-rich InGaN layers. Therefore, in future, the optimization of the solar cell structure design is required to improve device efficiency and to understand the influence of the different structural parameters on the cell performance.

References

1. S.P. Bremner, M.Y. Levy, C.B. Honsberg, Prog. Photovoltaics Res. Appl. **16**, 225 (2008)
2. Z. Xiao-Bin et al., Chin. Phys. B **20**(2), 028402 (2011)
3. J.W. Ager III, W. Walukiewicz, High efficiency, radiation-hard solar cells. Lawrence Berkeley National Laboratory, Paper LBNL 56326 (2004)
4. A.S. Bouazzi, H. Hamzaoui, B. Rezig, Theoretical possibilities of InGaN tandem PV structures. Sol. Energy Mater. Sol. Cells **87**, 595–603 (2004)
5. Indium Nitride and Gallium Nitride atoms per cm^3, 2 May 2007, http://www.ioffe.rssi.ru/SVA/NSM/Semicond/index.html
6. National Renewable Energy (NREL) Air Mass Zero (AM0) solar spectrum, 17 April 2007, http://rredc.nrel.gov/solar/spectra/am0/
7. J. Wu et al., Small band gap bowing in InGaN alloys. Appl. Phys. Lett. **80**(25), 4741–4743 (2002)
8. J. Wu, W. Walukiewicz, Band gaps of InN and group III nitride alloys. Superlattices Microstruct. **34**, 63–75 (2003)
9. R.E. Jones et al., Evidence for p-type doping of InN. Lawrence Berkeley National Laboratory, Paper LBNL 59255 (2005)
10. SILVACO Data Systems Inc.: Silvaco ATLAS User's Manual (2010)
11. R. Goldhahn et al., Dielectric function of "narrow" band gap InN. Mat. Res. Soc. Symp. Proc. **743**, L5.9.1–L5.9.6 (2003)
12. S. Valdueza-Felip, A. Mukhtarova, L. Grenet, C. Bougerol, C. Durand, J. Eymery, E. Monroy, Improved conversion efficiency of asgrown InGaN/GaN quantum-well solar cells for hybrid integration. Appl. Phys. Express **7**, 032301 (2014)

Feature Extraction of Hill Rain on the Basis of Reflectivity-Rain Rate Relationship

Madhura Chakraborty, Pooja Verma, Swastika Chakraborty,
Shreya Biswas, Smritikana Singha, Satya Prabha, Debarita Das,
and Sharmila Saha

Abstract An insight into the global hydrological cycle which is an important controlling factor of ecological system, requires the microphysical properties of rain to be understood very clearly. To estimate rain fall rate, Radar reflectivity (Z) and rainfall rate (R) relationship is a preferred choice. Therefore it is necessary to study the microphysical property of rain with a special focus in hills besides the plane land in tropical region as topography has a major impact on local weather. Here the place under investigation, Majhitar, Sikkim is having 200 metres of altitude and surrounded by hills. Rain events are separated into convective and stratiform type using Gamache-Houze method. Radar reflectivity is extracted from LASER Precipitation Monitor (LPM) data at Sikkim Manipal Institute of Technology (SMIT) site for one whole month (August, 2019) rain events. The Z–R relation for Sikkim is established and a clear deviation from Marshall—Palmer and Rosenfeld proposed model is reported. The variation in the pattern of convective and stratiform rain presented in the work is clear evidence of high altitude effect and the terrain of hill is introducing signature of orographic rain with the conventional rain also.

Keywords Z–R relationship · Rainfall rate · High altitude · Orographic rain

1 Introduction

Weather Radar system is a popular tool for measuring and forecasting very short time weather conditions. This system is having vast applications for measuring various hydrological parameters and can work efficiently for estimating rain fall rate.

Study of rain fall rate is the most important consideration for understanding the rain attenuation. Rain is having the most adverse effect on signal propagation along

M. Chakraborty (✉) · S. Biswas · S. Singha · S. Prabha
Department of Electronics and Communication Engineering, JIS College of Engineering, Kalyani, Nadia, India
e-mail: madhura.chakraborty@jiscollege.ac.in

P. Verma · S. Chakraborty · D. Das · S. Saha
Department of Electronics and Communication Engineering, SMIT, SMU, Majhitar, Sikkim, India

© Springer Nature Singapore Pte Ltd. 2020
R. Bera et al. (eds.), *Advances in Communication, Devices and Networking*,
Lecture Notes in Electrical Engineering 662,
https://doi.org/10.1007/978-981-15-4932-8_14

the path and most importantly causing high attenuation when frequency is greater than 10 GHz. For that detail study of characterisation of rain and estimation of rainfall rate has gained immense importance nowadays. Rainfall varies both in three-dimensional space and time and can be represented by multidimensional random process [1]. The Radar reflectivity (Z in mm^{-6} m^3) and rain fall rate (R in mm/hr) relation is known as Z–R relation. It is actually an empirically developed relation and is being studied from almost last 80 years [2]. The variations in Z–R are dependent on variation of Drop Size Distribution (DSD) [3]. According the formula (1) developed by Marshall and Palmer

$$Z = AR^b \tag{1}$$

The Z–R relation is expressed using the exponential DSD and found out the value of parameters, where $A = 200$ and $b = 1.6$.

Battan [4] presented total 69 numbers of Z–R relationships by collecting data from various parts of Globe. The relationships between Z and R [5–9] have helped a lot to identify various types of rain across the Globe. A standard deviation of rain rate with lesser than 1.5 mm/hr is considered to be stratiform rain [10].

For classification of rain, Gamache-Houze method (GH) [11] is a popular method. In this method 38 dBz has been considered as threshold value for distinguishing types of rain. Higher the threshold value is considered to be convective and below that is stratiform type of rain. According to Wilson [9] value of 'A' in Z–R is less in convective type than stratiform type and value of 'b' is more in convective type than stratiform type. On the other hand Rosenfield [12] has worked with data set based on tropical region and has established different value of A and b. The Eq. (2) is

$$Z = 250R^{1.2} \tag{2}$$

2 Experimental Set-up

A LASER Precipitation Monitor (LPM) (Model: V2.6x STD) has been used to measure DSD at Sikkim. The instrument set up (Fig. 1) is installed at Sikkim Manipal Institute of Technology (SMIT) at Majitar (27.1876° N, 88.4997° E) which is 200 metres high from sea level. The LPM is capable of sensing rain drop size diameter within range of 0.2–5.58 mm. At transmitter side infrared light with 785 nm wavelengths is produced by laser diode and at receiver side, photo diode with a lens which converts the light signal into electrical signal is used. Whenever rain drop passes through this LASER beam, signal amplitude reduces. Further digital signal processor produces the parameters in different units. With this set up, rain, hail, drizzle along with all parameters related to rain can be measured.

Fig. 1 Experimental set-up at SMIT

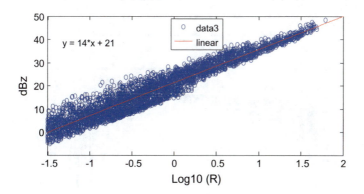

Fig. 2 Z (dBZ) versus $\log_{10}(R)$ plot

3 Experimental Results and Discussions

Figure 2 shows Z (dBZ) versus $\log_{10}$ (R) plot for 5306 DSD in minutes for the month of August, 2019 at SMIT. Total 26 rainy days data has been plotted together to find the $Z–R$ relation. Finally the linear regression technique is used to find the Z and R relation. The value of 'A' and 'b' is listed as Table 1. So it is found that the value of 'A' and 'b' is much different from what is established by Marshal-Palmar (MP). It may be the reason that MP model is recommended for in general stratiform precipitation. But in SMIT the rain is not stratiform type only. So the value of 'A' is heavily underestimated MP model developed value and same result is observed when compared with Rosenfield model too.

As per Gamache-Houze (GH) method 38 dBZ is taken as threshold value and below that it is stratiform type rain. In Fig. 3 implementing the same method underestimation is observed for both the value of 'A' and 'b' as per Table 2. Surprisingly before 38 dBZ and close to 35 dBZ rapid fluctuations of radar reflectivity is observed. This is pointing towards presence of lesser number of pure convective type of rain events and also nature is mixture of transition period and convective both (Fig. 4).

When for plotting convective rain above 38 dBz data has been taken into account and linearization on data shows underestimation in 'A' value and overestimation in 'b' value from GH method.

But the theory of Wilson that is value of 'A' in $Z–R$ is less in convective type than stratiform type and value of 'b' is more in convective type than stratiform type is

Table 1 $Z–R$ relationships for individual rain type and for the overall data set derived by linear regression of Z (dBZ) versus $\log_{10}$ (R)

Type	Value of 'A'	Value of 'b'
General result of SMIT	125.89	1.4
Marshall-Palmar (MP)	200	1.5
Rosenfield model	250	1.2

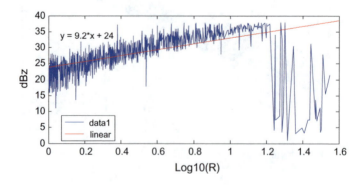

Fig. 3 Z (dBZ) versus $\log_{10}$ (R) plot till 38 dBZ

Table 2 Z–R relationships for individual rain type

Method	Type of rain	A	b
GH	Stratiform	273.57	1.35
Result at SMIT	Stratiform	251.188	0.92
GH	Convective	360.08	1.3
Result at SMIT	Convective	125.89	1.5

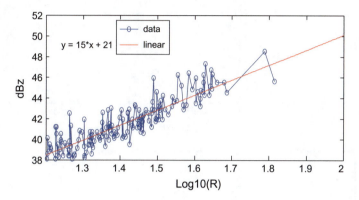

Fig. 4 Z (dBZ) versus $\log_{10}$ (R) plot beyond 38 dBZ

satisfying as per Table 2. Literature [13] shows that for orographic rain, the value of $A = 130$ and $b = 2.0$ which is very close to SMIT result to some extent especially for the value of A, pointing towards definite signature of orographic rain in the hill terrain.

For classification of rain, the whole 5306 number of samples are segregated event wise and total 54 numbers of rain events were taken into account. On 07/08/2019, highest rainfall rate is observed i.e. 65.2 mm/hr for a short duration. As per GH method rain rate higher than 10 mm/hr is convective type and lesser than that is stratiform type. On that basis, whole event is segregated, out of that 22 events are purely stratiform. Rest are mixed with stratiform, convective and transition in between. For every event of rain, the time taken to reach to the highest value is greater than the time taken to climb down from maximum value to ground. This is a very unique feature observed here and which is just in opposite of nature from what is described [14] for Malaysia, another tropical country.

In the Fig. 5a, b rain fall rate and corresponding Radar reflectivity was plotted for 13/08/2019–15/08/2019 as this is the highest duration when continuous rain is observed for almost for 3 days. As per GH method when rain reflectivity is greater than 38 dBz it is observed that rain fall rate is more than 30 mm/hr considering it as purely convective type of rain. Rest that is lower than 38 dBz, stratiform and transition period rain is observed.

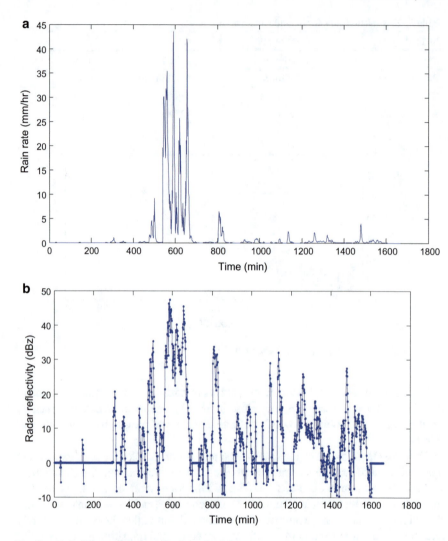

Fig. 5 **a** Rain fall rate, **b** radar reflectivity for 3 days

4 Conclusion

Z–R relationship is observed for 5306 samples for SMIT, Sikkim. With LPM installed in SMIT, rain rate intensity and radar reflectivity along with various parameters are measured and stored with data logger and signal is further processed with DSP unit. After deriving Z–R relationships, one of the results shows similarity of value of 'A' with orographic structured rain. On the other hand, the value of co-efficient 'A' and 'b' are following Wilson formula but differs from Marshall - Palmer and Rosenfeld proposed model, indicating difference in value due to tropical climate along with high altitude effect. But much investigation of its drop diameter and rain rate and drop velocity along with wind velocity is needed. DSD, standard derivation of rainfall rate and drop diameter will help to prove the orographic type rain presence further. During classification of rain with 54 events one important observation came into notice that leading rain slope duration is higher than trailing rain fall duration which is opposite in nature than another tropical location like Malaysia. So a new rain rate prediction model is needed to be developed to estimate rain fall rate as well as attenuation of signal for this location.

References

1. R. Cataneo, D.L. Vercellino, Estimating rainfall rate-radar reflectivity relationships for individual storms. J. Appl. Meteorol. **11**(1), 211–213 (1972)
2. J.S. Marshall, W.M. Palmer, The distribution of raindrops with size. J. Atmos. Sci. **5**, 165–166 (1948)
3. C.W. Ulbrich, Natural variation in the analytical form of the raindrop size distribution. J. Appl. Meteor. **22**(10), 1764–1775 (1983)
4. L.J. Battan, *Radar Observations of the Atmosphere* (Univ. Of Chicago Press, 1973), p. 323
5. D. Tokay, A. Short, Evidence from tropical raindrop spectra of the origin of rain from stratiform versus convective clouds. J. Appl. Meteor. **35**(3), 355–371 (1996)
6. D. Atlas, C.W. Ulbrich, F.D. Marks, E. Amitai, C.R. Williams, Systematic variation of drop size and radar Rainfall relations. J. Geophys. Res. **104**, 6155–6169 (1999)
7. D.A. Tokay, C.R. Williams, W.L. Ecklund, K.S. Gage, Tropical rainfall associated with convective and stratiform clouds: Intercomparison of disdrometer and profiler measurements. J. Appl. Meteor. **38**(3), 302–320 (1999)
8. M. Maki, T.D. Keenan, Y. Sasaki, K. Nakamura, Characteristics of the raindrop size distribution in tropical continental squall lines observed in Darwin, Australia. J. Appl. Meteor. **40**, 1393–1412 (2001)
9. C.L. Wilson, J. Tan, The characteristics of rainfall and melting layer in Singapore: Experimental results from radar and ground instruments, in *11th International Conference on Antennas and Propagation*, no. 480 (2001), pp. 852–856
10. V.N. Bringi, V. Chandrasekar, J. Hubbert, E. Gorgucci, W.L. Randeu, M. Schoenhuber, Raindrop size distribution in different climatic regimes from disdrometer and dual-polarized radar analysis. J. Appl. Meteorol. **60**(2), 354–365 (2003)
11. J.F. Gamache, A.R. Houze, Mesoscale air motions associated with a tropical squall line. Mon. Weather Rev. **110**, 118–135 (1982)
12. D. Rosenfeld, D.B. Wolff, D. Atlas, General probability-matched relations between radar reflectivity and rain rate. J. Appl. Meteorol. **32**(1), 50–72 (1993)
13. N.H.M. Sobli, A.F. Ismail, F.N.M. Isa, H. Mansor, Initial Assessment of radar reflectivity-rainfall rate, Z–R relationships for moderate rain events in malaysia, in *2013 IEEE Symposium on Wireless Technology and Applications (ISWTA)* (Malaysia, 2013)
14. L.S. Kumar, Y.H. Lee1, J X. Yeo, J.T. Ong, Tropical rain classification and estimation of rain from Z–R (reflectivity-rain rate) relationships, in *Progress in Electromagnetics Research B*, vol. 32 (2011), pp. 107–127

An Empirical Note on Positive Aspects of Online Direct to Consumer Promotion of Pharmaceutical Products

Jaya Rani, Samrat Kumar Mukherjee, Ajeya Jha, and Bibeth Sharma

Abstract Ethical and legal norms have been critical for marketing pharmaceutical products. Reaching out directly to patients for promoting drugs has been an anathema globally and is an illegal act even today barring two nations—USA and New Zealand. This rational policy, however, has been rendered futile with the advent of internet-based technologies. Patients today, in increasingly large numbers, are exploring online health information, including those served by pharmaceutical marketers. This development has been welcomed as well as has raised cudgels by the various stakeholders. Reason being the benefits and the challenges it brings in simultaneously. Direct to Consumer Promotion (DTCP) has positive aspects well acknowledged. Objective of this research is to quantify the patients' stated belief on positivity of DTCP. The null hypothesis has been stated as *Patients do not consider DTCP to be significantly positive*. Our result shows that patients are highly positive about DTCP. There are critical implications for the marketers, physicians and policy-makers on this front. This is essential in facilitating emergence of safe mechanisms that strengthen the healthcare system.

Keywords Online healthcare information · Positive aspects · Pharmaceutical products · Trust · Empowerment · Side effect awareness

1 Introduction

Healthcare has been central to social concerns and hence marketing of medicines has been subjected to strict control—both ethically and legally. This is a worldwide phenomena and is understandable because misuse, abuse and over use of medicines can have grave consequences. For this reason, promotion of drugs to patients directly is prohibited even today. In India The Drugs and Magic Remedies (Objectionable Advertisements) Act, 1954 prohibits '*promotion of diagnosis, mitigation, treatment*

J. Rani · S. K. Mukherjee · A. Jha (✉) · B. Sharma
Sikkim Manipal Institute of Technology, Sikkim Manipal University, Majitar, Rangpo, Sikkim, India
e-mail: ajeya611@yahoo.com

© Springer Nature Singapore Pte Ltd. 2020
R. Bera et al. (eds.), *Advances in Communication, Devices and Networking*,
Lecture Notes in Electrical Engineering 662,
https://doi.org/10.1007/978-981-15-4932-8_15

or prevention of any disease, disorder or condition stated in the Schedule, or any other disease, disorder or condition (by whatsoever name called) which may be specified in the instructions made under this Act to the general public'.

Violation of this may invite punishments including imprisonment up to 6 months, or a fine, or both. Repeat offences invite more serious punishments including imprisonment up to 1 year.

Internet has brought forth massive changes and which has rendered this law futile. Patients across the world are now receiving promotional materials of pharmaceutical products over the internet. The Governments have no control over this. DTCP, therefore, is not just unavoidable but is already happening.

Age research is now available, and which confirms this trend. Websites of major pharmaceutical companies provide enormous information directed towards patients. Post firms have created direct channels to communicate with patients through their corporate websites, disease websites and product (brand) websites.

2 Literature Review

In India, DTCP has invited the attention of many scholars (Khosla and Khosla [1], Pandey et al. [2], Vats [3] Goyal et al. [4], Pandey et al. [5], Rani and Jha [6], Jha and Pandey [7], Jha et al. [8], Pandey et al. [9]). DTC has been found to have positive consequences (Mukherjee et al. [10], Perri and Nelson [11], An [12], Bell [13], Bozic [14], Deshpande [15], Murray [16], Weissman [17], Bell [13], Burak [18], Choi [19], Datti [20], DeLorme [21], DeLorme [22], DeLorme [23], Herzenstein [24], Huh [25], Kon [26], Lee [27], Marinac [28], Menon [29], Joseph et al. [30], Turget [31], Rajani [32], Vats [3], Reast [33], Liang [34], Frosch et al. [35], Sumpradit et al. [36], Prigge et al. [37], Bélisle-Pipon and Williams-Jones [38], Ball et al. [39], Pechmann and Catlin [40], Adams and Gables [41]).

Positive impact of DTCP refers to their effects that are beneficial for the patient and significantly enhance the healthcare outcomes in terms of better, quicker and less distressful cure at an optimum cost. This could be real or perceptual. Even if perceptual, it still is significant because of corresponding placebo effect. Consequently, it needs to be measured in the context of patients.

How do patients in India view the positive aspects? This is a question that has been explored for answers in this work.

3 Methodology

The present research based on the data on the opinion communicated by patients on positive aspects of direct-to-consumer promotion of drugs. Accordingly, the objective of this work is to measure the patients' stated perception positive aspects of online health-related information. The null hypothesis has been stated as **Ho**: Patients do not consider DTCP to be significantly positive.

Variables for the study have been developed based on an exploratory study whereby patients were interviewed for their opinion on DTC-related issues. As, other than its impact, both negative and positive impacts emerged from this exploratory study they were further requested to provide specific reasons for considering it positive. These variables, thereafter, have been included in an interview schedule for measuring the quantitative belief of patients through a 5-point Likert scale.

The sample respondents of this research consist of patients taking recourse to allopathy. All 800 patients were approached for the screen survey. Responses submitted by 44 patients were found to be invalid and hence final sample size comprises of 754 patients. Out of these, 440 patients were not familiar with internet and hence were screened out in the first phase. The final survey sample therefore comprised of 314 patients. Cronbach's alpha has been used for measuring the reliability of data of 12-item scale and which has been found to be 0.843 and which is interpreted as being good. Values above 0.6 are considered acceptable for statistical analysis. The test statistics was checked and found to follow normal distribution (skewness and kurtosis were within 1 in most cases and within 2 in all cases). For the hypothesis testing the confidence limit is set at 95%. One-sample t-test was used to determine the significance of variations in the beliefs held by male and female patients. At 95% confidence limit for t-test is considered significant if its z-value is beyond ± 1.96 and corresponding significance (Sign.) less than 0.05.

4   Result and Discussion

Result has been shown in Table 1.

It is to be noted that all values are above 3 and which implies a general agreement with the suggestion that DTCP has positive aspects. Further t-values above 1.96 imply that this positivity is significant. This significance is further confirmed by p-values which should be below 0.05. In our case, we find that all the 12 variables have t-values more than 1.96 and the matching p-values below 0.05 (p-values can never be 0 but in our case any value below 0.001 is depicted as 0). Hence, our null hypothesis is rejected and it is inferred that DTCP is considered positive by the patients.

1. **Internet information is trustable**: How justified this trust is, obviously open to debate. Policies need to be in place to ensure that information being provided through DTCP is not misleading in any respect. Healthcare practitioner too can play a positive role in notifying erring DTCP campaigns as also by cautioning the patents. Perhaps a more cautious approach by patients, need to be evolved. It is in the interest of pharmaceutical markers too to remain honest and meticulous in this respect.
2. **Internet information is easier to understand than Physicians instructions**: We find that in terms of positivity, layman language provided in DTCP emerges

Table 1 [Patient advance t-test] positive aspect

S. No.	Statement	Mean	Std	t-value	p-value (sign.)
1	Internet information is trustable	4.04	0.77	23.8	0
2	Internet information is easier to understand than Physicians instructions	4.18	0.74	28.2	0
3	Internet information makes me feel empowered	3.52	1.04	8.9	0
4	Internet information facilitates more accurate diagnosis	3.39	8.98	7.7	0
5	Internet information provides better understanding of side effects of the drugs	4.07	0.78	24.2	0
6	Internet information forewarns of possible health issues	4.01	0.81	22	0
7	Internet information has resulted in better health for me	4.04	0.77	25.5	0
8	Without access to online information I feel helpless	4	0.78	22.4	0
9	Internet Information is highly educative	3.39	1.035	6.76	0
10	Internet information helps in understanding disease states better	3.44	0.898	7.66	0
11	Internet information leads to correct diagnosis	3.4	1.017	7	0
12	Internet information is very assuring	3.83	0.916	16.01	0

as its best attribute. This reflects a serious communication barrier between the physicians and patients which needs to be addressed appropriately by the physicians. Policy-makers may also take this into account and develop mechanisms to provide healthcare information in a language better understood in a multilingual country. Pharmaceutical companies must continue to provide required information in simple language with a view to strengthen the healthcare system.

3. **Internet information makes me feel empowered**: This information is important as empowering patients is in the interest of the entire healthcare environment. It enhances the decision choices for patients and makes them self-reliant.

4. **Internet information facilitates more accurate diagnosis**: Healthcare information though helpful in making preliminary diagnosis still harbours high potential for misleading diagnosis as well as under and over diagnosis. This doubt in the minds of patients is a positive indicator as it is physicians alone should legally be allowed to diagnose. Preliminary diagnosis, however, may help the physician to come to a conclusion with less effort and better conversation with the patient.

5. **Internet information provides better understanding of side effects of the drugs**: Therefore, it appears that DTCP helps patients by enhancing awareness about the side effects of the drugs. As side effects are common and at times can

be severe a forewarned patient naturally will find it easier to deal with them. This enhances the value of DTCP for the patients. It also reflects that without internet there was a paucity of this information from the patients' perspective.

6. **Internet information forewarns of possible health issues**: Patients response most often is reactive when faced with a healthcare issue and that also depending upon the severity of symptoms. Proactive response, triggered by DTCP, may be considered positive as it may lead to early, shorter and better treatment and avoid complications associated with delayed diagnosis.

7. **Internet information has resulted in better health for me**: It may be inferred from the finding that patients believe that overall impact of availability of online information results in a healthier life for them. In other words, they do associate this information with the fundamental need to suffer minimum because of health issues confronting them during day-to-day lives.

8. **Without access to online information I feel helpless**: This indicates that patients have come to a state when unavailability of health-related information makes them feel helpless. This implies that such an information is perceptually helpful to them. This perhaps is all the more reason to not to ban DTCP. Rather efforts need to be made to make it risk-free to the extent possible.

9. **Internet Information is highly educative**: One of the reasons DTCP is being criticized is that it is perceived as manipulative. It is suggested that attempts are being made to sell even prescription drugs directly to patient in the hope to increase sales. The DTCP promoters, however, insist that it is meant to an educative and not a commercial tool. As patients also find it educative its educational value appears to be established—at least in their minds. If it is educative, as the evidence suggest, it is distinctly a positive attribute.

10. **Internet information helps in understanding disease states better**: This may further be confirmed by level of significance which is near 0 and which is less than 0.05. An understanding of disease state may make the patients more appreciative about the risk posed as well as the significance of therapy. This naturally will get translated into better health outcomes and hence the perception that DTCP helps patient to understand their disease better may also be considered as a positive feature.

11. **Internet information leads to early diagnosis**: Patients appear to believe that internet helps them to make correct diagnosis. It has its positive aspects but has significant negative implications too. A general understanding of a disease is fine but correct diagnosis is a technical issue and perhaps ought to be left with the experts. Nevertheless, if information improves patient–physician communication then it has to be encouraged.

12. **Internet information is very assuring**: The patients find information on the internet highly assuring. Health issues are characterized by strong feeling of helplessness and which is not helped by the incomprehensibility and secrecy that generally is faced by the patients. This may put patients under unnecessary anxiety and apprehension. If DTCP is found to be assuring by the patients, it at least may bring down the trauma related to informational uncertainties and ambiguities.

5 Discussion

The study reveals that patients are highly positive regarding DTCP. They feel empowered due to health-related websites. They find it very assuring and feel helpless if it is banned. Patients trust the information available on internet; more than that they find it more understandable. Our finding indicates that patients are very positive about DTCP and believe that it helps in diagnosis of the disease and it also helps in being proactive. They believe that they are more aware of the side effects. According to them it is highly educative. Other researchers also found similar results. Peri and Nelson [11] reported that 82% of the patients believe that they get right kind of health information through internet. Singh and Smith [42] reported that customers claim that they are more aware and educated regarding health-related issues due to DTCP [16, 34]. We can say that in spite of their claim, they may not be fully aware of the benefits and risks of DTCP. Physicians should be aware of this condition that today's patient is seeking information through various sources (Parekh 2010), and being the trusted source of information encourage them to seek proper help. Unlike others few researchers claim that patients may not be fully aware of the risks involved into it and in an attempt to be rational it may lead to false expectation [43, 44]. Frosch et al. [35] found that quality and quantity of online information affect the understanding of risks and benefits. More recently, the positive aspects have been reported by Sumpradit et al. [36], Prigge et al. [37], Bélisle-Pipon and Williams-Jones [38], Pechmann and Catlin [40] and Adams and Gables [41]. In summary, it may be stated that DTCP is considered positively as it encourages patients to seek.

The implications for the policy-makers in this respect are that they must measure the actual positivity that is associated with DTCP. This may help in evolving mechanisms that can substantially strengthen the healthcare system. Practitioners need to introspect if their negative view on DTCP is sound and even-handed. They may perhaps require to overcome their natural apprehension and misgivings and not kill a phenomenon that enshrines several positive attributes that can strengthen our healthcare system.

6 Conclusion

This study was undertaken to understand the positive outcomes of DTCP. It is interesting to find that all the 12 variables on positivity of DTCP have been found to be so by the patients. The positivity is highest for the statement *I am healthier due to information I get on internet* and which reflects the overall mood of the patients in this regard. Pharmaceutical companies may also utilize these implications to fine tune their communication to patient directly for the overall welfare of the society and patients—their ultimate consumers. DTCP has come to stay and has positive outcomes and which may help to empower the patients. This is not to discount its negative outcomes which need to be regulated carefully.

References

1. P. Khosla, A. Khosla, Direct to consumer advertising of prescription drugs on internet: A Boon or a Curse. Indian J. Pharmacol. **43**(4), 483 (2011)
2. J. Pandey, D. Tiwari, A. Jha, Impact of on-line direct-to-consumer promotion of pharmaceutical products: a study of Indian patients, in *Communication, Cloud and Big Data: Proceedings of CCB 2014*
3. S. Vats, Impact of direct to consumer advertising through interactive internet media on working youth. Int. J. Bus. Adm. Res. Rev. **1**(2), 88–99 (2014)
4. M. Goyal, M. Bansal, S. Pottahil, S. Bedi, Direct to consumer advertising: a mixed blessing. Nat. J. Integr. Res. Med. **6**(3) (2015)
5. J. Pandey, M. Mishra, A. Jha, Negative impact of direct-to-consumer (DTC) promotion on Indian patients, in *Asian Business and Management Practices: Trends and Global Considerations* (IGI Global, 2015), pp. 92–106
6. J. Rani, A. Jha, Impact of age on online healthcare information search: a study on Indian patients. Asian J. Manag. **6**(1), 17–24 (2015)
7. A. Jha, J.R. Pandey, An empirical note on health information digital divide: a study of Indian patients. Int. J. Asian Bus. Inf. Manag. (IJABIM) **8**(2), 15–34 (2017)
8. A. Jha, J.R. Pandey, S.K. Mukherjee, An empirical note on perceptions of patients and physicians in direct-to-consumer promotion of pharmaceutical products: study of Indian patients and physicians, in *Management Strategies and Technology Fluidity in the Asian Business Sector* (IGI Global, 2018), pp. 65–87
9. J.R. Pandey, A. Jha, S.K. Saha, Impact of manipulative character of direct-to-consumer promotion, in *Dynamic Perspectives on Globalization and GlobalSustainable Business in Asia* (IGI, 2019), pp. 198–211
10. S.K. Mukherjee, J. Kumar, A.K. Jha, J.R. Rani, Role of social media promotion of prescription drugs on patient belief-system and behaviour. Int. J. e-Collab. (IJeC) **15**(2), 23–43 (2019)
11. M. Perri, A.A. Nelson, An exploratory analysis of consumer recognition of direct-to-consumer advertising of prescription medications. J. Health Care Market **7**, 9–17 (1987)
12. S. An, Antidepressant direct-to-consumer advertising and social perception of the prevalence of depression: application of the availability heuristic. Health Commun. **23**, 499–505 (2008)
13. R.A. Bell, R.L. Kravitz, M.S. Wilkes, Direct-to-consumer prescription drug advertising and the public. J. Gen. Intern. Med. **14**, 651–657 (1999)
14. K.J. Bozic, A.R. Smith, S. Hariri et al., The ABJS Marshall Urist Award—the impact of direct-to-consumer advertising in orthopaedics. Clin. Orthop. Rel. Res. **458**, 202–219 (2007)
15. A. Deshpande, A. Menon, M. Perri, G. Zinkhan, Direct-to-consumer advertising and its utility in health care decision making: a consumer perspective. J. Health Commun. Int. Perspect. **9**, 499–513 (2004)
16. E. Murray, B. Lo, L. Pollack, K. Donelan, K. Lee, Direct-to-consumer advertising: public perceptions of its effects on health behaviors, health care, and the doctor-patient relationship. J. Am. Board Fam. Pract. **17**, 6–18 (2004)
17. J.S. Weissman, D. Blumenthal, A.J. Silk, K. Zapert, M. Newman, R. Leitman, Consumers' reports on the health effects of direct-to-consumer drug advertising. Health Aff. **W3**, 82–95 (2003)
18. L.J. Burak, A. Damico, College students' use of widely advertised medications. J. Am. Coll. Health **49**, 118–121 (2000)
19. S.M. Choi, W.N. Lee, Understanding the impact of direct-to-consumer (DTC) pharmaceutical advertising on patient-physician interactions—adding the web to the mix. J. Advert. **36**, 137–149 (2007)
20. B. Datti, M.W. Carter, The effect of direct-to-consumer advertising on prescription drug use by older adults. Drugs Aging **23**, 71–81 (2006)
21. D.E. DeLorme, J. Huh, L.N. Reid, Age differences in how consumers behave following exposure to DTC advertising. Health Commun. **20**, 255–265 (2006)

22. D.E. DeLorme, J. Huh, L.N. Reid, A. Soontae, The state of public research on over-the-counter drug advertising. Int. J. Pharm. Healthc. Mark. **4**(3), 208–231 (2010)
23. D.E. DeLorme, J. Huh, Seniors' uncertainty management of direct-to-consumer prescription drug advertising usefulness. Health Commun. **24**, 494–503 (2009)
24. M. Herzenstein, S. Misra, S.S. Posavac, How consumers' attitudes toward direct-to-consumer advertising of prescription drugs influences ad effectiveness, and consumer and physician behavior. Mark. Lett. **15**(4), 201–212 (2004)
25. J. Huh, D.E. DeLorme, L.N. Reid, The third-person effect and its influence on behavioral outcomes in a product advertising context: the case of direct-to-consumer prescription drug advertising. Commun. Res. **31**, 568–599 (2004)
26. R.H. Kon, M.W. Russo, B. Ory, P. Mendys, R.J. Simpson, Misperception among physicians and patients regarding the risks and benefits of statin treatment: the potential role of direct-to-consumer advertising. J. Clin. Lipidol. **2**, 51–57 (2008)
27. B. Lee, C.T. Salmon, H.J. Paek, The effects of information sources on consumer reactions to direct-to-consumer (DTC) prescription drug advertising—a consumer socialization approach. J. Advert. **36**, 107–119 (2007)
28. J.S. Marinac, L.A. Godfrey, C. Buchinger, C. Sun, J. Wooten, S.K. Willsie, Attitudes of older Americans toward direct-to-consumer advertising: predictors of impact. Drug Inf. J. **38**, 301–311 (2004)
29. A.M. Menon, A.D. Deshpande, M. Perri, G.M. Zinkhan, Consumers' attention to the brief summary in print direct-to-consumer advertisements: perceived usefulness in patient-physician discussions. J. Public Policy Mark. **22**, 181–191 (2003)
30. M. Joseph, D.F. Spake, D.M. Godwin, Aging consumers and drug marketing: senior citizens' views on DTC advertising, the medicare prescription drug programme and pharmaceutical retailing. J. Med. Mark. **8**(3), 221–228 (2008)
31. T.C. Elif, Online health information seeking habits of middle aged and older people: a case study. Master's thesis (2010)
32. C.H. Rajani, A study to explore scope of direct to consumer advertisement (DTCA) of prescription drugs in India. Int. J. Mark. Hum. Resour. Manag. (IJMHRM) **3**(1) (2012)
33. J. Reast, D. Palihawadana, H. Shabbir, The ethical aspects of direct to consumer advertising of prescription drugs in the United Kingdom: physician versus consumer views. J. Advert. Res. **48**(3), 450–464 (2008)
34. B.A. Liang, T. Mackey, Direct-to-consumer advertising with interactive Internet media. Global regulation and public health issues. JAMA **305**(8), 824–825 (2011)
35. D.L. Frosch, D. Grande, D.M. Tarn, R.L. Kravitz, A decade of controversy: balancing policy with evidence in the regulation of prescription drug advertising. Am. J. Public Health **100**(1), 24–32 (2010)
36. N. Sumpradit, R.P. Bagozzi, F.J. Ascione, Give me happiness or take away my pain: explaining consumer responses to prescription drug advertising. Cogent Bus. Manag. 2(1) (2015). https://doi.org/10.1080/23311975.2015.1024926
37. J.K. Prigge, B. Dietz, C. Homburg, W.D. Hoyer, L. Burton Jr., Patient empowerment: a cross-disease exploration of antecedents and consequences. Int. J. Res. Mark. **32**(4), 375–386 (2015)
38. J.C. Bélisle-Pipon, B. Williams-Jones, Drug familiarization and therapeutic misconception via direct-to-consumer information. J. Bioeth. Inq. **12**(2), 259–267 (2015)
39. J.G. Ball, D. Manika, P. Stout, Causes and consequences of trust in direct-to-consumer prescription drug advertising. Int. J. Advert. **35**(2), 216–247 (2016)
40. C. Pechmann, J.R. Catlin, The effects of advertising and other marketing communications on health-related consumer behaviors. Current Opin. Psychol. **10**, 44–49 (2016)
41. C. Adams, F.L. Gables, Direct-to-consumer advertising of prescription drugs can inform the public and improve health. JAMA Oncol. **2**(11), 1395–1396 (2016)
42. T. Singh, D. Smith, Direct-to-consumer prescription drug advertising: a study of consumer attitudes and behavioural intentions. J. Consum. Mark. **22**(7), 369 (2005)

43. A. Mehta, S.C. Purvis, Consumer response to print prescription drug advertising. J. Advert. Res. **43**, 194–206 (2003)
44. J. Lexchin, B. Mintzes, Direct-to-consumer advertising of prescription drugs: the evidence says no. J. Public Policy Market **21**(2), 194–201 (2002)

Modelling Orographic Rainfall Using Regression Based Method—A Case Study

Pooja Verma, Swastika Chakraborty, Aishee Bhattacharya, Rupali Sinha, Joshna Kotha, Pragya Jaiswal, and Deepak Singh

Abstract The article attempts to develop regression based model for the prediction of stochastic—deterministic phenomena like orographic rain in North East Indian hills and valleys using historical thirty eight years data of rainfall over the three hill stations Majhitar, Silchar and Shillong. Considering the randomness, non-stationarity within the time series, the suitable model for prediction of rainfall has been determined. The performance of the prediction model is also calculated in terms of deviation from actual data. Results shows that for long term prediction of rainfall for the autoregressive integrated moving average ARIMA model (3, 1, 2) is suitable for obtaining the prediction of orographic rainfall of North eastern India.

Keywords ARMA model · ARIMA model · Seasonal ARIMA model · Rainfall prediction · Mean absolute deviation (MAD)

P. Verma · S. Chakraborty (✉) · A. Bhattacharya · R. Sinha · J. Kotha · P. Jaiswal · D. Singh
Electronics and Communication Engineering Department, Sikkim Manipal Institute of
Technology, Majitar, Rangpo, Sikkim, India
e-mail: swastika1971@gmail.com

P. Verma
e-mail: puja20verma@gmail.com

A. Bhattacharya
e-mail: aisheeb122000@gmail.com

R. Sinha
e-mail: 2706ahana@gmail.com

J. Kotha
e-mail: k.joshna1234@gmail.com

P. Jaiswal
e-mail: pragyajais.1110@gmail.com

D. Singh
e-mail: deepaksinghrajput2703@gmail.com

© Springer Nature Singapore Pte Ltd. 2020
R. Bera et al. (eds.), *Advances in Communication, Devices and Networking*,
Lecture Notes in Electrical Engineering 662,
https://doi.org/10.1007/978-981-15-4932-8_16

1   Introduction

1.1   Rainfall Modeling

Rainfall modelling is an important operational responsibility for the hill region like North East India where Indian summer monsoon is prevalent almost half of a calendar year. Rainfall, as it is advantageous for the agricultural country like India if proper forecasting can be done, in another way it has got devastating effect like landslide etc. for the hill area people of the society if rainfall cannot be predicted properly. Historical data of thirty years over Sikkim, a North East Indian state, shows that average annual rainfall is 3097 mm [1] and it is increasing at the final rate of 41.46 mm/decade. Long term prediction of rainfall may help to counter the devastating effect of rainfall as well as to restore rainfall for agricultural purpose. Therefore it is very important to find the suitable statistical prediction model of rain for those regions having such a high amount of rainfall like Sikkim. Atmospheric phenomenon like rainfall is highly chaotic in nature, which makes the performance of the numerical models for long range prediction as well as extreme rainfall prediction [2] poorer. To find the teleconnection, performance of the regression models in stochastic forecast is used in literature [3, 4]. Robustness, reliability in the reconstruction of time series considering highly spatio-temporal behaviour can be achieved using neural network as forecasting technique [5]. Due to orographic effect extreme rainfall in hill region varies strongly [6] over short spatial distance also. Rain gauge is extremely unavailable in high altitude where strong rainfall sometimes is affected by snow. Therefore there is a strong need of finding suitable statistical prediction model for the precipitation of North East Indian state like Sikkim, Assam, and Shillong where conventional theoretical model does not perform well in predicting rainfall. The aim of this work to explore suitable regression based model for the above said three stations to address the orographic precipitation with the help of regression based models.

2   Model Determination

2.1   The Prediction Model

To determine the stochastic-deterministic time series like orographic rainfall where non stationarity of the time series dominates stationarity, auto regressive moving average model (ARMA) is the most is widely used model by researchers. Primarily ARMA model is taking care of short term rainfall like hourly rainfall of the time series data where auto covariance of the hourly rainfall replicates auto covariance of low order ARMA model. When the inter-annual and inter-monthly variability comes into consideration in long term rainfall for the time series mapping so that Auto regressive integrated moving average model (ARIMA) is the preferred choice over ARMA model.

2.2 ARMA Model

ARMA model can be described mathematically by

$$y_t = \phi_1 y_{t-1} + \phi_2 y_{t-2} + \cdots + \phi_p y_{t-p} + u_t - \theta_1 u_{t-1} - \theta_2 u_{t-2} - \cdots - \theta_q u_{t-q}, \tag{1}$$

here, p is the autoregressive (AR) parameters, and
q is the moving average (MA) parameters,
$\phi_1, \phi_2, ..., \phi_p$ = autoregressive coefficients, and
the real parameters $\theta_j (j = 1, 2, ..., q)$ = moving average coefficients,
u_t = independent white noise sequence, i.e., $u_t \sim N(0, \sigma^2)$.
Usually the mean of $\{y_t\}$ = zero; if not, $y' = y_t - \mu$ is used.

The stationary processes like hourly rainfall where normal marginal distribution is there can be well described by the above equation. The parameters of the model like auto regressive and moving average coefficients can be derived from hourly data of rainfall.

2.3 ARIMA Model

Non stationarity of precipitation series dominates over stationarity when time period of observation moves from hourly to monthly and annually. Here non stationarity combined with cycles, trends and seasonality of past data creates impact on future data.

Lag operator (B) is then introduced in ARMA model and mathematically it is represented by

$$\phi(B) = 1 - \phi_1 B - \phi_2 B - \cdots - \phi_p B^p \tag{2}$$

$$\theta(B) = 1 - \theta_1 B - \theta_2 B^2 - \cdots - \theta_q B^q, \tag{3}$$

where, $\phi(B)$ = autoregressive operator and
$\theta(B)$ = moving-average operator.
Then the model can be simplified as

$$\phi(B)y_t = \phi(B)u_t \tag{4}$$

If $\{y_t, t = 1, 2, 3, \ldots, n\}$ is non stationary, we obtain the z_t (stationarized sequence) by means of difference, i.e.,

$$z_t = (1 - B)^d y_t = \nabla^d y_t, \tag{5}$$

where, d = number of regular differencing,

Then the corresponding ARIMA model (p, d, q) for y_t can be made [7], where, d = number of differencing passes by which the non-stationary time series might be described as a stationary ARMA process.

2.4 Seasonal ARIMA Model

Periodicity of the time series like orographic rainfall when is not proper or in other words when seasonality has to be taken into consideration, ARIMA model is mathematically modified with the seasonal effect i.e. seasonal autoregressive, seasonal moving average and seasonal differencing parameter are:

$$\phi_p(B)\Phi_p(B^s)(1 - B)^d y_t = \theta_q(B)\Theta_q(B^s)u_t \tag{6}$$

where, p = number of seasonal autoregressive parameter, and q = number of seasonal moving average order, s = the period length (in month in this work), d = number of differencing passes.

2.5 Climate of Study Area

Climate of Silchar (Assam) is tropical monsoon climate having rainfall in most of the months of the year with a very short dry period. As the monsoon moves in the region during April, very hot and humid weather with heavy thunderstorms occurs almost every afternoon until the mid of October. Average annual rainfall in Silchar is 3144 mm with average temperature 24.9 °C. In Majhitar (Sikkim) also, the climate is warm tropical climate with temperate inclination. The rain is comparatively higher in summer than the winter. Average annual rainfall in Majhitar is 3148 mm with average temperature 18.0 °C. The variation in the temperature throughout the year is 11.1 °C. Pleasant and pollution-free climate is the beauty of Shillong. In the summer, the temperature varies from 23 °C (73 °F). In Shillong, the summers are cool and very rainy, whereas winters are cool and dry. Average annual rainfall in Shillong is 3385 mm with average temperature 17.0 °C throughout the year. According to Koppen climate classification Shillong is classified as dry climate where as the other two stations are tropical climate.

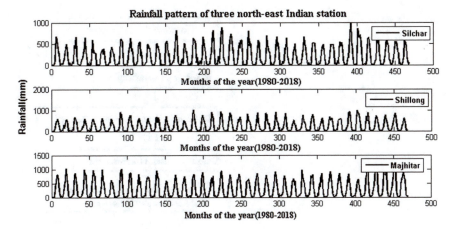

Fig. 1 Rainfall pattern of three station

2.6 Data Source

Giovani is the web data source for providing user the National Aeronautics and Space Administration (NASA)'s geophysical gridded data from various satellites and surface observations to analyse to derive important scientific conclusions. It facilitates researchers with the capability to analyse examine data on atmospheric parameters. Historical rainfall data of thirty eight years (1980–2018) for three North East (NE) Indian hill station having very slight climatic variation is considered in this study.

Figure 1 shows the rainfall pattern of three north-east Indian station i.e. for Silchar, Shillong and Majhitar in the state of Assam, Meghalaya and Sikkim respectively. So, all the three stations shows the different variations of rainfall. Because of this we actually knows that how rainfall pattern can be seen also it's easy to compare for one station to the other stations.

2.7 Methodology

The method of modelling time series is started with the testing of stationarity of the time series data. After testing the data is processed to eliminate the trending effect and seasonal effect. Then suitable regression model is identified for each of the three stations. Iteration has been carried out to identify the most suitable model parameter. After suitable model identification white noise test is carried out to identify the forecasting model for orographic precipitation. The methodology is described by following flow chart (Fig. 2).

Fig. 2 Flow chart for
forecasting model

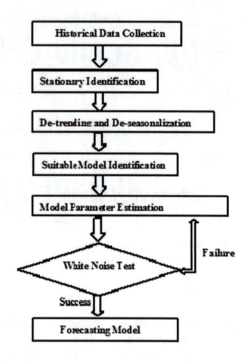

Fig. 2 Flow chart for
forecasting model

2.8   Results and Discussion

As a first step for identifying the stationarity of data and auto-correlation of the time
series data for three stations are plotted in Fig. 3. From the figure there is no serial
dependence of time series is found. Auto-correlation of the data clearly indicates non
stationarity of the time series with distinct randomness. It is positive at some point,
negative at some other point with some point where autocorrelation vanishes.

From Fig. 3 seasonality of the data is coming out which is to be addressed in
model identification.

2.9   Proposed Model Description

To identify the time series of each of the three stations taking the advantage of
autocorrelation, autoregressive model (AR) is tried as a first step. After defining AR
parameter and taking the advantage of Holt-Winters' method [8] to smooth the data,
the trend, and the seasonal index are taken care. For monthly observation, yearly
moving average, MA (12), eliminate or averages out with the seasonal effect.

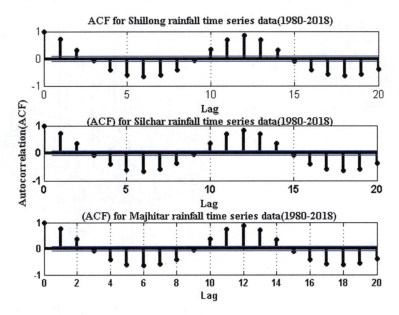

Fig. 3 ACF of rainfall time series for the three stations

3 ARIMA Model Parameter Identification

To combine the above two effects ARMA model is considered for defining the time series of the three stations. The most important things about the ARMA model is to know its specification. After that pre-fit identification and post-fit testing is to be done with the projected ARMA model before forecasting the time series with it. As a first step of finding ARMA model de-trending of the data is done and the effect of seasonality is eliminated. Considering stationarity and invertibility within the data series the fitting percentage of the model with the original data is considered. Simultaneously the error percentage of fitting AR model and MA model is observed by changing the order of the model. Looking at error statistics for each iteration of changing order of AR and MA model for each of the three stations it is found that ARIMA (3, 1, 2) model performs well for Shillong and Majhitar and Silchar.

3.1 Inclusion of Seasonal Effect in ARIMA (SARIMA) Model

Actual time series data of thirty-nine years is seasonally changing to get more insight into the inter-monthly variation of the data series. Seasonal consideration of ARIMA model is taking care of both inter- annual and inter-monthly variation of the data series. The graphical representation of ARIMA model (3, 1, 2) for predicted values of rainfall as well as actual rainfall shows the seasonal effect in Fig. 4.

Fig. 4 a ARIMA (3, 1, 2) model predicted rainfall along with actual rainfall. **b** ARIMA (3, 1, 2) model predicted rainfall along with actual rainfall taking care of seasonal effect

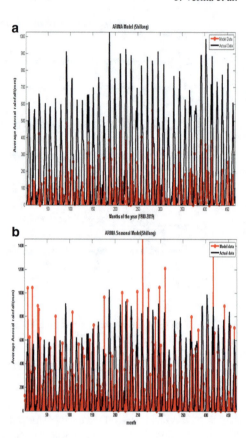

3.2 Comparison of Proposed SARIMA (3, 1, 2) Model with Existing Conventional Model

To get the best result we calculate the variance of ARIMA Model (3, 1, 2) for all the three stations: Silchar, Shillong and Majhitar yearly. By this we found the maximum and minimum variance which shows that in which month highest rainfall can be occurred (Table 1).

The ARIMA model (3, 1, 2) catches the correct trend overall and predicts the monthly precipitation in most months with high accuracy. It predicts highly accurately for the dry seasons, such as January, February, March, November and December.

However it overestimates the precipitation of July and August.

Table 1 Variance of ARIMA model (3, 1, 2) for three stations

S. No	Months	Silchar			Shillong			Majhitar		
		SARIMA	ARIMA	AR	SARIMA	ARIMA	AR	SARIMA	ARIMA	AR
1	January	0.296887	0.886272	1.07133	0.57149	0.82535	0.219394	0.196031	4.11454	1.3163
2	February	0.125526	1.78012	1.47058	0.08794	1.66273	0.52345	0.07063	2.39291	2.13682
3	March	0.176845	21.265	2.336464	0.23878	16.116	0.622284	0.72257	6.83858	1.505944
4	April	0.359884	16.1115	1.049684	0.3184	32.5682	0.425378	0.846815	78.6497	3.29185
5	May	0.022175	220.964	3.574593	0.17876	124.401	0.941582	0.073199	227.162	2.344497
6	June	0.07828	168.255	3.52269	0.07427	233.04	0.948771	0.537939	255.515	2.668497
7	July	0.37972	162.412	2.511329	0.095110	261.582	0.025418	0.324292	147.621	1.483851
8	August	0.007453	211.03	2.065458	0.065050	244.325	0.186437	0.19999	318.58.3	1.466391
9	September	0.351662	105.346	1.732247	0.490517	110.385	0.732247	0.225695	239.529	1.474019
10	October	0.34445	90.4437	1.06885	0.362954	164.26	1.07353	0.067380	41.4005	1.292821
11	November	0.177344	25.0602	1.08837	0.292212	7.85162	1.08577	0.046570	13.84	1.906867
12	December	0.0084113	2.69602	1.53726	0.955688	3.05856	1.53728	0.202253	1.54230	1.00679

3.3   Mean Absolute Deviation (MAD)

It is the total forecasting error for a model. It is calculated by taking sum of the absolute values of the forecasting error and divided by the number of data periods. Following Fig. 5 shows the plot of mean absolute deviation for each of the three stations.

$$\text{MAD} = \frac{\sum |\text{Actual Value} - \text{Forecasting Value}|}{n} \tag{7}$$

where:

n = number of data periods.

From the above facts and figures it is obvious that regression based model performs well in forecasting nature of orographic precipitation but not very well in predicting values of precipitation showing the error in prediction as much as 200%.

However inclusion of seasonally differenced time series may improve the prediction accuracy to a significant value.

4   Motivation

The motivation of the present study is to determine an appropriate univariate representation model of regression to account for a very special type of rainfall, named orographic rain in North-East Himalayan hills considering the effect of non-linearity randomness of the long term time series.

5   Speciality

Updraft of moist air along the side of the hills causes a rainfall (orographic rain) which is not very intense in terms of rain rate (below 20 mm/h), but having a longer duration during most of the time of the day from May to October of each year at North-Eastern Himalayan region. The rain accounts 75% of total duration causes severe effect on society causing landslide and associated problems needs to be predicted beforehand to take any precaution before its happening.

6   Conclusions

Thirty Eight years rainfall data is used in this work to find the suitable regression based model for orographic precipitation after the identification and estimation of the proper model for each of the NE (North-East) hill station, diagnostic checking

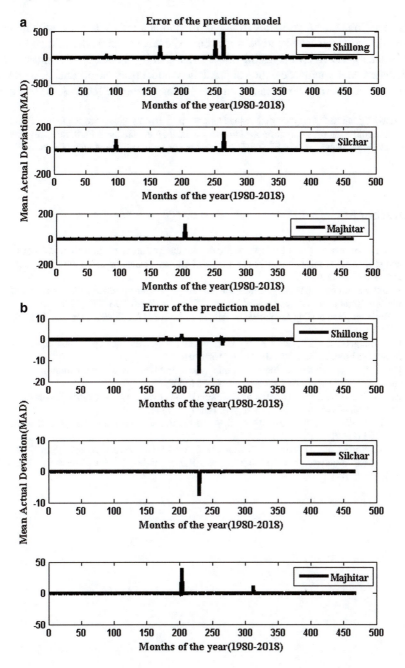

Fig. 5 a MAD values for non-seasonal ARIMA of the three stations, b MAD values for seasonal ARIMA of the three stations

has been done for the residuals from the fitted model for white noise with iterative process. Errors are calculated between original data and the modelled data for each of the station. Then for the testing of the success of the model, error performance is calculated. Mean actual deviation of prediction model from the original data series is found to decrease to a significant level considering seasonal effect in ARIMA model.

Acknowledgements Analyses and visualizations used in this study were produced with the Giovanni online data system, developed and maintained by the Goddard Earth Sciences Data and Information Services Center (GES DISC), National Aeronautics and Space Administration (NASA).

References

1. H. Rehman, An analysis of past three decade weather phenomenon in the mid-hills of Sikkim and strategies for mitigating possible impact of climate change on agriculture. http://sikenvis.nic.in/writereaddata/
2. P. Guhathhakurta, Long-range monsoon rainfall prediction of 200 for the districts and sub-division Kerala with artificial neural network. Curr. Sci. **90**(6), 773–779 (2006)
3. P. Burlando, Forecasting of short-term rainfall using ARMA models. J. Hydrol. **144**, 193–211 (1993)
4. H.R. Wang, C. Wang, X. Lin, J. Kang, An improved ARIMA model for precipitation simulations. Nonlinear Process. Geophys. **21**, 1159–1168 (2014)
5. P.S. Lucio, Spatiotemporal monthly rainfall reconstruction via artificial neural network-case: study: South of Brazil. Adv. Geosci. **10**, 67–76 (2007)
6. Z. Yun-Jian, Changes in extreme precipitation events over the Hindu Kush Himalayan region during 1961–2012. Adv. Clim. Change **8**, 166–175 (2017)
7. S. Makridakis, M. Hibon, ARMA models and the Box-Jenkins methodology. J. Forecast. **16**, 147–163 (1997)
8. C.P.D. Veiga, C.R.P.D. Veiga, A. Catapan, U. Tortato, Silva, Demand forecasting in food retail: a comparison between the HoltWinters and ARIMA models. WSEAS Trans. Bus. Econ. **11**, 608–614 (2014)

Electric Vehicles in the Indian Scenario

Noimisha Hazarika, Pratyasha Tamuli, and Amit Kumar Singh

Abstract Over the last few years, electric vehicles have been gaining popularity in and around the world with their growing adoption in countries like China, Europe and the USA. Several countries are aiming to make the vehicles fully electric in the coming years. In India, electric vehicles seem to replace the conventional vehicles soon due to the increasing prices of fuel along with the growing environmental pollution. Researchers throughout the globe are focusing on accelerating the adoption of electric vehicles by the masses. This paper focuses on the advancements in the field of electric vehicles and its feasibility in the long run. In the first part, the challenges encountered by the EVs in the market of India have been discussed along with the Indian EV market scenario with the plans presented by the government. In the subsequent section, the recent studies on EV in India are discussed trailed by a discussion on wireless power transfer.

Keywords Electric-vehicle · Charging · Cost · Market · Wireless charging

1 Introduction

With rapid urbanization in developing countries, the growing CO_2 emissions because of the transport sector pose a threat to the environment. In India, the transport sector contributes to 11.23% CO_2 emissions (cars, trains, two-wheelers, etc.) of total fuel combustion in accordance with 2014 data [1]. In accordance with the 2019 database of the World Health Organization, 22 of the top 30 maximum polluted cities all over the world are found to be in India, with Gurugram topping the list [2]. Consequently,

N. Hazarika · P. Tamuli · A. K. Singh (✉)
Department of Electrical and Electronics Engineering, Sikkim Manipal Institute of Technology, Sikkim Manipal University, Rangpo, Sikkim, India
e-mail: amit.s@smit.smu.edu.in

N. Hazarika
e-mail: noimisha_201600097@smit.smu.edu.in

P. Tamuli
e-mail: pratyasha_201600699@smit.smu.edu.in

© Springer Nature Singapore Pte Ltd. 2020
R. Bera et al. (eds.), *Advances in Communication, Devices and Networking*,
Lecture Notes in Electrical Engineering 662,
https://doi.org/10.1007/978-981-15-4932-8_17

the health issues resulting from air pollution cannot be ignored. In 2013, the Global Burden of Disease Study for 2010 stated that diseases resulting from air pollution caused 62,000 deaths becoming the fifth largest killer in India [3]. The major reasons for the growing pollution from automobiles in India include the rising population of the country and the resulting demand for private vehicles, fuel adulteration, traffic congestion, bad road conditions, faulty traffic management systems, etc. [4].

In the twenty-first century, the increasing focus on renewable energy has led to the popularity of EVs [5] which prove to be cleaner than most of the conventional vehicles. In India, the government has been promoting electric vehicles as moving toward a clean and pollution-free environment is the need of the hour for the country with one of the major automobile industries in the world. In 2013, the Government of India launched the National Electric Mobility Mission Plan (NEMMP) 2020 with a view to bring a shift in the automobile industry of the country by introducing programmes and policies for about 6–7 million EV by the year 2020 [6]. In 2017, the National Institution for Transforming India (NITI) Aayog, a policy think tank of the Government of India published a report which stated that the use of shared and electric vehicles could save up to $34 billion in petrol and diesel costs and reduce up to 1 gigaton of carbon emissions by the year 2030 [7]. The Government of India launched a scheme named the FAME India, which provides incentives on the adoption of EVs up to Rs. 13,000 for bikes and Rs. 1.13 lakh for cars [8].

1.1 Challenges in the Implementation of Electric Vehicles in India

(1) The cost of EVs is a major reason of the sluggish demand of EV market in India. The price of an electric car happens to be around 2.5 times that of a conventional fuel car [9], which is usually due to the expensive Li-ion batteries which are imported from other countries [10]. The battery manufacturing companies like Panasonic, BYD, Tesla are not based in India, which makes it necessary to import batteries. This is not feasible as it would put India back to the market of petrol and diesel. The only way to make EVs affordable is by the production of batteries in India and to test them to cope with Indian roads and weather conditions [11]. However, in 2017, Suzuki MotoCorp has announced that it has entered into a joint venture with Toshiba and Denso Corp, two Japanese firms, to produce lithium-ion batteries in India [12].

(2) The need for a robust charging station network in India is like the heart of EVs today. It is one amongst the major factors behind the lesser vending of EVs in India. It becomes difficult to convince the customers to buy the electric vehicles as even a fast charger may take 2 h to charge an electric car [12]. None of the electricity distribution utility is willing to set up charging infrastructure without the increase of sales of electric vehicles in India. According to observations made from a report released in 2017, India lags in the availability of enough

charging stations with only around 300 in the country. In order to reach the target of 2020, the government needs to invest around Rs. 445–1000 crore in the establishment of charging infrastructure and power generation capacity of 125–225 MW [13]. Ather Energy, a Bangalore based start-up claims to be the largest EV charging infrastructure in the country [13] Ola associated with Indian Oil to launch the first EV charging station in Nagpur in 2017 [14] While the government has been aiming at the increase in setting up of charging setup across the country, the process appears to be quite sluggish.

(3) Lack of public awareness is a key reason behind the slow adoption of EVs in India. Public should be made aware of the benefits of electric vehicles including less dependence on oil, decrease in creation of green-house gases and improving the air quality, which makes it more likely for people to shift towards EVs.

(4) Time consumption: Shifting to EVs will undoubtedly bring employment opportunities along with other benefits in the country. But this shift from conventional fuel cars to EVs is expected to take at least a decade or so. Factors such as policies, convincing the customers, cost, etc., play a significant part in the progression [12]. The introduction of the FAME II scheme in 2019 is likely to contribute in the faster implementation of EVs in the country, with its purpose to bring ample electrification in the transport sector by 2030.

(5) Technical challenges: Technical challenges in EVs include dealing with major problems like enhancement in battery technologies, charging time reduction, getting enough driving range, etc. One of the major barriers in achieving higher driving ranges is the low energy-density of EV batteries like the lead–acid batteries which are very bulky in nature and hence increases the weight of the vehicle in general [15].

(6) Lack of incentive-based Government policies: Introducing incentives in the EV market not only reduces the cost of Hybrid and EVs, but also encourages their growth in the market. The National Electric Mobility Mission Plan (NEMMP) was set up in 2012, under which an incentive-based scheme titled Faster Adoption and Manufacturing of Hybrid and Electric Vehicle (FAME) was launched in 2015 which gives an aid on the marketing price of cars. Moreover, the Central Government of India and some State Governments like the National Capital Territory of Delhi (NCT of Delhi) provide tax incentives to consumers who preferably treat Hybrid and EVs over conventional technologies [15].

1.2 Steps Taken for the Adoption of EVs in India

National Electric Mobility Mission Plan (NEMMP) 2020:

- Targets at introducing 5–7 million EVs in India by the year 2020 [16].
- Particularly emphasizes in the importance of government incentives in improving the EVs' market and in the coordination between the industries and academia [16].

- Aims at achieving national fuel security by encouraging the implementation of Hybrid and EVs in India. It has, in fact, set an ambiguous goal of selling around 6–7 million EVs per year from the year 2020 onwards [17].
- An overall asset of Rs. 20,000–23,000 Cr (USD 3 billion approx.) is required [16].

E-rickshaw implementation through DeenDayal Scheme in June 2014:

- Aims at helping in the funding and procurement of e-rickshaws in India.
- The Motor Vehicles Amendment Bill was passed in March 2015 that established e-rickshaws as a valid means of transport in the country.
- The number of e-rickshaws in the transport eco-system of Delhi had increased from 4,000 in 2010 to more than 1,00,000 in 2014.

Rapid Adoption and Production of Electric and Hybrid Vehicles in India (FAME India Scheme):

- The FAME India organization was administered by the Department of Heavy Industry since 01 April 2015.
- Under this scheme, 11 cities around the country will be given funding for introducing electric buses, taxis and three-wheelers. The nine big cities will get 15 buses each while the other two cities under the special category will get 15 buses each. Subsidies for taxis are given to four cities while subsidies for three-wheelers are given to three cities [16].

(7) Lack of mineral reserves in India: India is entirely reliant on imports for almost all the major minerals required in the process of manufacturing the batteries. Lack of availability of mineral reserves in the homeland becomes a key factor against the adoption of EVs in India [18].

(8) Urban Planning: Electric mobility is expected to have an impact on the planning and design of future Indian cities. The planning must ensure allocation of large spaces for the implementation of large-scale charging stations for EVs that can accommodate many vehicles at a time for charging. The problem of urban planning is especially true to the Indian scenario because usually open urban spaces and proper parking facilities are a rarity in the country. Large-scale charging stations will have an impact on the residential neighbourhoods, workplaces, shopping centres and other entertainment facilities [19].

2   Indian EV Market Scenario

2.1   Size of the Indian EV Market

The current Indian Electric-Vehicle Market is at an incipient phase encompassing only 1% of the entire automobile sales in the country. Though 95% of the EV market of India is conquered by two-wheelers and three-wheelers, the overall share in the market is negligible [20]. Electric two-wheelers and three-wheelers were first launched in the Indian automobile market back in the 1990s but were soon discontinued because they did not prove to be as successful in the market. A breakthrough was observed in the early 2000s when Mahindra Electric launched its first electric car named Reva in the Indian automobile market in the year 2001. Mahindra Electric highlighted the development of EVs at a rational price. An improved model of this car was launched in the year 2013 which was powered by lithium-ion batteries and had a driving range of around 100 km [15]. The company strategizes to manufacture 34,000 EVs yearly from the year 2020 [20]. Later, Toyota launched the Prius Hybrid model in the year 2010 which was followed by Camry Hybrid in 2013 [15]. The Indian EV market had about 25,000 units during the period 2016–17. Later, Tata Motors has introduced its EV named Tata Tigor in 2018. In 2019, the Indian unit of South Korea's Hyundai Motor Co. Ltd launched its EV in the country. Maruti Suzuki will enter the scene in 2020 [20]. According to a study conducted by P&S Intelligence, the Indian EV market is likely to reach US $427.4 million by the year 2025 [21]. The market is likely to grow in the coming years and successfully stimulate the growth of EV implementation in the country.

2.2   Maharashtra Presides Over the Indian EV Market

Maharashtra had the maximum sales in electric cars in the year 2017 accounting for a market share of 17.9% in the Indian Electric-Vehicle market. Maharashtra is likely to continue its foremost contribution in the EV market as the state government's EV Policy 2018 with an aim to help in the growth of the Indian EV market by supporting EV manufacturing, development of charging stations and providing subsidies to the consumers of EVs (the subsidies amounted up to US $1275 (INR 100,000) per electric vehicle). However, the state of Tamil Nadu is likely to beat all other Indian States in the EV market with a CAGR of 26.5% over the years [21].

2.3   Policy to Push Electric Vehicles

After Karnataka and Andhra Pradesh, Maharashtra is also planning to implement the policy to push electric vehicles by setting up more charging stations, offering subsidies to the buyers on the cost price of the electric vehicles and even refunding taxes to the companies that manufacture battery-powered electric vehicles [22]. The points that the policy has to offer are mentioned below:

- 25–100% of reimbursement on land and loan agreements for the makers of electric vehicles.
- Return of state GST if the car is sold within the state.
- Offering inducement to the battery makers as well as other component makers.

- Offering a subsidy of 25% or Rs. 10 Lakh (whichever is lesser) on the first 225 charging stations constructed near bus depots, petrol pumps or any other public parking areas.
- Subsidies up to Rs. 10 lakhs for buyers (based on the type of vehicle).
- Exemption from road taxes and registration fees [22] (Table 1).

Table 1 Some of the recent research works on the study of EVs in India

Paper title	Objectives of research	Salient features of research	Conclusion
Hybridization of light commercial vehicles in India [27]	To provide the essentials, profits and future for growth of light Hybrid EVs in India	• Advantages as well as obstacles in the employment of hybrid technology in light marketable vehicles, types of hybrid system, etc., are analyzed	• Applying hybrid-technology in lightweight marketable vehicles can uplift the automobile sector of India
Creating a platform to foster the growth of electric vehicles in Indian market by increasing the electric vehicle public charging stations [28]	To provide a technique to charge electric vehicles from electric poles while keeping the charging rate limited	• Developing public charging stations (Type 2) near electric poles with the help of NEMA 5–15 connector and household socket • Developing of Type 1 charging stations by stepping down the grid voltage from 137 V AC to 120 V AC	• Setting up of charging infrastructure can aid the increase of ownership of EVs in India
Safety considerations for EV charging in India [29]	To provide information regarding regulations and safety requirements in the maintenance and installation of charging infrastructure in India	• A comparative study on the present and international standards for setting up of charging infrastructure • Safety considerations for wireless charging as well as battery swapping for EV	• Standardization of charging infrastructure considering electrical safety can be achieved keeping in view with international standards
India's charging infrastructure-biggest single point impediment in EV adaptation in India [30]	To present a study of the challenges of charging setup in India, charging standards and the development needed to overcome the barriers	• Study of factors disturbing the progress of charging setup in India • Needs of development and improvement in charging standards and other sectors to scale the growth of charging infrastructure	• Necessary improvements in the required fields can speed up the charging infrastructure in India

(continued)

Table 1 (continued)

Paper title	Objectives of research	Salient features of research	Conclusion
Affordable hybrid topology for PV end LDVs in prospering India [31]	To present a case study of 23 V (P)HEV system benefits, comparative study of P2 vs P4 approaches using simulations using 23 V system on a diesel-fueled car engine	• 0-D Simulation of hybrid components on the New European Driving Cycle (NEDC) and real-world cycles using a 1, 1L 3 cylindrical diesel engine from FEV database • Simulation of the data and evaluation of the potential of 23 V system in the vehicle performance and emission • Cost analysis of the preferred architecture for the Indian market	• P2 architecture is more efficient than P4 architecture. However, with better control strategies P4 architecture can be as efficient as P2 architecture
Route to electrification for trucks and buses in India [32]	To present measures for cooling circuits for cost-effective cars and buses in hot climate countries	• CAE-simulations and calculations of measures on cooling methods on e-Axle and battery level • Explanation of integrated cooling systems such as integrated e-Axle and modulated battery systems • Presenting a solution for durability and cooling of BECV battery systems and to keep scalability and modularity of battery cells and modules	• Development of thermal solutions for highly integrated systems can increase durability of battery commercial vehicles • Results of simulations of control strategies can be helpful to get information on energy and power demands of the battery
Challenges of electric vehicles from lab to road [33]	To present a study on the challenges met by the automobile manufacturers in bringing EVs to market in India	• Detailed analysis of the challenges faced by the vehicle manufacturers at different levels	• Proper regulations and testing standard need to be implemented for the growth of EVs
Electric vehicles in India: current status, future trend and environmental impact [34]	To focus on the current scenario, prospects of EVs in India in future with emphasis on the effect on environment and economy	• A detailed study on the present situation of EVs in India, methods of improvement in future as well as the impact on environment	• Upgradation of batteries coupled with their reuse and recharging can help in the reduction of pollution during their manufacturing • Support from the government in setting up charging stations and overcoming other obstacles can prove beneficial to the EV market

(continued)

Table 1 (continued)

Paper title	Objectives of research	Salient features of research	Conclusion
Design and analysis of a parallel hybrid electric vehicle for Indian conditions [35]	A parallel Hybrid EV design and the regulating policies for both traction and braking functions	• Determining the various characteristics required to meet the drive requirements to design the drive system • Numerical simulation considering a real-time speed plot and the Modified Indian Drive Cycle to analyze the benefits of a hybrid power train compared to a typical Indian vehicle	• The maximum power output of the engine in hybrid mode was reduced up to 27% than that of conventional Internal Combustion Engine (ICE) drives • In full parallel hybrid configuration, the net fuel economy was improved up to 32% than that of conventional Internal Combustion Engine (ICE) drive
Challenges in transition from internal combustion vehicles to electric vehicles in India by 2030 [36]	To analyze the challenges that can be met in the evolution from conventional cars to EVs in India and to provide possible solutions with the help of the Indian government	• Study of the challenges faced by the electric vehicle market in India at different levels • Provide possible solutions to the challenges	• Proper analysis of the obstacles to the implementation of EVs in the country can speed up the process
Optimal placement of electric, hybrid and plug-in hybrid electric vehicles (xEVs) in Indian power market [37]	To propose an outline based on switching-mechanism from one place of trade to another to lessen the charging (G2V) charge and exploit the discharging (V2G) charge	• A case study with a fleet of 200 eV is accomplished to validate the proposed framework • Four types of EVs are considered to analyze the influence of various types of xEVs • Analysis of the uncertainty in load demand as well as selling and procuring charge	• The aggregator paid 4.22%, 9.40% and 4.37% lesser energy cost as associated to DA, bilateral and DISCOM-based trading platform, respectively. under the proposed methodology
Selection of propulsion motor and suitable gear ratio for driving electric vehicle on Indian city roads [38]	To present selection of appropriate motor along with base and peak power rating, design of suitable gear ratio in view of overloading and underloading performance for effective mileage and safety	• MATLAB simulation of vehicle dynamics following Indian drive cycle after considering the suitability of Induction motor for EVs • Determination of number of teeth at 3-stages of gear box considering the specified gear ratio	• Performance of the electric vehicle is found to be satisfactory with well-designed gear box, calculated gear ratio and quantity of teeth for suitable load and torque-speed characteristics of the motor

(continued)

Table 1 (continued)

Paper title	Objectives of research	Salient features of research	Conclusion
A proposal for meeting power demands of electric vehicle transportation in India [39]	To present a proposal to meet the power demands of EV transportation in India with the help of the Indian River Grid (IRG)	• Analysis of the scope of the Indian River Grid (IRG) in meeting the electric vehicle transportation power demands, keeping in view with the current scenario	• Utilizing the hydro potential can be beneficial for EV transportation in a lot of countries
Modelling the barriers for mass adoption of electric vehicles in Indian automotive sector: an Interpretive Structural Modelling (ISM) approach [40]	To identify the hindrances in adaptation of EVs in India and modelling them using the technique of Interpretive Structural Modelling (ISM)	• Categorizing 33 barriers into 17 barriers and analyzing them appreciating their scopes by developing an ISM-model	• Incentives provided by the Government and customer characteristics are the major concerned areas in the improvement of EV penetration in India
Impact of vehicle to grid on voltage stability: Indian scenario [41]	To emphasize the importance of voltage to grid technology on bringing voltage stability in power grid and hence lessen the load on other sources of energy by supplying power back to the grid on peak hours	• A Simulink model of 15 MW diesel power generator, a wind-farm supplying 4.5 MW of the peak electricity an 8 MW solar farm is considered • Different cases of electric vehicles owners are taken, considering the presence of charging stations at homes • Designing a V2G algorithm and testing it on Nissan Leaf and Tesla Model S under city-conditions	• Integration of a greater number of smart grids can help in the stability during voltage fluctuations, during bad weather conditions or during peak load hours • Charging the vehicles during off-peak hours can be beneficial to both the customers as well as maintaining balance in the grid
End of life considerations for EV batteries- ISO and Indian business perspectives [42]	To highlight the importance of adopting the International Organization for Standardization (ISO) practices for environmental management systems to mitigate the risks involving electric vehicle batteries	• Analyzing the impact of EV batteries, the need for environmental management systems along with the strengths, weaknesses and opportunities	• Adoption or integration of environmental management systems should be given importance to benefit the customers and stakeholders

3   Wireless Charging for Electric Vehicles

Low cost of the electric vehicles coupled with full automation can bring the electric vehicle market to a new level. These can be achieved with the help of wireless charging [23]. Not stopping for recharging can extend the time of service and can also reduce the number of vehicles needed for passenger demand [23]. Wireless charging can also reduce the size of the batteries and hence the cost of the vehicles. However, wireless charging technology to be fully implemented, challenges associated with it such as cost, performance and safety need to be overcome [23].

Near field wireless power transfer (WPT) systems are of two types:

(1) Inductive WPT system: This system uses magnetic field coupling between con-ducting coils and it is preferred for medium-range applications. By supplying alternating current to a coil, electric field is generated, and electricity is induced in the coil kept in the magnetic field. High frequency passes through one coil and electricity is induced in the adjacent coil. The number of windings and gap between the coils determine the strength of electricity [24]. However, the need for ferrite cores for magnetic flux guidance and shielding makes them expensive and bulky [23]. The frequencies in these systems are limited to 100 kHz to limit losses in the ferrite cores, resulting in low power transfer density and high cost [23].

(2) Capacitive WPT system: This system uses electric field coupling between con-ducting coils to transfer energy [23]. This system uses two plates—one on the road and the other in the bottom of the car. High-frequency power is trans-ferred from the charging station to the battery through electrostatic effect [24]. This does not need electromagnetic shield, making them advantageous over the inductive systems. Ferrite cores are not required in this system which makes it possible to operate at higher frequencies, making them less costly and bulky [23]. But operating at high frequencies makes the design challenging because of the very small capacitance between the roads and vehicle plates. The other limitations associated with the capacitive system are the low efficiency due to carbon fillers and insufficient transfer of power due to less area.

Charging of vehicles without the need to stop requires proper designing of the properties of the transmitter and receiver, such as orientation and position with respect to one another [25]. So, constantly changing these parameters or keeping both the transmitter and receiver still makes the system design complex [25]. Research has been constantly going on to design techniques to deliver power at high efficiency to moving cars. In 2017, Israel government collaborated with start-up firm ElectRoad announced to have tested a similar system on an 80 feet track [26]. Qualcomm announced that a 100 m track has been constructed successfully in Paris, capable of charging vehicles at 20 KW driving along the track at highway speeds. It also works when two vehicles are travelling in opposite directions along the track. The first electrified-road in the world has been opened in Sweden that can revitalize the batteries and cars driving on it [26].

The benefits of wireless power transfer include smaller batteries, less weight, less cost and to be able to run indefinitely without the need to stop for recharging. Installing the inductive charging hardware in the roads also seems to be a safer option than overhead power lines to power electric vehicles. However, the disadvantages associated with the system cannot be overlooked. The roads must be destroyed to lay the electromagnets along the length, which is costly and disruptive. Secondly, the system is expensive on its own [26]. These disadvantages question the possibility of implementation of such a system in a country like India, where electric vehicles are on the struggling phase to gain popularity among the masses.

4 Conclusion

The literature review on the current scenario and studies on electric vehicles in India reveal that electric vehicles can prove to be a key technology to reduce the alarming pollution in the country. Costliness and lack of charging infrastructure happen to be the major reasons behind slow growth of the EV market in India. The fall in battery prices and improvement in battery technology can result in affordable electric vehicles. Policies and incentives from the government are recognized to be major drivers in the growth of EV market in the country. Developing a suitable strategy and implementing it can help in the realization of targets set by the government.

References

1. The World Bank Data, https://data.worldbank.org/indicator/EN.CO2.TRAN.ZS
2. Article on list of most-polluted cities by particulate matter concentration, https://www.ndtv.com/gurgaon-news/india-air-pollution-gurugram-worlds-most-polluted-city-6-others-in-india-in-top-10-study-2002404
3. Article on air pollution in India, https://en.wikipedia.org/wiki/Air_pollution_in_India
4. Article on automobiles and pollution in India, https://community.data.gov.in/automobiles-and-pollution-in-india/
5. Article on history of the electric vehicle, https://en.wikipedia.org/wiki/History_of_the_electric_vehicle
6. Article on National Electric Mobility Mission Plan, https://www.dhi.nic.in/UserView/index?mid=1322
7. Article on NITI Aayog report, www.pluginindia.com/blogs/niti-aayog-report-indian-government-is-fully-committed-to-electric-vehicles
8. Article on Govt eyes made-in-India Lithium ion batteries to lower cost of electric vehicles, https://www.hindustantimes.com/business-news/govt-eyes-made-in-india-lithium-ion-batteries-to-lower-cost-of-electric-vehicles/story-d2zDR7DKJQVjmaFv4eInaJ.html
9. K.V. Muralidhar Sharma, M.R. Kulkarni, G.P. Veerendra, N. Karthik, Trends and challenges in electric vehicles. Int. J. Innov. Res. Sci. Eng. Technol. 5(5) (2016)
10. Article on opinion-India's electric vehicle challenge, https://www.gaadi.com/car-news/opinion-indias-electric-vehicle-challenge-108433
11. Article on opinion on electric vehicles India challenge, https://www.outlookindia.com/website/story/opinion-indias-electric-vehicle-challenge/320355

12. Article on the 7 challenges of electric vehicles market in India, https://electricvehicles.in/the-7-challenges-of-electric-vehicles-market-in-india/
13. Article on Grant Thornton CII report, https://www.grantthornton.in/press/press-releases-2017/establishing-the-charging-infrastructure-is-a-significant-challenge-in-adoption-of-electric-vehicles-in-india-grant-thornton-cii-report/
14. Article on electric vehicle charging station Nagpur, https://inc17.com/buzz/electric-vehicle-charging-station-nagpur/
15. Khan, W., Ahmad, F., Ahmad, A., M.S. Alam, A. Ahuja; Electric vehicle charging infrastructure in india: viability analysis, in *ISGW 2017: Compendium of Technical Papers* (Springer, Singapore, 2017)
16. Article on global Ni-MH Battery Market Growth 2019–2024. Report ID: 343189 (2019), https://www.fiormarkets.com/report/global-ni-mh-battery-market-growth-2019-2024-343189.html
17. M. Brain, How electric cars work, http://auto.howstuffworks.com/electric-car2.htm. Accessed 29 Jan 2010
18. A. Pandey, Article on rise of the electric vehicle market in India (2018), https://www.ciol.com/rise-electric-vehicle-market-india/
19. Article on use of electric vehicles to transform mass transportation in India. PricewaterhouseCoopers Private Limited (PwCPL)] (2018), https://www.pwc.in/assets/pdfs/publications/2018/use-of-electric-vehicles-to-transform-mass-transportation-in-india.pdf
20. Article on rise of electric vehicle market in India, DQINDIA Online, https://www.dqindia.com/rise-electric-vehicle-market-india/. Accessed 22 Oct 2018
21. Article on India electric car market to reach $427.4 million by 2025 (2018), https://www.psmarketresearch.com/press-release/india-electric-car-market
22. Article on India State with most cars plans Electric Vehicle Push (2018), https://www.bloombergquint.com/business/indian-state-with-most-cars-plans-electric-vehicle-push
23. Article on wireless charging of electric vehicles, https://www.powerelectronics.com/automotive/wireless-charging-electric-vehicles
24. Article on wireless charging for electric care: how convenient, https://getelectricvehicle.com/wireless-charging-for-electric-cars/
25. Article on wireless power to moving vehicles closer to reality, https://spectrum.ieee.org/cars-that-think/transportation/advanced-cars/wireless-power-to-moving-electric-vehicles-closer-to-reality
26. Article on the case for building roads that can charge electric cars on the go, https://www.technologyreview.com/s/347902/the-case-for-building-roads-that-can-charge-electric-cars-on-the-go/
27. P. Desai, A.M. Deshpande, Hybridization of light commercial vehicles in India, in *2015 IEEE International Transportation Electrification Conference (ITEC)*, Chennai, 2015, pp. 1–6
28. N. Joshi, D. Naik, Creating a platform to foster the growth of electric vehicles in Indian market by increasing the electric vehicle public charging stations. Quest J. J. Res. Mech. Eng. **2**(7), 01–03 (2015). ISSN(Online) 2321-8185
29. Gambhir, Safety considerations for EV charging in India: overview of global and Indian regulatory landscape with respect to electrical safety, in *2017 IEEE Transportation Electrification Conference (ITEC-India)*, Pune, 2017, pp. 1–5
30. S. Nair, N. Rao, S. Mishra, A. Patil, India's charging infrastructure—biggest single point impediment in EV adaptation in India, in *2017 IEEE Transportation Electrification Conference (ITEC-India)*, Pune, 2017, pp. 1–6
31. A. Kumaran et al., Affordable hybrid topology for PV and LDV's in prospering India: Case study of 23 V (P)HEV system benefits, in *2017 IEEE Transportation Electrification Conference (ITEC-India)*, Pune, 2017, pp. 1–6
32. M. Ackerl, M. Kordon, H. Schreier, H. Petutschnig, M. Huetter, Route to electrification for trucks & busses in India, in *2017 IEEE Transportation Electrification Conference (ITEC-India)*, Pune, 2017, pp. 1–6
33. Y. Somayaji, N. K. Mutthu, H. Rajan, S. Ampolu, N. Manickam, Challenges of electric vehicles from lab to road, in *2017 IEEE Transportation Electrification Conference (ITEC-India)*, Pune, 2017, pp. 1–5

34. P. Rodge, K. Joshi, Electric vehicles in India: current status, future trend and environmental impact, in *2018 International Conference on Smart Electric Drives and Power System (ICSEDPS)*, Nagpur, 2018, pp. 22–27

35. C.S.N. Kumar, and S. C. Subramanian, Design and analysis of a parallel hybrid electric vehicle for Indian conditions, in *2015 IEEE International Transportation Electrification Conference (ITEC)*, Chennai, 2015, pp. 1–12

36. S. Parulekar, R.M. Holmukhe, S. Mehta, A. Raj, R. Raj, P.B. Karandikar, Challenges in transition from internal combustion vehicles to electric vehicles in India by 2030, in *2017 International Conference on Energy, Communication, Data Analytics and Soft Computing (ICECDS)*, Chennai, 2017, pp. 779–784

37. F. Ahmad, M.S. Alam, M. Shahidehpour, Optimal placement of electric, hybrid and plug-in hybrid electric vehicles (xEVs) in Indian power market, in *2017 Saudi Arabia Smart Grid (SASG)*, Jeddah, 2017, pp. 1–7

38. P. Mishra, S. Saha, H.P. Ikkurti, Selection of propulsion motor and suitable gear ratio for driving electric vehicle on Indian city roads, in *2013 International Conference on Energy Efficient Technologies for Sustainability*, Nagercoil, 2013, pp. 412–418

39. K.V. Rupchand, A proposal for meeting power demands of electric vehicle transportation in India, in *2011 IEEE Power and Energy Society General Meeting*, Detroit, MI, USA, 2011, pp. 1–5

40. S. Prakash, M. Dwivedy, S.S. Poudel, D.R. Shrestha, Modelling the barriers for mass adoption of electric vehicles in Indian automotive sector: an Interpretive Structural Modeling (ISM) approach, in *2018 5th International Conference on Industrial Engineering and Applications (ICIEA)*, Singapore, 2018, pp. 208–212

41. M. Gupta, S. Giri, S.P. Karthikeyan, Impact of vehicle-to-grid on voltage stability—Indian scenario, in *2018 National Power Engineering Conference (NPEC)*, Madurai, 2018, pp. 1–5

42. P. Gupta, End of life considerations for EV batteries—ISO and Indian business perspectives, in *6th Hybrid and Electric Vehicles Conference (HEVC 2016)*, London, 2016, pp. 1–5

Self-Medication and Direct-to-Consumer Promotion of Drugs

Jaya Rani Pandey, Saibal Kumar Saha, Samrat Kumar Mukherjee, and Ajeya Jha

Abstract Self-medication is one of the major healthcare concern globally and World Health Organization (WHO) has underlined the importance of appropriately probing and controlling it. Self-medication has been found to be associated with negative consequences like resource wastage, misdiagnosis, increased resistance to pathogens, extended period of use and use of undue drug dosage. Increase in self-medication practices has been linked to enhanced patient awareness of the products because of online information. This study is undertaken to quantify the self-expressed beliefs of physicians regarding extent of self-medication in India. Allopathic physicians who are currently practicing in India were the sample respondents for this study. Primary data collected through a tool comprising of seven items on a 5-point Likert scale. Random sampling has been employed for the research. In all 200 physicians were approached for the survey. The results indicate that physicians are of the opinion that self-medication is encouraged by DTCP.

Keywords Pharmaceutical products · Self-medication · Direct-to-consumer promotion · Unnecessary medication · Over-diagnosis

J. R. Pandey · S. K. Saha · S. K. Mukherjee · A. Jha (✉)
Sikkim Manipal Institute of Technology, Sikkim Manipal UniversitySMIT, Majitar, Sikkim, India
e-mail: ajeya611@yahoo.com

J. R. Pandey
e-mail: jayaranim@rediffmail.com

S. K. Saha
e-mail: saibal115@gmail.com

S. K. Mukherjee
e-mail: samrat.k@smit.smu.edu.in

© Springer Nature Singapore Pte Ltd. 2020
R. Bera et al. (eds.), *Advances in Communication, Devices and Networking*,
Lecture Notes in Electrical Engineering 662,
https://doi.org/10.1007/978-981-15-4932-8_18

1 Introduction

As per the WHO, self-medication is defined as "use of pharmaceutical or medic-
inal products by the consumer to treat self-recognized disorders or symptoms, the
intermittent or continued use of a medication previously prescribed by a physician
for chronic or recurring disease or symptom, or the use of medication recommended
by lay sources or health workers not entitled to prescribe medicine". It is one of
the major healthcare concern globally and WHO has underlined the importance of
appropriately probing and controlling it. In India it has become a critical issue as
self-medication has gone from 31% in 1997 to 71% in 2011 [10]. Self-medication
has been found to be associated with negative consequences like resource wastage,
misdiagnosis, increased resistance to pathogens, extended period of use and use of
undue drug dosage [11]. Increase in self-medication practices has been linked to
improved patient and consumer awareness about product availability [19].

Direct-to-consumer marketing has remained regulated till recently and direct-to-
patient promotion is illegal across the world (Jha et al. [8]; Mathur et al. [12]; Goyal
et al. [6]). Internet, however, has rendered this law virtually ineffective as patient
across the world has access to online healthcare information and they are searching
for it frequently [15]

2 Literature Review

With the onset of e-commerce pharmaceutical industry has substantially deployed
online healthcare information so-much-so that online healthcare search by common
people has become an integral part of their existence. Self-medication has emerged
as one of the foremost outcome of drug promotion (Mintzes et al. [13]; Donahue
et al. [3]; Folayan et al. [4]; [18]; Kłak et al. [9]; Hamzehei et al. [7]; Ahsan et al.
[1]; Atanasova et al. [2]; TG [20]). As already stated that despite some benefits of
self-medication it actually poses a challenge to the healthcare outcomes and hence
it is important to measure the extent of it. View of physicians are important in this
respect as the physicians are aware of the self-medication cases and they also have
knowledge of its ill-effects. Also it may also affect patient-physician relationship,
particularly if a legal issue arises out of self-medication instance.

This study, therefore is undertaken to quantify the self-expressed beliefs of physicians regarding extent of self-medication in India. Why the beliefs of physicians is
important on what essentially is a patients' behavior? The answer lies in the fact that
physicians legally and ethically are responsible for favorable health-outcomes and
which has a high possibility of compromised because of this patient behavior. The
second reason is the patient-physician relationship and which is considered sacro-
sanct (Tucker et al. [21]; Pellegrini [17]; Weeger and Farin [22]; Fuertes et al. [5];
Pavelić et al. [16]; Nie et al. [14]). The patent behavior under study is significant
because of its potential in adversely affecting patient-physician relationship. Lastly,

the consequences of elf-medication eventually arrive at the door-steps of physicians and hence they are aware of the extent and gravity.

3 Methodology

Study is empirical and survey based. The study's objective is to measure if the physicians believe that significant self-medication amongst patients is resulting due to the access to online information. The alternate hypothesis is that *Physicians significantly believe that DTCP promotes self-medication.*

Allopathic physicians who are currently practicing in India were the sample respondents for this study. For conducting the survey 200 physicians were approached. Four responses were invalid and therefore the final sample size was 196. The tool comprised of seven items on a 5-point Likert scale. Random sampling has been employed for the research. After obtaining the list of physicians for a city, respondents were chosen on systematic random basis. They researcher explained the purpose of the survey and handed over the tool. After guarantying the confidentiality of the information, the filled questionnaires were collected.

4 Result and Discussion

Results and discussion (Table 1) in this respect are as follows.

1. **Patient with online information prescribe medicine to others**: For this variable mean value is 3.75 which refers that there is general parity in the beliefs of physicians' that patient with online information prescribe medicine to others. The t value is more than 1.96 (8.921). Hence, the null hypothesis is rejected. The conclusion is confirmed as the significance value is less than 0.05 (0.0). Therefore, there is general conformity in the belief of physicians that patient with online information have prescribed medicine to others. This accentuates the fears that medications are being informally prescribed by non-physicians. It is an unhealthy trend even if only a few physicians believe it. Self-styled experts are not uncommon cross professions but in case of medicine it assumes a dangerous proportion. Medicine is a profession in true sense and which implies only the ones who have undergone long and arduous training should be allowed to practice it. This is also the legal position. It is to be noted that other healthcare professionals like clinical psychologists, pharmacists, physiotherapists and nurses are also not permitted to prescribe medicine.

2. **Request for advertised medicines by internet savvy patients'**. For this variable mean value is 3.62 which refers that there is general parity in the beliefs of physicians' that request for advertised medicines is made by internet savvy patients'. The t value is more than 1.96 (5.876). Hence, the null hypothesis is rejected. The

Table 1 Physician [t-test]: self-medication

Sl. no	Statement	Mean	SD	t. value	Sig.
1	Patient with online information prescribe medicine to others	3.79	1.205	8.921	0
2	Request for advertised medicines by internet savvy patients'	3.62	1.059	5.876	0
3	Patient inspiration of use costly medicines is based on online information	3.77	1.1	9.701	0
4	Patient inspiration of use unnecessary medication is based on online information	3.75	1.103	9.701	0
5	Patient inspiration of use self-medication is based on online health information	3.86	1.043	12.213	0
6	Brand endorsement by patients with sound knowledge of drugs and diseases on the basis of online information	2.51	1.074	–	–
7	Over diagnosis of condition by patient based on online information	3.6	1.112	7.916	0

Physicians believe that DTCP does not promote self-medication

conclusion is confirmed as the significance value is less than 0.05 (0.0). This accentuates the fears that patients are looking for brands they see online. Even if there is few occurrences of such a behavior it may presage perilous portends. When patients direct a physician to prescribe a certain medicine, they imply a mistrust on his/her abilities. Such a breach of trust could have adverse effects on patient-physician relationships. Also in pharmaceutical products chemical-equivalence does not directly result in bio-equivalence. Finally, medicines are available in excessively varying cost and branded drugs are many more times expensive than generic drugs. Economics, therefore, also makes self-medication a costly affair.

3. **Patient inspiration of use costly medicines is based on online information**: From the table, we observe that the mean value is 3.77 which indicated an agreement in the beliefs of the physician that the online information encourages the patients to seek out expensive medications. In Table 1, we discover the value of t as 9.701, which is greater than 1.96. We, therefore, reject the null hypothesis. Confirmation of this is obtained from the corresponding level of significance which is 0.000—much less than 0.05. A country poor as India can hardly afford to waste its resources in this manner. For a strong and resilient health system

medication has to be cost-effective and which is being undermined by online information.

4. **Patient inspiration of use unnecessary medication is based on online information**: In Table 1, we observe the mean value as 3.75 which signify general agreement in the beliefs of physicians that patient inspiration of use unnecessary medication is often based on online information. This is further confirmed by the value of t 9.701—far above 1.96. Hence, null hypothesis is rejected. As significance is less than 0.05 (0.000) the conclusion is further confirmed.

5. **Patient inspiration of use self-medication is based on online health information**: From the table, we see that the mean value is 3.86. It may further be interpreted that there is an agreement in the belief that patient inspiration of use self-medication is based on online health information. The t value is 12.213 which is much above than tabulated t-value 1.96 so we reject the null hypothesis. It is further confirmed by the level of significance which is 0.501 more than the standard 0.05. This is positive that Physicians believe that DTCP does not promote self-medication.

6. **Brand endorsement by patients with sound knowledge of drugs and diseases on the basis of online information**: From the table we can say that the mean value is 2.51, which says that there is disagreement in the belief of the physicians that patients having good knowledge of disease and drugs based on online information act as ambassador of a brand. Hence, we accept the null hypothesis. This possibility may have been discarded but could certainly be a possibility. The stake-holders certainly need to guard against such behavior that may result from DTCP campaigns.

7. **Over diagnosis of condition by patient based on online information**: For this variable mean value is 3.6 which indicates that there is general covenant in the belief that often there is over diagnosis of condition by patient based on online information. The t value is more than 1.96 (7.916). Hence, the null hypothesis is rejected. The conclusion is confirmed as the significance value is less than 0.05 (0.0). Yet again we find that physicians agree to such a view. Over-diagnosis may result in self-medication on one hand and may lead to unnecessary medication and corresponding expenditure which patients can ill-afford.

5 Discussion

From the findings it indicates that physicians are of the opinion that DTCP inspires self-medication. It even leads to brand endorsement by patients. Self-medication leads to prescribing medications to others. In the context of India, we know that it is happening but the extent to which it is due to DTCP should to be measured. This study implies that there is a direct link between online healthcare information and self-medication practice at least as per the beliefs expressed by physicians. The implications for the study for regulatory bodies are that thy need to enforce responsible online information with proper cautionary advises regarding consultation of physicians before acting upon the information provided by online sources. Regulators must ban questionable websites and initiate mechanisms to regularly identify websites that cross the ethical lines. Physicians have the additional task to guide the patients regarding misinterpretations of information patients gather online. They must be in a position to guide the patients regarding the risks involved, particularly that of self-medication. Patients find self-medication economic and time-saving but this is a myth. It comes to them at a high cost. They must learn to be cautious about self-medication. A big onus is on pharmaceutical marketers. The balance in healthcare is extremely delicate. Whether the self-medication driven by online information is intentional on part of pharmaceutical firms (to drive profits) or unintentional is a moot point because consequences are unacceptable. Pharmaceutical marketers must be careful while providing online information and discourage any action on part of patients. Without due consultation of a registered physician. Are the pharmaceutical companies doing it? A glimpse of online healthcare information by US based firms reveal that they contain following cautions:

- "DTC is intended for the citizens of US alone." This is because DTC is legal in US. These instructions are not valid as websites are visible to people from all over the world.
- "Website is just for educational purpose and patients must consult a physician before consuming them."

The advertised brands are allowed to be purchased from websites.

It is not mentioned in the websites of European companies that the promotion is only for a particular country. Also, the caution that "information is educational" or "a physician must be consulted before consuming them" is not used. These type of communications are referred to as patient information. Product comparison is often done by European companies. For example, manuals are provided for different ailments, cures and their prevention by Merck (A German firm). An aggressively ethical approach on part of pharmaceutical marketers alone can keep online relatively safe in the context of self-medication.

6 Conclusion

This study has been undertaken to understand the extent (In view of the physician) of self-medication that is being driven by online promotion of drugs. It is obvious that various manifestations of self-medication are believed to be linked to availability of online healthcare information. This can have serious consequences and hence all the stake-holders need to get together and discourage the negative consequences.

Refernces

1. M. Ahsan, R. Batool, U. Farheen, S. Kumar, S. Javed, S. Khan, S. Hussain, Prevalence and attitude towards health related web searches. Int. J. Res. **5**(20), 1508–1529 (2018)
2. S. Atanasova, T. Kamin, G. Petrič, The benefits and challenges of online professional-patient interaction: comparing views between users and health professional moderators in an online health community. Comput. Hum. Behav. **83**, 106–118 (2018)
3. J.M. Donahue, M. Cevasco, M.B. Rosenthal, A decade of direct-to-consumer advertising of prescription drugs. N. Engl. J. Med. **357**, 673–681 (2007)
4. O.F. Folayan, F.O. Adeosun, O.T. Adeosun, B.O. Adedeji, The influence of the internet on health seeking behaviour of nursing mothers in ekiti state, Nigeria. Int. J. Adv. Res. Publ. **1**(5), 1–6 (2017)
5. J.N. Fuertes, A. Toporovsky, M. Reyes, J.B. Osborne, The physician-patient working alliance: theory, research, and future possibilities. Patient Educ. Couns. **100**(4), 610–615 (2017)
6. M. Goyal, M. Bansal, S. Pottahil, S. Bedi, Direct to consumer advertising: a mixed blessing. Nat. J. Integr. Res. Med. **6**(3) (2015)
7. R. Hamzehei, M. Kazerani, M. Shekofteh, M. Karami, Online health information seeking behavior among Iranian pregnant women: a case study. Libr. Philos. Pract. (2018)
8. A. Jha, J.R. Pandey, S.K. Mukherjee, An empirical note on perceptions of patients and physicians in direct-to-consumer promotion of pharmaceutical products: study of indian patients and physicians, in *Management Strategies and Technology Fluidity in the Asian Business Sector*, (IGI Global, 2018), pp. 65–87
9. A. Kłak, E. Gawińska, B. Samoliński, F. Raciborski, Dr Google as the source of health information–the results of pilot qualitative study. Pol. Ann. Med. **24**(2), 188–193 (2017)
10. V. Kumar, A. Mangal, G. Yadav, D. Raut, S. Singh, Prevalence and pattern of self-medication practices in an urban area of Delhi, India. Med. J. Dr. DY Patil Univ. **8**(1), 16 (2015)
11. M. Locquet, G. Honvo, V. Rabenda, T. Van Hees, J. Petermans, J.Y. Reginster, O. Bruyere, Adverse health events related to self-medication practices among elderly: a systematic review. Drugs Aging **34**(5), 359–365 (2017)
12. M.B. Mathur, M. Gould, N. Khazeni, Direct-to-consumer drug advertisements can paradoxically increase intentions to adopt lifestyle changes. Front. psychol. **7**, 1533 (2016)
13. B. Mintzes, M.L. Barer, R.L. Kravitz, A. Kazanjian, K. Bassett, J. Lexchin, S.A. Marion, influence of direct to consumer pharmaceutical advertising and patients' requests on prescribing decisions: two site cross sectional survey. Br. Med. J. **324**(7332), 278–279 (2002). https://doi.org/10.1136/bmj.324.7332.278. and patients' requests on prescribing decisions: two site cross sectional survey. BMJ 2002; 324:278-9
14. J.B. Nie, J.D. Tucker, W. Zhu, Y. Cheng, B. Wong, A.M. Kleinman, Rebuilding patient-physician trust in China, developing a trust-oriented bioethics, (2017)
15. J.R. Pandey, A. Jha, S.K. Mukherjee, S.K. Saha, An empirical note on comparative perceptions of indian patients and physicians in direct-to-consumer promotion of pharmaceutical products, in *Dynamic Perspectives on Globalization and Sustainable Business in Asia* (IGI Global, 2019), pp. 128–153
16. K. Pavelić, S.K. Pavelić, T. Martinović, E. Teklić, J. Reberšek-Gorišek, Patient-physician relationship in personalized medicine in *Personalized Medicine in Healthcare Systems*, (Springer, Cham, 2019), pp. 217–226
17. C.A. Pellegrini, Trust: the keystone of the patient-physician relationship. J. Am. Coll. Surg. **224**(2), 95–102 (2017)
18. S. Shakeel, S. Nesar, N. Rahim, W. Iffat, H.F. Ahmed, M. Rizvi, S. Jamshed, Utilization and impact of electronic and print media on the patients' health status: Physicians' perspectives. J. Pharm. Bioallied Sci. **9**(4), 266 (2017)
19. S. Singh, A. Banerjee, Internet and doctor–patient relationship: cross-sectional study of patients' perceptions and practices. Indian J. Public Health **63**(3), 215 (2019)
20. R.S. TG, Mapping of online pharmaceutical marketing practices: review of literature. IMPACT: Int. J. Res. Humanit. Arts Lit. **5**(11), 15–26 (2017)

21. J.D. Tucker, Y. Cheng, B. Wong, N. Gong, J.B. Nie, W. Zhu, ... W.C. Wong, Patient–physician mistrust and violence against physicians in Guangdong Province, China: a qualitative study. BMJ Open **5**(10), e008221 (2015)
22. S. Weeger, E. Farin, The effect of the patient–physician relationship on health-related quality of life after cardiac rehabilitation. Disabil. Rehabil. **39**(5), 468–476 (2017)

Performance Index Modeling for Urban Water System Using Hierarchical Fuzzy Inference Approach

Pooja Shrivastava, M. K. Verma, Meena Murmu, and Ishtiyaq Ahmad

Abstract In developing country like India, water management is growing exploding and all these water resources should maintain in the developing countries. There are many failures in the urban water system occur because of lack of assessment of present and future conditions. This needs a performance assessment method to assess performance and inform any irregularity for better performance of the urban water system. In this paper, we find performance index value through an integrated hierarchical fuzzy inference approach for urban water system. There are eight important parameters used as input for the proposed integrated hierarchical fuzzy inference model. This integrated modeling minimizes the different conditions in the rule base. For a very large set, reduction in rule is significant because of simplicity and interpretation. Additional, it is very important to design many rules because that requires skilled expertise; otherwise, it will lead to unexpected incidents. As compared to traditional fuzzy inference with integrated hierarchical fuzzy inference, a design that needs only fewer rules. For the proposed model, we take only 165 rules and for the similar eight variables, the traditional inference approach of fuzzy requires 250,000 rules. For this study, we select eight variables as input because these taken from the survey of a city. This proposed method will also used for the system with several parameters. In traditional fuzzy inference, there are several rules exponentially increases, also with the additional of new parameters. Integrated hierarchical

P. Shrivastava (✉)
National Institute of Technology, Raipur, India
e-mail: pooja.shrivastava04@gmail.com

M. K. Verma · M. Murmu · I. Ahmad
Civil Engineering Department, National Institute of Technology, Raipur, India
e-mail: mkseem670@gmail.com

M. Murmu
e-mail: mmurmu.ce@nitrr.in

I. Ahmad
e-mail: iahmad.ce@nitrr.ac.in

M. K. Verma
CSVTU, Bhilai, India

© Springer Nature Singapore Pte Ltd. 2020
R. Bera et al. (eds.), *Advances in Communication, Devices and Networking*,
Lecture Notes in Electrical Engineering 662,
https://doi.org/10.1007/978-981-15-4932-8_19

fuzzy inference approach used for computing the performance index of the whole system. The index values traces the possibility of system failure and results shows the performance values under different circumstances.

Keywords Hierarchy, fuzzy logic · Performance index · Urban water system · Integrate model

1 Introduction

In, any multidimensional system when several inputs are very large then it may become idle to apply a rule based fuzzy controller. To avoid this problem, a hierarchical fuzzy inference approach proposed for the present work. It is very time overwhelming activity to design a fuzzy controller with defined control structure and their rules along with knowledge acquisition. Previous there is one of the major troubles in the fuzzy inference is how to decrease the maximum numbers of rules and their calculations. In the traditional fuzzy inference, there are rules increases exponentially along with the addition of parameters in the fuzzy controller.

The conventional inference approach defined mn rules for n number of parameters used as input and m as a membership function for each variable to build a complete fuzzy controller. With increasing of input variables, the rules base will fully loaded the memory and make difficult to implement the fuzzy controller. The problem of increasing exponentially with addition of input parameters called dimensionality curse that is not exclusive to a fuzzy inference approach. Hence, to handle the problem of dimensionality curse and rules expansion, the hierarchical fuzzy inference approach introduced. These fuzzy hierarchical inferences have the advantages to tackle the numbers of rules linearly with the addition of input parameters that reduce the dimensionality curse and their rules.

Fuzzy inference approach has been widely applied in many areas of engineering to assess and analyses performance of the system. In this present work, we use the application of hierarchical fuzzy inference for then integrated urban water system by design a fuzzy model to reduce the dimensionality curse. For study, we considered following important parameters as input that affects the performance of the urban water system of Raipur i.e. water availability, groundwater stress, reliability, resilience, leakage, health and hygiene conditions, water management and vulnerability.

2 Literature Review

Altunkaynak et al. [1] illustrated a fuzzy approach for prediction of subsequent monthly water consumption for three different existed water consumption capacities, which defined as an independent parameter. This model applied for Istanbul city in

Turkey for monthly water consumptions variation calculation. Bagheri et al. [5] studied the fuzzy inference method for aggregating performance indicators for an urban water system to the performance of the entire system. Fuzzy inference system design a model by planning the mapping from the inputs to the outputs according to if-then statements called rules. It involved three important sections i.e. membership function, if–then rules and fuzzy logic operators.

Adriaenssensa et al. [2] stated that fuzzy logic could trade in with several uncertain, linguistic, variable and vague data or knowledge. It can allow a coherent and transparent stream of information from data information to decision-making by uses of data. Chen and Chang [7] explored the fuzzy multi-objective analysis having greater flexibility by using four fuzzy operators for representing complexity. These multi-objectives included the following constraints are physical, chemical, socio-economic, management and technical factors that reflecting adaptive needed of water resources execution. Jelleli and Alimi [11] presented the concept of hierarchical rule basis gives better readability of the rules with guaranteed interpretability. The hierarchy system considerably minimizes the rules for the highly dimension system and ensures the flexibility for removing or replacing variable changing no rule base. Almedia et al. [3] stated that the fuzzy inference method used as a multi-criteria decision-making tool and a theoretical concept-based support system. Potential of water reprocess identify through the fuzzy inference approach also described.

Xu and Qin [17] explained uncertainty data associated with water requirements, capacities of water treatment plant, leakage rate and water sharing price described by trapezoidal membership function and rooted into fuzzy programming. Complex uncertainties in the urban water distribution system are tackling by using fuzzy programming and decision analysis framework. Zhang et al. [18] explained a compressive evaluation method using fuzzy in order to evaluate the risk index value under the unwanted accidents. He also observed water hammer under various conditions to control the burst in the water supply line, which is more severe. Malinowska and Hejmanowski [14] presented breakdown hazard in a water pipeline of mining area by critical analysis used mamdani fuzzy inference model coupled with a Geographic information system (GIS). Li and Yao [13] described fuzzy approach similarity along with triangle fuzzy number for an extended water transmission pipeline to quantify risk assessment based on fuzzy logic. This approach was a design for a common operation of water resources project. Fayaz et al. [8] described the blended hierarchical fuzzy logic model coupled with GIS for assessment of water supply pipelines risk index. He also explained that hierarchical fuzzy logic required minimum rules as compared to the traditional fuzzy logic inference for data sets.

3 Study Area

Raipur is the capital of the Indian state of Chhattisgarh, established as capital after the separation of Chhattisgarh from the Madhya Pradesh state on 1 November 2000. City is located between 82° 6' to 81° 38'E Longitude and 22° 33'N to 21° 14'N

Latitude in the fertile plain of state. Raipur is encircle by part of the Orissa state and Raigarh in East and west side by Durg, while Bilaspur, Bastar and part of Orissa state touched north and south face of city. District thus splits into two main geographic regions, i.e. flat and hilly regions. Raipur Municipal Cooperation (RMC) has 13 Tehsil and 15 revenue blocks in the administrative district. The city's slope is typically northward. The Raipur's main river is the Mahanadi River. Kharun River, Groundwater and Gangrel Dam, which is constructed on the Mahanadi River, are the main sources of water supply in the city. All river Mahanadi and kharun is Non-perennial River in the region. Kharun River is flowing by the side of the city and Mahanadi is situated 60 km away from the city. Water supplied in the city from a small storage facility, i.e. bhatagaon anicut, built across the Kharun River and feed by Pandit Ravishankar Reservoir in Dhamtari district From bhatagaon water is supply to the water treatment plant at Rawanbhatta, treated water from where is supplied in the city. Approximately 67.5 MLD from the Gangrel Dam taken for the city where approximately 13.5 MLD is lost in the percolation, evaporation and illegal activity of farmers which is about 20%. It circulated in the wards using booster pump after treating water up to the softening point (Fig. 1).

4 Methodology

Fuzzy logic maps various input parameters to an output through the fuzzy inference approach based on rules reduction method, which comprised of classified if-then rules, each parameter membership function and operation of fuzzy logic. There are three fuzzy inference methods available for modelling i.e. Mamdani fuzzy, Tsukamoto Fuzzy, Sugeno fuzzy inference system. All these approaches divided into two parts; first part is fuzzifying the crisp values of input parameter into membership values according to existing fuzzy sets and this process is same for all approaches. There is only difference in the second part when the results of all rules integrated into a single value for accurate output. The linguistic variables in Mamdani fuzzy inference are in both the antecedent and consequent parts of the rules. So we can define Mamdani as MISO i.e. Multi-input with a single output system and defined if-then rules in the following type;

If Z_1 is A_1 and Z_n is A_n then W is B.

Where Z_i and W are input and output linguistic variables, respectively. A_i and B are linguistic values. The block layout diagram for the Mamdani representation is showing in Fig. 2. There are four major components of the Mamdani inference model: Fuzzification, knowledge base, inference engine and defuzzifier. Crisp values are converted into linguistic values in the Fuzzification application.

The knowledge base portion includes the database, i.e. descriptions of fuzzy set and membership function parameters and rule base of parameters, i.e. if-them rules sets. For rules and input information, the inference engine performs the reasoning, fuzzy reasoning and fuzzy logic. Convert the crisp meaning to linguistic values for

Fig. 1 Satellite image of Raipur city. *Source* Bhuvan

Fig. 2 Block structure
diagram for Mamdani's
fuzzy inference system

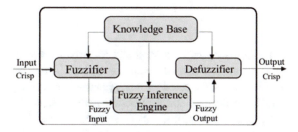

tests at the last defuzzifier. Sometimes it named a linguistic or descriptive inference
method because of this Mamdani inference model.

The process of Mamdani fuzzy inferences consists of following steps; (i) determine fuzzy rules that explain how the fuzzy logic inference should be decision making
to classify the input or controlling the output, (ii) fuzzify input variables using the
membership functions of input variables. Within interval it defines between 0 and 1

Fig. 3 Graph for triangular
membership functions

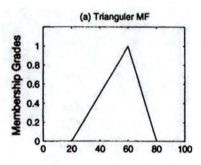

(a) Trianguler MF

the fuzzy linguistic sets as degree of membership for each variable, (iii) combining
the all fuzzy input rules by using the fuzzy operator "and" and "or" sometimes "not".

For "and" ^{u} A∩B=T($u_A(x)$, u_B (x))
For "or" ^{u} A∪B=T($u_A(x)$, u_B (x))

(iv) After the vigor and the membership of the output are combined, the consequences of the rule will be found (v) integrate the output consequences of all fuzzy
rules to a single output distribution, (vi) defuzzification applied for output distribution
either of centre of gravity method or by means of maximum method.

In the study, we used three factors (a, b, c) defined triangular membership functions. Using a, b and c factors, triangular membership functions are defined by the
following equations. Triangular membership graph functions as illustrated in Fig. 3.

$$triangle(x; a, b, c) \begin{cases} 0, & x \leq a, \\ \dfrac{x-a}{b-a}, & a \leq x \leq b, \\ \dfrac{c-x}{c-b}, & b \leq x \leq c, \\ 0, & c \leq x. \end{cases}$$

Performance index assessment for urban water system is a challenging assignment because it involves various aspects in the system's analysis. For accurate
measurements, we consider many parameters. The input layer comprises eight variables i.e. water availability, groundwater stress, reliability, resilience, leakage, health
and hygiene conditions, water management and vulnerability. Figure 4 shows the
operation diagram for proposed integrated hierarchical fuzzy inference.

In proposed method, we used seven fuzzy logics are S1_FIS, S2_FIS, S3_FIS,
S4_FIS, S5_FIS, S6_FIS, and the integrated fuzzy inference system (PI_FIS). There
are five fuzzy sets defined for water availability, groundwater stress, reliability,
resilience, leakage, water management and four fuzzy sets for health and hygiene
conditions and vulnerability. In this fuzzy model outputs for S1_FIS, S2_FIS,
S3_FIS, S4_FIS, S5_FIS and S6_FIS are represented by y_1, y_2, y_3, y_4, z_1 and z_2
respectively.

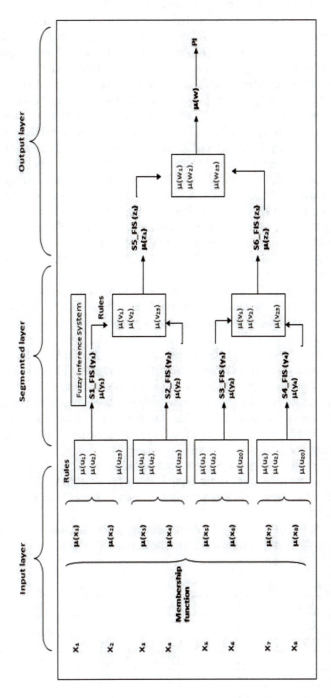

Fig. 4 Operation diagram of the integrated hierarchical fuzzy inference

The S1_FIS has two inputs parameters, water availability and groundwater stress. Assigned the fuzzy sets of water availability as Very small (VS), Small (S), Average (Av), Large (La), Very Large (VLa). Likewise, five fuzzy sets are assigned for the groundwater stress parameter: Low (Lo), Average (Av), Marginal (Mar), Large (La), Severe (Sev). The second segmented layer also has two inputs parameters, reliability and Resilience. Assigned five fuzzy sets for reliability: Unacceptable (Un), Poor (P), Acceptable (Acc), Good (G), Excellent (E). Likewise, five fuzzy sets are for resilience: Very less (VLe), Less (Le), Medium (Med), High (Hi), Very high (VHi). Designated five fuzzy sets for Leakage parameter: no-leakage (NL), very low (VLo), low (Lo), medium (Me), high (Hi). Assigned five fuzzy sets for Health and hygiene conditions parameter: Inconsiderable (Incons), Moderate (Mod), Considerable (Cons), Best (best). Likewise allotted five fuzzy sets for water management: Worst (w), Bad (B), Good (G), Better (Bet), Best (Best). Assigned five fuzzy sets for Vulnerability: Less (Le), Good (G), Superior (Sup), Ultimate (Ul). The S1_FIS logic, and S2_FIS logic, has one output parameter i.e. S5_FIS logic the segmented Urban water system Performance index. The S3_FIS logic and S4_FIS logic has another one output parameter i.e. S6_FIS logic the segmented Urban water system performance index.

For both S5_FIS and S6_FIS fuzzy inference system, there are five fuzzy sets: Very Low (VL), Low (L), Medium (M), High (H), and Very High (VH). Additional inputs to the integrated fuzzy inference system are the S5_FIS logic and S6_FIS logic parameters.

PI_FIS has one output parameter for the performance index of urban water system (UWSPI). The final performance index result of the integrated hierarchical fuzzy logic model is the output of the integrated fuzzy inference system.

We classified a different number of membership functions for each input parameter depending on the statistical variation. In the proposed model, we define different rules based on the number of fuzzy sets input and output. For, the water availability and groundwater stress parameter we delineate five fuzzy sets. It is therefore possible to define the number of rules for S1_FIS by 5 x 5 = 25. We carried out the rules as shown in Table 1. Rules for S2_FIS, S3_FIS, S4_FIS, S5_FIS and S6_FIS Fuzzy logic Fuzzy logic as shown in Tables 2, 3, 4, 5, 6 and 7 are also laid down.

The outputs of S5_FIS and S6_FIS are the inputs of the integrated fuzzy logic, and the output is the final performance index for the urban water system. For every

Table 1 Rules of fuzzy in S1FIS

X2\X1	V S	S	Av	La	VLa
Lo	M	M	H	H	VH
Av	M	M	M	H	H
Mar	L	L	M	M	H
La	VL	VL	L	M	M
Sev	VL	VL	L	M	M

Table 2 Rules of fuzzy in S2FIS

X4\X3	Un	P	Ac c	G	E
VLe	VL	L	L	L	M
Le	VL	L	L	L	M
Med	L	L	M	M	H
Hi	M	M	H	H	VH
VHi	M	M	H	H	VH

Table 3 Rules of fuzzy in S3FIS

X6\X5	N L	VLo	Lo	Me	Hi
Incons	V L	VL	V L	V L	V L
Mod	L	L	L	L	V L
Cons	H	H	M	M	L
Best	V H	V H	H	H	M

Table 4 Rules of fuzzy in S4FIS

X8\X7	W	B	G	Bet	Best
Le	V L	V L	L	M	M
G	L	L	M	M	H
Sup	M	M	H	H	V H
Ul	M	M	H	V H	V H

Table 5 Rules of fuzzy in S5FIS

S2_FIS\S1_FIS	V L	L	M	H	VH
VL	V L	V L	L	M	M
L	L	L	L	M	M
M	L	L	M	M	H
H	M	M	M	H	H
VH	M	M	H	VH	VH

Table 6 Rules of fuzzy in S6FIS

S4_FIS\S3_FIS	V L	L	M	H	VH
VL	V L	VL	L	M	M
L	L	L	L	M	M
M	L	L	M	M	H
H	M	M	M	H	H
VH	M	M	H	VH	VH

Table 7 Rules of fuzzy in P.I._FIS

S6_FIS\S5_FIS	VL	L	M	H	VH
VL	1	2	3	4	5
L	2	2	3	4	5
M	4	3	4	5	6
H	6	5	7	8	9
VH	7	8	8	9	10

input variable, we described five fuzzy sets. The numbers of rules for integrated FIS was determined $5 \times 5 = 25$. We illustrate the rule designing for integrated fuzzy logic in Table 7.

5 Results and Discussion

On MATLAB 7.14.0739, carrying out the proposed integrated fuzzy inference model. In the proposed integrated hierarchical fuzzy inference model, according to the data, each input and output is level and specifies membership functions of a different number and a different scale for each parameter.

Linguistic terms for an output parameter are Very Low (VL), Low (L), Medium (M), High (H), and Very High (VH) same for six FIS. Rules are designed for all six-output i.e. S1-FIS, S2-FIS, S3-FIS, S4-FIS, S5-FIS, S6-FIS. The rule viewer defined the fuzzy process of each output parameter. The figure includes three plots representing the antecedent and consequent of the rule. Figure 5 illustrates the integrated FIS triangular membership functions. The linguistic terms allocated to the UWS_PI membership functions are 0, 1, 2, 3, 4, 5, 6, 7, 8, 9, 10. Figure 6 shows the FIS editor and rule viewer for UWS_PI with S5_FIS and S6_FIS output.

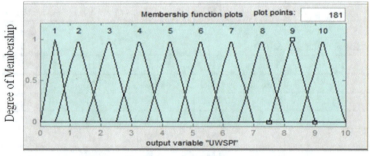

Fig. 5 Membership function for UWSPI

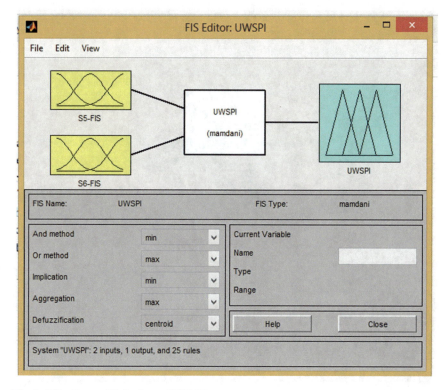

Fig. 6 FIS editor and Rule viewer UWS_PI

All performance indices of Raipur's urban water system are aggregated into an overall index in the proposed integrated fuzzy inference model. We show the outline of this aggregate performance index by using surface viewer Fig. 7.

We designed the proposed integrated hierarchical fuzzy inference to reducing the dimensionality curse, i.e. reducing the number of rules in the rule base. There are six sub-fuzzy inference logics in the proposed model. It is possible to define the likely number of rules for the integrated hierarchical fuzzy inference model (RI) as;

$$RI = (5 \times 5) + (4 \times 5) + (5 \times 5) + (4 \times 5) + (5 \times 5) + (5 \times 5) + (5 \times 5) = 165$$

Describe all probable rules in traditional (RC) fuzzy inference logic as below.

$$RC = (5 \times 5 \times 4 \times 5 \times 5 \times 5 \times 5 \times 4) = 250,000$$

It is very tough to design many considerable rules and curse of dimensionality damage the simplicity of rules reduction. In addition, it is important to designing rule with more attentiveness. With fewer rules, an expert may be careful, but in the case of thousands of rules, there may be an error in the rules leading to unexpected incidents. Therefore, for accurate rule designing for several rules we proposed here integrated hierarchical fuzzy inference.

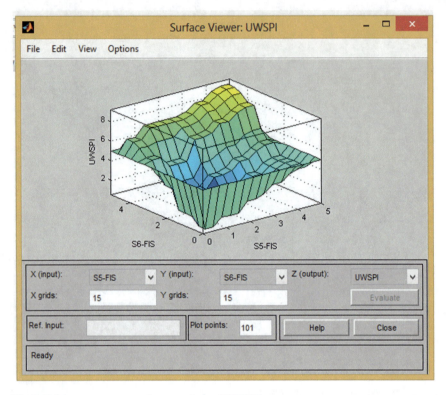

Fig. 7 Urban water system performance index (UWSPI)

6   Conclusion

The performance of an urban water system takes into account eight variables in the framework. A performance assessment method proposed to develop the indices to determine the capability of the urban water system in Raipur, India, to get its sustainable objectives. Such parameters are being established to trace the urban water system recitation of the region. The proposed integrated hierarchical fuzzy inference was taking several aspects of the Raipur city from the previous year's draft report. The approach can lessen the number of rules in the process and solve the problems of higher-dimensional hierarchical fuzzy logic. We found that it would minimize the number of rules as defined in the modeling for an eight variables. As, a compare to traditional fuzzy logic inference with hierarchical fuzzy inference model was very simple in the structure, and it do not augment rule exponentially as a new parameter go through the system. In traditional fuzzy inference, only one parameter has a design rule at one time only, which increase complexity in the system and increase the chance of errors while designing rules.

The model was using to aggregate each index value of segmented fuzzy inference system into overall performance indices. The model can also be used to assess a water system's performance in other urban areas and other water utilities. The present model considered only the triangle membership functions. For future research work, we would apply a different distribution membership functions in order to improve accuracy. For more accuracy, more parameters will be a study for urban water system performance.

References

1. A. Altunkaynak, M. Özger, M. Çakmakci, Water consumption prediction of istanbul city by using fuzzy logic approach. Water Resour. Manage **19**(5), 641–654 (2005). https://doi.org/10.1007/s11269-005-7371-1
2. V. Adriaenssensa, B.D. Baetsb, P.L.M. Goethalsa, N.D. Pauwa, Fuzzy rule-based models for decision support in ecosystem management. Sci. Total Environ. **319**, 1–12 (2009)
3. G. Almeida, J. Vieira, A.S. Marques, A. Kiperstok, A. Cardoso, Estimating the potential water reuse based on fuzzy reasoning. J. Environ. Manage. **128**, 883–892 (2013). https://doi.org/10.1016/j.jenvman.2013.06.048
4. Z. Bien, W.-K. Song, Blend of soft computing techniques for effective human–machine interaction in service robotic systems. Fuzzy Sets Syst. **134**(1), 5–25 (2003). https://doi.org/10.1016/s0165-0114(02)00227-0
5. A. Bagheri, A. Asgary, J. Levy, M. Rafieian, A performance index for assessing urban water systems: a fuzzy inference approach. J. Am. Water Works Assoc. **98**(11), 84–92 (2006). https://doi.org/10.1002/j.1551-8833.2006.tb07807.x
6. O. Castillo, F. Valdez, P. Melin, Hierarchical genetic algorithms for topology optimization in fuzzy control systems. Int. J. Gen Syst **36**(5), 575–591 (2007). https://doi.org/10.1080/03081070701321860
7. H.-W. Chen, N.-B. Chang, Using fuzzy operators to address the complexity in decision making of water resources redistribution in two neighboring river basins. Adv. Water Resour. **33**(6), 652–666 (2010). https://doi.org/10.1016/j.advwatres.2010.03.007
8. M. Fayaz, S. Ahmad, I. Ullah, D. Kim, A Blended Risk Index Modeling and Visualization Based on Hierarchical Fuzzy Logic for Water Supply Pipelines Assessment and Management. Processes **2018**(6), 61 (2018)
9. M. Fayaz, S. Ahmad, L. Hang, D. Kim, Water supply pipeline risk index assessment based on cohesive hierarchical fuzzy inference system. Processes **7**(4), 182 (2019). https://doi.org/10.3390/pr7040182
10. Y. Icaga, Fuzzy evaluation of water quality classification. Ecol. Ind. **7**(3), 710–718 (2007). https://doi.org/10.1016/j.ecolind.2006.08.002
11. T.M. Jelleli, A.M. Alimi, Automatic design of a least complicated hierarchical fuzzy system, in *Proceedings of the International Conference on Fuzzy Systems*, (Barcelona, Spain, 18–23 July 2010), pp. 1–7
12. L.A. Zadeh, Fuzzy sets. Inf. Control **8**, 338–353 (1965)
13. D. Li, W. Yao, Risk assessment of long- distance water transmission pipeline based on fuzzy similarity evaluation approach, in *Proceedings of the 2016 12th International Conference on Natural Computation, Fuzzy Systems and Knowledge Discovery (ICNC-FSKD)*, (Changsha, China, 13–15 August 2016), IEEE: Piscataway, (NJ, USA, 2016), pp. 1096–1102
14. A.A. Malinowska, R. Hejmanowski, Fuzzy-logic assessment of failure hazard in pipelines due to mining activity. Proc. IAHS **372**(105–109), 2015 (2015). https://doi.org/10.5194/piahs-372-105-
15. T. Ross, *Fuzzy Logic With Engineering Applications* (McGraw-Hill, New York, 1995)
16. W. Li, Z. Huicheng, Urban water demand forecasting based on hp filter and fuzzy neural network. J. Hydroinformatics **12**(2), 172–184 (2010). https://doi.org/10.2166/hydro.2009.082

17. T.Y. Xu, X.S. Qin, Integrating decision analysis with fuzzy programming: application in Urban water distribution system operation. J. Water Resour. Plann. Manag. **140**(5), 638–648 (2014). https://doi.org/10.1061/(asce)wr.1943-5452.0000363
18. J. Zhang, J. Gao, M. Diao, W. Wu, T. Wang, S. Qi, A case study on risk assessment of long distance water supply system. Procedia Eng. **70**, 1762–1771 (2014)

Design and Simulation of an Efficient FPGA Based Epileptic Seizure Detection System

Abrar Ahmad Banday, Abdul Imran Rasheed, and Ugra Mohan Roy

Abstract Epilepsy is a neurological disorder. It effects almost 50 million people worldwide. Presently noise removal for accurate epilepsy detection has been potential area of research. In this paper a Lifting Scheme based Discrete Wavelet Transform (LDWT) and Inverse Lifting Scheme based Discrete Wavelet Transform (IDWT) is used to denoise the EEG signal. Least Mean Square (LMS) adaptive filter has been used to compute minimum Mean Square Error. The epileptic Electroencephalography (EEG) signal has higher spikes and sharp curve density than the normal EEG signal. Therefore, the error output and total error energy of the Epileptic EEG signal is higher than the normal. The calculated "Total Error Energy" of the EEG signal is passed through a decision-making block that decides if the subject is epileptic or not. The developed RTL architecture has been implemented on Artix-7 FPGA. The maximum operating frequency of the architecture is 100 MHz and the maximum delay is 10 ns. The developed architecture utilizes 2223 slice LUT's, consuming 0.56 watts of total power. The accuracy and sensitivity of the design is 92.1% and 95.2% respectively.

Keywords Seizure · EEG · LDWT · LMS adaptive filter · Total error Energy

1 Introduction

Seizure is defined as abnormal or synchronous neural activity in the brain. This unre-strained electrical activity of the brain has various manifestations that result in physical convulsions, inability to think and react, sensory hallucinations or combination

A. A. Banday (✉) · A. I. Rasheed · U. M. Roy
M. S. Ramaih University of Applied Sciences, Bengaluru, India
e-mail: astroabrar@gmail.com

A. I. Rasheed
e-mail: imran.ec.et@msruas.ac.in

U. M. Roy
e-mail: mohanroy.ec.et@msruas.ac.in

© Springer Nature Singapore Pte Ltd. 2020
R. Bera et al. (eds.), *Advances in Communication, Devices and Networking*,
Lecture Notes in Electrical Engineering 662,
https://doi.org/10.1007/978-981-15-4932-8_20

Fig. 1 Overview of epileptic seizure detection system

of all symptoms. The condition is term as 'Epilepsy when the patient has recurring seizures. Seizures may be categorized as focal or generalized. Focal seizures originate in one hemisphere of brain while as generalized seizures affect the entire brain. Electroencephalography (EEG) is the most efficient tool in evaluating the condition of patients with epilepsy. Epilepsy is a chronic neurological disorder that affects more than 50 million people worldwide. The patients suffering from epilepsy undergo unpredictable and persistent seizures, which limits the independence of an individual, increases the risk of serious injury [1]. Presently noise removal for accurate epilepsy detection has been potential area of research. The ultimate need is to develop a reliable Epileptic Seizure detection systems shown in Fig. 1 that filters the smallest noise in EEG signal and estimates the error energy to detect epilepsy in the patient. In this paper, LDWT is used to remove the noise from the signal. 1D, 5 level decomposition of the signal is carried out using LDWT and the signal is decomposed into 5 physiological EEG bands. The noise associated with high-frequency details coefficients is removed through hard thresholding. IDWT is used to reconstruct the signal without any loss of information [2, 3]. Windowing helps to convert non-stationary quasi-infinite signals into stationary signals. The stationary denoised signals are given to LMS adaptive filter. LMS adaptive algorithms used in the filter adapt the weights by iteratively approaching the Mean Square Error (MSE) minimum. Epileptic EEG signals have sudden changes and spikes associated with them while non-epileptic EEG signals are smother. LMS adaptive algorithm cannot track these changes resulting in larger Error output, which results in larger "Total Error Energy". The calculated "Total Error Energy" of the EEG signal is passed through a decision-making block that decides, whether the subject is epileptic or not.

2   EEG Data Set

EEG database is developed by the Department of Epileptology, University of Bonn, Germany [4]. There are four EEG data subsets in the database, denoted as A1, B2, C3, and D4. Each data set contains 50.txt files. The duration of each discrete time is 23.6 s. The sampling frequency of the recorded dataset with 128 channel EEG recording system is 173.6 Hz [5]. The subject description for each subset is as follows: A1: Healthy volunteers without epilepsy with eyes open and recorded from outside the skull. B2: Healthy volunteers without epilepsy with eyes closed. C3: Patients with epilepsy during interictal period. D4: Patients with epilepsy during ictal period [6].

3   Denoising of EEG Signal Using IDWT

The signal is made reliable by reducing the noises of the signal. The more common pre-processing procedure is using a bandpass filter, but in a complex signal like EEG signal, the sharp features like spikes and bumps, which carry information are also removed by these filters. LDWT noise removal technique solves this problem. It reduces the noise without degrading the signal. The important features of the signal remain intact with the signal. LDWT denoising require involves three steps to denoising a signal. The first step involves the decomposition of the signal into multilevel wavelets. Multilevel decomposed of quasi-infinite EEG signal is performed using wavelet transform. At every level, high-frequency coefficients and low-frequency coefficients known as details coefficients and approximates coefficients respectively are obtained. The filtering coefficients are also downsampled by factor 2 at each level. The cutoff frequency of different level frequency bands is given by Eq. 1. After scaling, translation is performed, which occurs at integer level. This kind of sampling eliminates redundancy in coefficients. The output of the transforms yields same number of outputs coefficients as the length of input signal. Therefore, it requires less memory. Noise in the EEG signal results in smaller DWT coefficients and are removed through thresholding.

$$f_{\text{cutoff}} = f_{\text{max}}/2^i \tag{1}$$

where i $= 1, 2, 3\ldots$ and represents different levels.

Table 1 shows the decomposition of EEG signal into five physiological EEG bands.

The second step is identifying the threshold. There are two thresholding techniques soft threshold and hard threshold. Hard thresholding is used in paper. In hard thresholding output coefficients at each level are compare with threshold and the coefficients less than threshold are shrunk to zero while the coefficients greater than threshold remain the same. Universal threshold is given by Eq. 2 [7].

$$T = \sigma\sqrt{2 \cdot \log(n)} \tag{2}$$

$\sigma = $ median(ab(Di))/0.6745

where, Di is detail coefficients.

The third step is reconstruction of the signal. IDWT is used to reconstruct the signal and at each level, the coefficients are sampled up by 2.

Table 1   Different EEG bands

Level	1	2	3	4	5
Frequency band	Gamma	Beta	Alpha	Theta	Delta
Frequency range (Hz)	30–60	13–30	8–13	4–8	0–4

4 LMS Adaptive Filter and Total Energy Estimation

The EEG signal is quasi-infinite signal and it makes the EEG signal non-stationary [8]. Rectangular windowing technique is used to make the signal stationary, the windowing technique is also known as the boxcar or Dirichlet windowing. The EEG signal includes 3696 samples and samples are divided into 400 long epochs. The least mean square or LMS adaptive algorithm adapts the weights by iteratively approaching the MSE minimum. MSE is estimated by using Eq. 3. The goal is to find y(n) in such a way that minimizes the error between the true value d(n) and the estimated value y(n) [9].

$$e(n) = d(n) - y(n) \tag{3}$$

To update the weights in LMS adaptive filter, Stochastic Gradient Descent method is used. The LMS algorithm for a pth order algorithm is given by Eqs. 4, 5 and 6, where, Initial value of $\hat{h}(0) = 0$.

$$Y(n) = [y(n), y(n-1), y(n-2), \ldots, y(n-p+1)] \tag{4}$$

$$e(n) = d(n) - \hat{h}(n)Y(n) \tag{5}$$

$$\hat{h}(n+1) = \hat{h}(n) + \mu\, e(n)\, Y(n) \tag{6}$$

Parameters: P = filter order, μ = step size.

Computation: For n = 0, 1, 2, ...; e(n) = error; $\hat{h}(n+1)$ = updated coefficients

Y(n) is the EEG signal. d(n) is the desired signal, e(n) is difference between desired signal and estimated signal and μ is step size. Filter algorithm cannot track sudden changes like spikes and the error energy increases on that instant. Therefore, the error and error energy of the EEG signal of an epileptic patient is higher than normal EEG signals. Total error energy of each epoch (E_k) is given by Eq. 7 [10]. The comparator threshold unit compares total error energy with the threshold. If the Error energy is more than threshold, then the segment of EEG signal is epileptic otherwise EEG signal is non-epileptic. With the majority vote, the signal is categorized as epileptic ornon-epileptic signal.

$$E_k = \sum N\,[(n)]^2 \tag{7}$$

5 Architecture

The RTL design is developed, simulated and implemented on XILINX Artix-7 platform, target device is xc7a100tcg324-1. As shown in Fig. 2, the top module of the design consists of a ROM sub-block, 5 lifting scheme based DWT sub-blocks, five Inverse scheme based DWTs, RAM block for the temporary storage of data, one LMS adaptive filter and Error Energy calculation sub-block to calculate error and total error energy of each segment. The inputs to the signal are clock, reset, enable and data. The EEG data to be processed is stored in the ROM. The EEG signal is first converted into ".coe" file, only then it is stored into ROM. Each of the sample data stored has 17-bit width and each sample is stored in respective address locations. LDWT is computed by successive low pass and high pass filtering in the discrete-time domain. There are four adders, two multipliers and some flip flops in DWT sub-block. Five DWT blocks are used for five-level decomposition. At each level, there are 2 outputs namely, the details coefficient and approximates coefficient. From Fig. 2 it is observed that the details coefficients are passed through threshold comparator to remove noise.

Five IDWT sub-blocks are used to reconstruct the signal. At each level, there is an upsampling of coefficients by a factor 2. Every IDWT block consists of 2 adders, 2 subtractors, 2 multipliers, and some flip-flops. The denoised signal is then temporarily stored in RAM for further processing. The denoised data enters the LMS adaptive filter sub-block. Other inputs to the LMS filter block is clock, reset and enable. The outputs from the LMS adaptive filter sub-block are "Error output" and "filtering done". There are 1336 LUTs in the LMS filter sub-block. EEG signal of Epileptic patients results in more error as compared to normal subjects. Finally, the comparator with a threshold set by inspecting the statical distribution of Error Energy density is used to detect the seizure. If epilepsy is detected in the subject, logic "1" is read, else it readslogic "0".

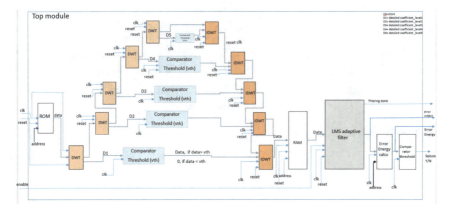

Fig. 2 Epileptic seizure detection system design in FPGA

6 Simulations and Experimental Results

The software implementation of the Epileptic Seizure Detection System is carried out on MATLAB R2018a and hardware implementation on XilinxArtix-7. The noisy and denoised signal is shown in Fig. 3. The estimated Peak Signal to Noise Ratio(PSNR) is 35.31 dB. Total Error energy of the epileptic segments (ictal period) is higher due to spikes that are present during epileptic seizures. The reason being that the filter algorithm cannot track sudden changes like spikes and the error energy increases on that instant, resulting in higher "Error output" in LMS adaptive filter. Figure 4 shows the EEG recording of a healthy and epileptic patient and error energy output of same subject at every 400 samples of data signal respectively. Figure 5 shows that denoised EEG signal is much more smoother than noisy data_in EEG signal. The signal used includes 3696 samples and the samples are divided into 400 long epochs. Hence, Each EEG signal is divided into nine 400 sample segments. Each segment provides its Error Energy output and when compared with the threshold in the Threshold comparator block.

If the total Error Energy is more than the threshold then Seizure Y/N reads logic "1" otherwise logic "0" is read.

Seizures Y/N is a 1-bit output that tells us if the seizure is "true" or "false". Figures 6 and 7 show the simulations of a healthy and epileptic patient respectively. The success rates of the seizure detection system are tabulated and comparison between software and hardware implementation is also given in Table 2. The device hardware utilization is shown in Table 3.

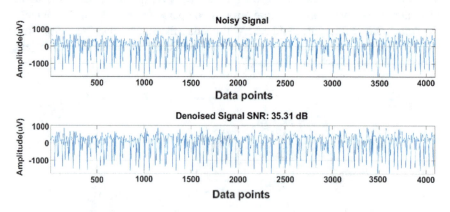

Fig. 3 Noisy and denoised signal

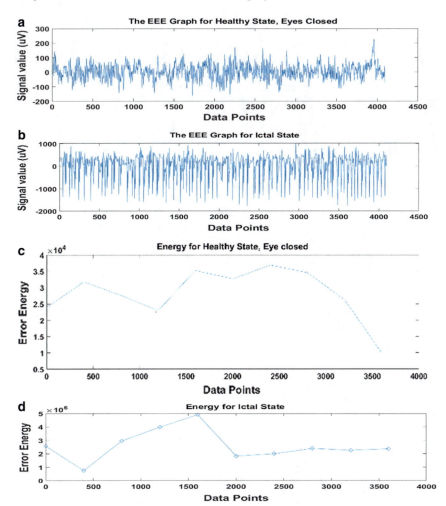

Fig. 4 **a** EEG recording of a healthy subject, **b** EEG recording of a patient in ictal state, **c** modeling error of healthy patient, **d** modeling error of a patient in ictal state

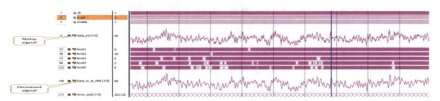

Fig. 5 Simulation results of denoised signal using XilinxArtix-7 and Vivado

Fig. 6 Simulation results of healthy volunteers

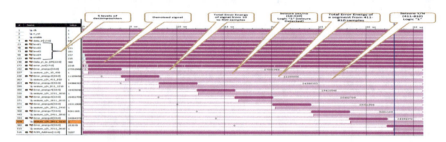

Fig. 7 Simulation results of patients with epilepsy

Table 2 Recording details and success rates

S.no	Group	Healthy/epileptic	Success		
			Software implementation (%)	Hardware implementation (%)	
1	A, B	Healthy eyes open, healthy eyes closed	90.5, 93.8	89, 96	
2	C, D	Epileptic inter-ictal, epileptic ictal state	92, 95.22	90.2, 98	

Table 3 Hardware utilization

Logic utilization	Used	Available	Utilization (%)
Number of slice registers	2365	126800	1.87
Number of Slice LUTs	2223	63400	3.51
Number of LUT-FF pairs	1897	126800	1.50

7   Conclusion

We propose an Epileptic Seizure detection system. The developed RTL architecture has been implemented on Artix-7 FPGA. The maximum operating frequency of the architecture is 100 MHz and the maximum delay is 10 ns. The whole design consists of 181 pins, 54 DSP blocks, one block ROM and one block RAM. The developed architecture utilizes 2223 slice LUT's. The accuracy and sensitivity of the design is 92.1% and 95.2% respectively.

References

1. A.H. Shoeb, *Application of Machine Learning to Epileptic Seizure Onset Detection and Treatment*, (Doctoral Dissertation, Massachusetts Institute of Technology, 2009)
2. P.R. Pal, R. panda, Classification of EEG signals for epileptic seizure evaluation, in *Proceedings of the 2010 IEEE Students' Technology Symposium*, vol. 25, pp. 72–76, 2010
3. M. Bahoura, H. Ezzaidi, FPGA-implementation of discrete wavelet transform with application to signal denoising. Circuits Syst. Signal Proc. **31**(3), 987–1015 (2012)
4. D. Lal, A.K. Ruppert, H. Trucks, H. Schulz, C.G. de Kovel, D.K.N. Trenité, A.C. Sonsma, B.P. Koeleman, D. Lindhout, Y.G. Weber, H. Lerche, Burden analysis of rare microdeletions suggests a strong impact of neurodevelopmental genes in genetic generalised epilepsies. PLoS Genet. **11**(5), e1005226 (2015)
5. G. Ekim, N. Ikizler, A. Atasoy, The effects of different wavelet degrees on epileptic seizure detection from EEG signals, in 2017 IEEE International Conference on Innovations in Intelligent Systems and Applications (INISTA), (July 2017), (pp. 316–321). IEEE
6. J.T. Oliva, J.L.G. Rosa, How an epileptic EEG segment, used as reference, can influence a cross-correlation classifier? Appl. Intell **47**(1), 178–196 (2017)
7. M. Aqil, A. Jbari, A. Bourouhou, ECG signal denoising by discrete wavelet transform. Int. J. Online Eng. **13**(9) (2017)
8. H. Adeli, Z. Zhou, N. Dadmehr, Analysis of EEG records in an epileptic patient using wavelet transform. J. Neurosci. Methods **123**(1), 69–87 (2003)
9. R. Martinek, R. Kahankova, H. Nazeran, J. Konecny, J. Jezewski, P. Janku, P. Bilik, J. Zidek, J. Nedoma, M. Fajkus, Non-invasive fetal monitoring: a maternal surface ECG electrode placement-based novel approach for optimization of adaptive filter control parameters using the LMS and RLS algorithms. Sensors **17**(5), 1154 (2017)
10. T. Lajnef, S. Chaibi, A. Kachouri, M. Samet, Epileptic seizure detection using linear prediction filter, in *12th International Conference on Sciences and Techniques of Automatic Control And Computer Engineering*, 2010

Design of a Novel Reconfigurable Microstrip Bandpass Filter with Electronic Switching

Hashinur Islam, Sagnik Sarkar, Tanushree Bose, and Saumya Das

Abstract A novel multifunctional reconfigurable microstrip bandpass filter with two electronic switching is presented. In this paper, four possible combinations are explored using two electronic switches and four different bandpass filters with multiband characteristics are presented. This type of reconfigurable filter structure is beneficial for multifunction wireless communication systems.

Keywords Bandpass filter (BPF) · Reconfigurable filter · Multiband · Switchable filter

1 Introduction

In the multifunction communication system, the importance of designing a reconfigurable bandpass filter (BPF) is noticeable. One of the important facts of a reconfigurable bandpass filter is low cost, simple structure, low power consumption, and small size [1, 2]. Because of all these exciting characteristics, the reconfigurable bandpass filter is expected to play an interesting role for the next-generation multiband and multimode wireless communication technologies, and the activity of growing research is focused on them [3, 4].

Over the years, a large number of frequency-adaptive RF filters in terms of center frequency, bandwidth, transmission zeros (TZs), and different type of filtering profile

H. Islam (✉) · S. Sarkar · T. Bose · S. Das
Department of Electronics and Communication Engineering, Sikkim Manipal Institute of Technology, Sikkim Manipal University, Sikkim, India
e-mail: hashinur0001@gmail.com

S. Sarkar
e-mail: sagniksarkar0102@gmail.com

T. Bose
e-mail: tanushree.contact@gmail.com

S. Das
e-mail: saumya.das.1976@gmail.com

© Springer Nature Singapore Pte Ltd. 2020
R. Bera et al. (eds.), *Advances in Communication, Devices and Networking*,
Lecture Notes in Electrical Engineering 662,
https://doi.org/10.1007/978-981-15-4932-8_21

have been reported [5, 6]. The frequency bands are required in practical applications in GSM, UMTS, ISM, WiMAX, and WLAN [7–9]. Numerous researchers have proposed various configurations for reducing filter size and improving filter performance [10].

In this work, a reconfigurable microstrip bandpass filter is designed using meandered but symmetrical structure. The reconfigurability is achieved using two electronic switches. The proposed microstrip bandpass filter achieved four different switchable states. The return loss and the insertion loss are satisfactory for all the four states.

2 Filter Configuration

The geometry of the proposed reconfigurable filter is developed on FR4 substrate ($\varepsilon_r = 4.3$, loss tangent $\tan\delta = 0.025$) with thickness of 0.8 mm which is shown in Fig. 1. The overall size of the filter is 20 mm x 20 mm and the dimensions of each and every part of the meandered structure are mentioned in Fig. 1. HFFS, CST-MWS, and ADS simulation softwares are used to simulate the proposed filter.

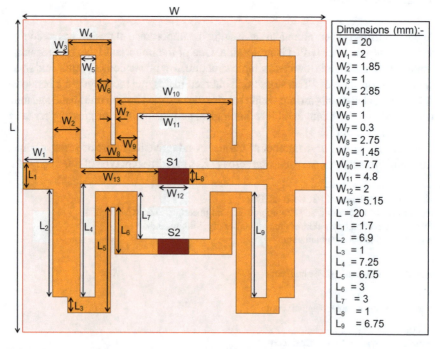

Fig. 1 Geometry of the proposed reconfigurable bandpass filter

To design the proposed filter, two switches are incorporated and different operational states of the proposed filter are as follows.

State 1

When both switches are in OFF state, the designed filter gives the frequency range 2.93–3.54 GHz, which is applicable for WiMAX and WLAN band applications.

State 2

When the switch S1 is in OFF state and switch S2 is in ON state, then the designed filter gives the frequency ranges 1.69–2.1 and 3.2–3.52 GHz. The multiband obtained corresponds to DCS, GSM, PCS, DCET, UMTS, WiMAX, and WLAN band applications, respectively.

State 3

When the switch S1 is in ON state and switch S2 is in OFF state, then the proposed filter gives the frequency range 1.49–1.71 GHz for the use of GPS band application.

State 4

When both switches are in ON state, the proposed filter gives the frequency ranges of 2.33–3.35 and 5.58–5.87 GHz. The multiband obtained for this frequency range corresponds to ISM, WLAN, and WiMAX band applications.

Fractional bandwidth (FBW) and percentage bandwidth (% BW) of the bandpass filter can be given by the following formulas [11]:

$$\text{FBW} = \frac{f_U - f_L}{f_C} \tag{1}$$

$$\%\text{BW} = \frac{f_U - f_L}{f_C} \times 100\% \tag{2}$$

where f_U = upper frequency, f_L = lower frequency, and f_C = center frequency which is defined by

$$f_C = \frac{f_U + f_L}{2} \tag{3}$$

Table 1 Switching states of the proposed reconfigurable microstrip filter

States	S1	S2	Frequency range (GHz)	Operating frequency (GHz)	3dB BW	Applications
1	OFF	OFF	2.93–3.54	3.2, 3.5	28%, 26%	WiMAX and WLAN
2	OFF	ON	1.69–2.1, 3.2–3.52	1.8,1.88,1.9,1.95, 3.3, 3.5	77%, 74%, 73%, 71%, 24%, 22%	DCS, GSM, PCS, DCET, UMTS, WiMAX and WLAN
3	ON	OFF	1.49–1.71	1.575	35%	GPS
4	ON	ON	2.33–3.35, 5.58–5.87	2.4, 2.5, 3.2, 5.7, 5.8	66%, 63%, 50%, 12%, 11.5%	ISM, WLAN and WiMAX

3   Results and Discussion

The designed microstrip BPF gives four different states using two electronic switches shown in Table 1.

The simulated S-parameters of the filter are validated using three different electro-magnetic simulation softwares, HFSS, CST-MWS, and ADS. The simulated return loss is achieved above 10 dB in all bands for State 1 and insertion loss in all the bands is less than 2.3 dB which is shown in Fig. 2. The 3 dB percentage bandwidth achieved in each band is 28% for WiMAX and 26% for WLAN bands, respectively, and the 10 dB percentage bandwidths achieved in all bands are 19% for WiMAX and 17.5% for WLAN, respectively, using Eq. (2).

The simulated insertion loss for lower bands is less than 1.5 dB and for upper bands is less than 2.4 dB and return loss is achieved above 10 dB in all bands for State 2 as shown in Fig. 3. The 3 dB percentage bandwidth achieved in each band is 77% for DCS, 77% for GSM, 74% for PCS, 73% for DCET, 71% for UMTS, 24% for WiMAX, and 22% for WLAN bands, respectively, using Eq. (2). The 10 dB percentage bandwidths achieved in all bands are 23% for DCS, 23% for GSM, 22% for PCS, 21.5% for DCET, 21% for UMTS, 9% for WiMAX, and 10% for WLAN bands, respectively, using Eq. (2).

Figure 4 shows the simulated return loss and insertion loss for State 3 which are achieved above 10 dB and less than 1.5 dB, respectively. The 3 dB and 10 dB percentage bandwidths achieved for GPS band are 35% and 14%, respectively, using Eq. (2).

The simulated insertion loss for lower bands is less than 1.8 dB and for upper bands is less than 2.5 dB and return loss is achieved above 10 dB in all bands for State 4 as shown in Fig. 5. The 3 dB percentage bandwidths achieved for lower bands are 66% for ISM, 63% for WLAN, and 50% for WiMAX and for upper bands are 12% for ISM, 11.5% for WiMAX and Wi-Fi band applications, respectively, using

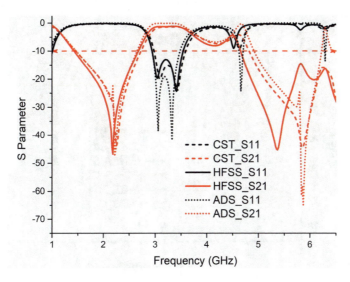

Fig. 2 Simulated S-parameters when both switches are in OFF state

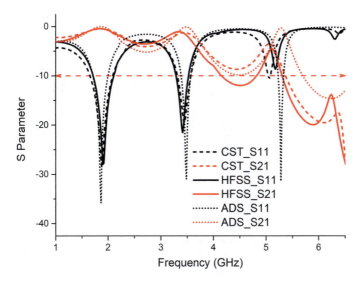

Fig. 3 Simulated S-parameters when switch S1 is in OFF state and switch S2 is in ON state

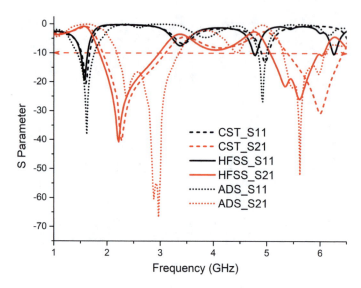

Fig. 4 Simulated S-parameters when S1 is in ON state and S2 is in OFF state

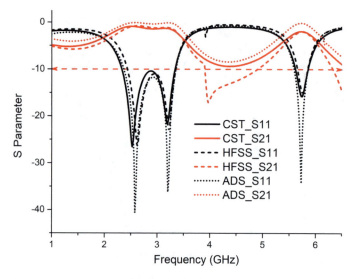

Fig. 5 Simulated S-parameters when both switches are in ON state

Eq. (2). The 10 dB percentage bandwidths achieved for lower bands are 42.5% for ISM, 41% for WLAN, and 32% for WiMAX and for upper bands are 5.2% for ISM, 5% for WiMAX and Wi-Fi band applications, respectively, using Eq. (2).

4 Conclusion

In this article, a reconfigurable microstrip bandpass filter using meandered but symmetrical structure is proposed. Four possible combinations are explored using two electronic switches and four different bandpass filters with multiband characteristics are achieved. The microstrip reconfigurable bandpass filter provides return loss always more than 10 dB and insertion loss below 3 dB in all the states. This proposed reconfigurable filter is very useful for multifunction communications like DCS, GSM, PCS, DCET, UMTS, ISM, WiMAX, and WLAN.

In the present work, only simulation results are shown and verified using different softwares (HFSS, CST-MWS, and ADS). The prototype model of the same can be developed and validated.

Acknowledgements The authors would like to thank Sikkim Manipal University, Sikkim, India for providing TMA Pai University Research Fund (Grant/Award Number: 118/SMU/REG/UOO/104/ 2019) for this research work.

References

1. P. Kumar Singh, A. Kumar Tiwary, N. Gupta, Ultra-compact switchable microstrip band-pass filter-low-pass filter with improved characteristics. Microwave Opt. Technol. Lett. **59**(1), 197–201 (2017)
2. H.A. Mohamed, H.B. El-Shaarawy, E.A. Abdallah, H.M. El-Hennawy, Frequency-reconfigurable microstrip filter with dual-mode resonators using RF pin diodes and DGS. Int. J. Microwave Wireless Technol. **7**(6), 661–669 (2015)
3. S. Kingsly, M. Kanagasabai, G.N.A. Mohammed, S. Subbaraj, Y. Panneer Selvam, R. Natarajan, Multi-band reconfigurable microwave filter using dual concentric resonators. Int. J. RF Microwave Comput. Aided Eng. **28**(6), e21,290 (2018)
4. P. Vryonides, S. Nikolaou, S. Kim, M.M. Tentzeris, Reconfigurable dual-mode band-pass filter with switchable bandwidth using pin diodes. Int. J. Microwave Wireless Technol. **7**(6), 655–660 (2015)
5. W. Feng, Y. Shang, W. Che, R. Gómez-García, Q. Xue, Multifunctional reconfigurable filter using transversal signal-interaction concepts. IEEE Microwave Wireless Compon. Lett. **27**(11), 980–982 (2017)
6. C. Shi, W. Feng, R. Gómez-García, X. Zhang, Y. Zhang, W. Che, Ultra-wideband reconfigurable filter with electronically-switchable bandpass/bandstop states. In: 2018 2nd URSI Atlantic Radio Science Meeting (AT-RASC), pp. 1–4. IEEE (2018)
7. B. Lui, F. Wei, X. Shi, Switchable bandpass filter with two-state frequency responses. Electron. Lett. **47**(1), 40–41 (2011)
8. J. Mazloum, A. Jalali, M. Ojaroudi, Miniaturized reconfigurable band-pass filter with electronically controllable for wimax/wlan applications. Microwave Opt. Technol. Lett. **56**(2), 509–512 (2014)
9. J. Sahay, D. Goutham, S. Kumar, A novel compact ultrawide band filter for reconfigurable notches. Microwave Opt. Technol. Lett. **57**(1), 88–91 (2015)
10. P.K. Singh, A.K. Tiwary, Design of switchable microstrip low pass-band stop-band pass filter. Microwave Opt. Technol. Lett. **59**(2), 257–260 (2017)
11. J.S.G. Hong, M.J. Lancaster, *Microstrip Filters for RF/Microwave Applications*, vol. 167 (Wiley, 2004)

Guided Filter Based Colour Image Rain Streaks Removal Using L_0 Gradient Minimization Method

Manas Sarkar, Priyanka Rakshit Sarkar, Ujjwal Mondal, and Debashis Nandi

Abstract In recent time, removal of rain from videos and images is becoming a popular research interest in computer vision world. Outdoor visibility degrades rapidly due to rain and it gives direct affect in object detection, surveillance systems and many other image processing applications. Different research works on image rain removal have already been proposed which perform their operations based on dictionary learning and sparse coding, different image decomposition techniques, probabilistic model, entropy maximization and many more. In this paper, a novel rain removal technique is proposed where high frequency rain component of a rainy image is guided by low frequency non-rain component to remove the rain streaks by guided filter. As the guided filter is not able to remove satisfactory number of rain streaks, therefore L_0 gradient minimization technique is applied on the image obtained after guided filtering. Overall sharpness of the output image is enhanced by unsharp masking (USM) operation. Finally contrast limited adaptive histogram equalization (CLAHE) technique generates the contrast enhance desired output. Qualitative and quantitative comparisons are performed to show the superiority of the proposed technique.

Keywords Gaussian filtering · Guided filtering · L_0 gradient minimization · Image smoothing · Rain removal

M. Sarkar (✉) · P. R. Sarkar
Haldia Institute of Technology, Haldia, India
e-mail: manasm.sarkar@gmail.com

P. R. Sarkar
e-mail: priyanka.rakshitsarkar@gmail.com

U. Mondal
Calcutta University, Kolkata, India
e-mail: ujjwalmondal18@gmail.com

D. Nandi
National Institute of Technology, Durgapur, India
e-mail: debashisn2@gmail.com

© Springer Nature Singapore Pte Ltd. 2020   199
R. Bera et al. (eds.), *Advances in Communication, Devices and Networking*,
Lecture Notes in Electrical Engineering 662,
https://doi.org/10.1007/978-981-15-4932-8_22

1   Introduction

In many of the applications in computer vision, rain gives an insignificant and poor contribution. Detection of objects, tracking and surveillance systems and many other systems are badly affected by rain. Moreover different image parameters like contrast, sharpness, intensity are also changed in a rainy image. Rain also causes huge degradation of visibility. Therefore, to improve overall image quality, an appropriate rain removal technique should be applied.

Rain is nothing but the collection of water droplets which are spherical in shape and random in nature [1]. In a rainy image, pixels covered by rain have got different intensity values from the pixels not covered by rain. Therefore deraining is a method which restores the image scene behind rain by the elimination of rain streaks or by changing the intensity values of rain affected pixels. A lot of rain removal approaches have already been proposed. A probabilistic approach is introduced by Tripathy et al. [2] which successfully differentiate the rain and non-rain pixels by the automatic adjustment of threshold value. It can handle dynamic scene but fails to remove heavy rain. Kang et al. [3] proposed a novel deraining method by performing dictionary learning and sparse coding based on morphological component analysis (MCA). It successfully eliminates rain but works only in gray image. Zheng et al. [4] proposed a technique where high frequency image component with rain is guided by low frequency component to remove rain from overall image. But some other image information are also lost due to smoothing operation. As a rainy image can be considered as the combination of rain and background layer, Li et al. [5] proposed a rain removal method based on that concept. Rain patch prior is formed based on the Gaussian mixture model. This method has its limitation in size and location of the patches. Hence, after knowing all the merits and demerits of some of the rain removal techniques, we are inspired to introduce a new method which produces excellent outputs. Our proposed algorithm removes rain streaks from an image by the use of guided filter where high frequency rain component is guided by low frequency non rain component and L_0 gradient minimization technique.

2   Review of Previous Works

Kang et al. [3] proposed a novel single image rain removal technique by separating rain and non-rain image components. Rainy image is decomposed into low and high frequency component using bilateral filter. Using MCA based dictionary learning and sparse coding method, rain component is extracted separately from the high frequency component of the rainy image. Non-rain structure with low frequency component of the rainy image jointly produces the desired output. But this method works only for grayscale images.

Chen et al. [6] developed a colour image rain removal method which is operated by guided filter based image decomposition technique and dictionary learning with sparse coding based rain detection and its removal technique. Moreover, a hybrid feature set is introduced and applied on high frequency image part for better detection and removal of rain component. But the method does not give satisfactory result in heavy rain.

Chen et al. [7] introduced a motion segmentation of a dynamic scene based image rain removal technique. To enhance the visibility in heavy rain, this method is proposed. This technique overcomes the problems occurred during the removal of rain in dynamic scene and the scene which are in motion. By fixing a threshold value, the variations of pixel intensity in a rainy image is restricted and desired result is obtained.

Wang et al. [8] proposed a rain removal technique where the mapping of rain pixels are done and using rain characteristics, rain streaks are detected. A linear model $p = \alpha s + \beta$ is used to measure the intensity of the rain pixels where p denotes the observe intensity of rain affected pixels. s shows the intensity values of non-rain pixels. A cost function is generated after forming L_2 norm for all rain detected pixels inside a window. By measuring the cost function and obtaining the α and β values, the rain removal pixel intensity is calculated.

3 Proposed Algorithm

Based on the limitations found in some of the proposed algorithms, we developed a new technique to obtain the optimum result. As heavy rain causes severe problem in visibility, therefore we are trying to solve this problem to recover information behind the rain. We also take care of some other problem occurred due to rain in a rainy image. The algorithm introduced by us can be splitted into two separate units such as: (i) main processing unit and (ii) post processing unit. Rain streaks elimination is done in main unit by the application of guided filter [4, 9] and L_0 gradient minimization method [1, 10] whereas overall visibility and image quality is improved in post unit through USM [11] and CLAHE [12] method. A block diagram of our proposed technique is shown in Fig. 1.

A rainy image can also be mathematically modeled and represented as:

$$I_{in} = I_{bi} + I_{ri} \tag{1}$$

This model is formed by observing the textural difference between background (I_{bi}) and rain (I_{ri}) of a rainy image. In this method, low (I_{LF}) and high frequency (I_{HF}) components are obtained by decomposing the input rainy image after applying Gaussian filter [13]. The low frequency component does not contain rain whereas the rain streaks are found in high frequency component.

$$I_{in} = I_{LF} + I_{HF} \tag{2}$$

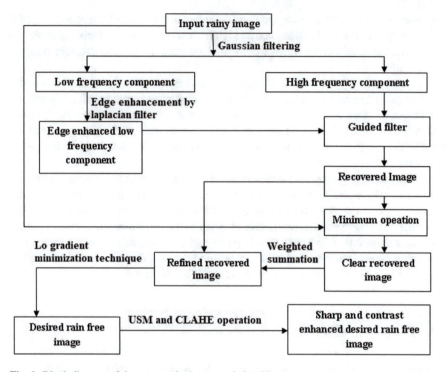

Fig. 1 Block diagram of the proposed rain removal algorithm

By observing the textural difference between rain and other background edges, a background is estimated from the low frequency image part. Thereafter, laplacian filter is used to enhance the edges of low frequency component near to the edges of the input image. Edge enhancement operation is done by following the given equation:

$$I_{\mathrm{LFE}} = I_{\mathrm{LF}} + \omega \nabla I_{\mathrm{LF}} \tag{3}$$

where I_{LF} has a gradient ∇I_{LF} and ω is an adjustment factor chosen as 0.1 for our work. This edge enhanced image does not contain rain streaks but other edges are prominent. For this reason, I_{LFE} is used as a guidance image of guided filter to guide the input guided image I_{HF} and to remove the rain streaks. I_r is the image obtained after guided filtering [9]. The recovered image I_r has got the high frequency component with minimum rain streaks but this image is a bit blurred. To overcome this problem, a new output I_{cr} is achieved by the given equation:

$$I_{\mathrm{cr}} = \min(I_r, I_{\mathrm{in}}) \tag{4}$$

Weighted summation of I_{cr} and I_r gives more refined result and is shown by I_{ref}.

$$I_{\text{ref}} = \beta I_{\text{cr}} + (1 - \beta)I_r \qquad (5)$$

where β is selected by the entropy value of the input rainy image. It is observed that the recovered image I_{ref} has also some rain edges. Therefore to remove those highest contrast edges, and L_0 gradient minimization method [10] is applied. This method smoothens the image by finding out the global non zero gradients. So, most of the rain streaks can be removed from the refined image and information can be extracted which are hidden behind the rain. Lastly, USM and CLAHE techniques are used to enhance the visibility by enhancing the sharpness and contrast of the image obtained from L_0 gradient operation. Scaling constant value of USM is chosen as 1.4. For CLAHE method, tile size is set to be [8.8] and contrast enhancement limit is chosen as 0.005.

4 Simulation Results and Performance Analysis

Simulation results of our proposed technique are compared with some of the recent techniques for quantitative and qualitative analysis. It is observed that performance of the proposed method is better than other rain removal algorithms. By the use of some quality metrics, the performance of different rain removal techniques are observed and quantitative analysis is accomplished. These quality metrics are (i) percentage of rain removal (PRR) [1], (ii) colour information entropy (CIE) [14], (iii) contrast gain (CG) [15].

PRR metric plays a vital role to judge the performance of different rain removal algorithms. By considering high frequency component of a rainy image and by estimating background edges, the number of rain streaks is calculated. This calculation is performed for input rainy image as well as output rain removed image to obtain the PRR value. More value of PRR means more rains are removed from an image.

The value of CG is calculated by finding the difference between the mean contrast values of input rainy image and output rain removed image. Higher value of CG implies better performance of the output.

$$CG = C_{I,\text{der}} - C_{I,r} \qquad (6)$$

where $C_{I,\text{der}}$ represents the mean contrast of rain removed image and $C_{I,r}$ denotes the same for rainy image.

Colour is one of the important factors of an image. Proper selection of RGB colour ratio makes an image better in terms of visibility and quality. CIE measures the colour ratio in an image. Larger the value of CIE gives more improved image quality.

$$CIE = -\sum_{k=0}^{L-1} P_k log_2(P_k) \qquad (7)$$

where the number of intensity levels are defined by L, probability which is related to intensity level k is indicated by P_k.

Qualitative analysis is accomplished through visual comparison of different output images obtained from different algorithms. The comparison of each result of different algorithms is shown in Figs. 2, 3, 4 and 5.

For the pictorial comparison of different algorithms, various intensity of rain is used. Heavy rain is found in 'Image 3' and 'Image 4' whereas moderate rain is observed in 'Image 1' and 'Image 2'. We feel that for all of the above test images, the proposed outputs (Figs. 2d, 3d, 4d and 5d) give more clear view compared to other methods. We also observe that proposed result contains more image information with respect to the results.

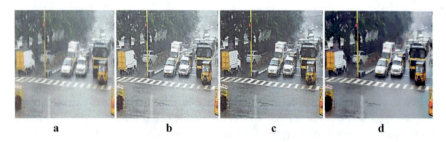

Fig. 2 'Image 1': **a** rainy image, **b** Zheng et al. (2013), **c** Luo et al. (2015), **d** proposed result

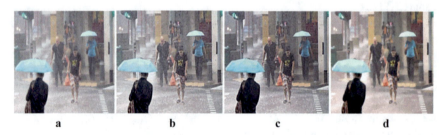

Fig. 3 'Image 2': **a** rainy image, **b** Zheng et al. (2013), **c** Luo et al. (2015), **d** proposed result

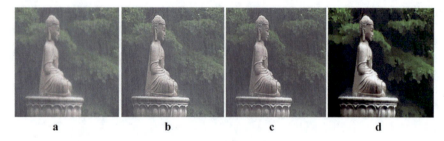

Fig. 4 'Image 3': **a** rainy image, **b** Zheng et al. (2013), **c** Luo et al. (2015), **d** proposed result

Fig. 5 'Image 4': **a** rainy image, **b** Zheng et al. (2013), **c** Luo et al. (2015), **d** proposed result

Table 1 Comparison of CG and PRR of proposed method with the previous methods

Method	Zheng et al. [4]		Luo et al. [16]		Proposed	
Metrics	CG	PRR%	CG	PRR%	CG	PRR%
'Image 1'	0.085	56	0.113	68	0.182	82
'Image 2'	0.094	35	0.122	39	0.139	51
'Image 3'	0.089	37	0.091	43	0.145	70
'Image 4'	0.129	39	0.132	41	0.178	62

Table 2 Comparison of CIE of proposed method with the previous methods

Method	Zheng et al. [4]	Luo et al. [16]	Proposed
Metrics	CIE	CIE	CIE
'Image'	7.402	7.453	7.623
'Image'	7.625	7.616	7.688
'Image'	6.123	5.976	6.731
'Image'	7.761	7.617	7.813

Values of different quality metrics of different rain removal techniques are evaluated and shown in Tables 1 and 2.

It is observed from Tables 1 and 2 that the different metric values for different test images of our proposed technique are better compared to other techniques in terms of contrast, colour quality and percentage of rain.

5 Conclusions

This article introduces a unique rain removal technique which extracts image information behind the rain. The proposed technique preserves all edge information from an image except high intensity rain edges. The method is operated based on the application of guided filter to the low and high frequency image components and the application of L_0 gradient minimization technique to the output of guided filter. Gaussian filter is used to separate the low frequency non-rain component and high frequency rain component. The overall image clarity and visibility is improved by enhancing the sharpness and contrast through USM and CLAHE operation respectively. Superiority of our proposed algorithm is shown pictorially and quantitatively by the use of some quality metrics through the comparison with some of the recent techniques.

References

1. B.N Manu, Rain removal from still images using L0 gradient minimization technique, in *7th International Conference on Information Technology and Electrical Engineering (ICITEE)*, Thailand (2015)
2. A.K. Tripathi, S. Mukhopadhyay, A probabilistic approach for detection and removal of rain from videos. IETE J. Res. **57**(1), 82–91 (2011)
3. L.W. Kang, C.W. Lin, Y.H. Fu, Automatic single-image-based rain streaks removal via image decomposition. IEEE Image Process. **21**, 4 (2012)
4. X. Zheng, Y. Liao, W. Guo, X. Fu, X. Ding, *Single-Image-Based Rain and Snow Removal Using Multi-Guided Filter. In Neural Information Processing* (Springer, Berlin Heidelberg, 2013), pp. 258–265
5. Y. Li, R.T. Tan, X. Guo, J. Lu, M.S. Brown, Rain streak removal using layer priors, in *Proceedings of the IEEE Conference on Computer Vision Pattern Recognition (CVPR)*, pp. 2736–2744 (2016)
6. D.Y. Chen, C.C. Chen, L.W. Kang, Visual depth guided color image rain streaks removal using sparse coding. IEEE Trans. Circ. Syst. Video Technol. **24**(8), 1430–1455 (2014)
7. J. Chen, L.-P. Chau, A rain pixel recovery algorithm for videos with highly dynamic scenes. IEEE Image Process. **23**, 3 (2014)
8. Y. Wang, S. Liu, B. Zeng, Removing rain streaks by a linear model. IEEE Trans. Image Process. (2018)
9. J. Xu, W. Zhao, P. Liu, X. Tang, Removing rain and snow in a single image using guided filter. Proc. CSAE **2**, 304–307 (2012)
10. L. Xu, C. Lu, Y. Xu, J. Jia, Image smoothing via L0 gradient minimization. ACM Trans. Graph. (TOG) **30**, 6 (2011)
11. L. Ying, M.N. Tek, L.B. Beng, A wavelet based image sharpening algorithm, in *International Conference on Computer Science and Software Engineering*, pp 1053–1056 (2008)
12. K. Zuiderveld, *Contrast Limited Adaptive Histogram Equalization. Graphic Gems IV* (Academic Press Professional, San Diego, 1994), pp. 474–485
13. I. Agustina, F. Nasir, A. Setiawan, The implementation of image smoothing to reduce noise using Gaussian filter. Int. J. Comput. Appl. **177**(5), 15–19 (2017)
14. Y. Wang, C. Fan, Single image defogging by multiscale depth fusion. IEEE Trans. Image Process. **23**(11), 4826–4837 (2014)
15. A.K. Tripathi, S. Mukhopadhyay, Single image fog removal using anisotropic diffusion. IET Image Proc. **6**(7), 966–975 (2012)
16. Y. Luo, Y. Xu, H. Ji, Removing rain from a single image via discriminative sparse coding, in *Proceedings of the IEEE International Conference on Computer Vision (ICCV)*, pp. 3397–3405 (2015)

Image Contrast Enhancement Using Histogram Equalization-Based Grey Wolf Optimizer (GWO)

Saorabh Kumar Mondal, Arpitam Chatterjee, and Bipan Tudu

Abstract Image contrast enhancement is a common image enhancement method in various image applications. Histogram equalization (HE) is a popular conventional and improved way to enhance the image contrast. This conventional HE and its updated versions try to maintain the original image characteristics especially in original image brightness distribution during contrast enhancement. But it suffers various restrictions including false contouring, lack of original image features, etc. Optimization can be a possible solution to overcome such problems and to maintain desired image characteristics as well as enhanced image contrast. Grey wolf optimizer (GWO) is a newly high potential algorithm in the computational intelligence field to solve this type of problem. This paper presents an application of HE-based contrast enhancement using GWO. The objective function has been designed using different parameters to preserve original image characteristics. The results have been taken using the standard database of both gray and colour images. The results have also been compared against various renowned techniques by standard image quality evaluation metrics. The visual as well as parametric presentation indicates the distinguished prospective of GWO for image contrast enhancement.

Keywords Contrast enhancement · Histogram equalization · Grey wolf optimizer

S. K. Mondal (✉)
Haldia Institute of Technology, Haldia, India
e-mail: sm_751@yahoo.com

A. Chatterjee · B. Tudu
Jadavpur University, Kolkata, India

© Springer Nature Singapore Pte Ltd. 2020
R. Bera et al. (eds.), *Advances in Communication, Devices and Networking*,
Lecture Notes in Electrical Engineering 662,
https://doi.org/10.1007/978-981-15-4932-8_23

1 Introduction

Image enhancement is an imperative process in different image processing applications [1]. Contrast enhancement is one of the most popular steps of image enhancement systems which will improve the contrast of the image for visual appearance or further processing such as segmentation, edge extraction, etc. [2].

In a poor quality image or low contrast image, the entire dynamic range available there is not employed. Conventional histogram equalization (CHE) is a process, which probabilistically remaps the existing image intensity levels to the available intensity levels to improve image contrast [3]. Although CHE is popular method, it cannot preserve the brightness distribution of original image and thus many further methods were proposed, such as brightness preserving bi-histogram equalization (BBHE) [4], dynamic histogram equalization (DHE) [5], etc. Depending on input mean value, BBHE separates the input image histogram into two parts. After separation, each part is equalized independently. DHE generates the specified histogram dynamically from the input image. These methods are tried to preserve the original image brightness mean and enhancing the image contrast.

Apart from HE techniques, some other methods are also implemented such as exact histogram specification (EHS) [6] and adaptive gamma correction weighted distribution (AGCWD) [7] for image contrast enhancement. EHS method helps to reschedule the pixel mapping to match a specified histogram. AGCWD optimizes the gamma parameter based on the weighted distribution function with the help of PDF (probability distribution function) and CDF (cumulative distribution function).

However, these techniques also provide some limitations like whitening of the image, over enhancement, false contouring, etc. HE-based optimization may be a possible alternative to deal with such problems with conventional techniques.

Metaheuristic optimization techniques are the most popular optimization techniques inspired by physical phenomena, biological behaviours, or evolutionary concepts [8]. One of the rapidly growing fields of metaheuristics optimization techniques is swarm intelligence (SI) techniques. These techniques are mostly inspired from natural colonies, herds and flock. Some of the most popular SI techniques are genetic algorithm (GA) [9], particle swarm optimization (PSO) [10], artificial bee colony (ABC) [11], etc. All those algorithms start with a randomly generated population and update the agent information following the searching behaviour of natural swarms. For example, ABC imitates the food searching behaviour of honeybees while PSO uses a number of agents (particles) that comprise a swarm moving around in the search area looking for the best solution. GWO [12] is comparatively new and efficient optimization technique based on the characteristics of grey wolves.

2   Grey Wolf Optimizer (GWO)

2.1   Dynamics of GWO

According to GWO techniques, grey wolves maintain a social hierarchy where the leaders of this hierarchy are called alpha (α). Alphas are the decision-maker. The decisions of alphas are followed by the other members in the group. For this reason, the alpha wolf is also called the dominant wolf and belong to the top position of their group. The second position in the hierarchy is beta (β). The beta helps the alpha for managing the group. The beta wolves follow the alphas, but give the instructions to the other lower position wolves as well. The third position in the group is called delta (δ). Delta wolves have to follow the instructions of alphas and betas, but they can give the instructions to the lowest position of grey wolves in the group. The other members of this group include scouts, sentinels, elders, hunters and caretakers. These members are responsible for different activities for their life lead. The lowest position grey wolves are called omega (ω). Apart from this social hierarchy, grey wolves also perform hunting operation. The main three steps of grey wolf hunting are (a) tracking, chasing and approaching the prey; (b) pursuing, encircling and harassing the prey until it stops moving and finally (c) attacking the prey. Encircling the prey can be mathematically represented as Eqs. 1, 2 [12] where $\vec{X}$ is the position vector and $\vec{X}_p$ is the position of prey at current iteration t.

$$\vec{D} = |\vec{C} \cdot \vec{X}_p(t) - \vec{X}(t)| \tag{1}$$

$$\vec{X}(t+1) = \vec{X}_p(t) - \vec{A} \cdot \vec{D} \tag{2}$$

where A and C are coefficient vectors that are calculated as Eqs. 3 and 4, respectively.

$$\vec{A} = 2 \cdot \vec{a}\,\vec{r_1} - \vec{a} \tag{3}$$

$$\vec{C} = 2 \cdot \vec{r_2} \tag{4}$$

where components of a are linearly decreased from 2 to 0 throughout the iterations and r_1 and r_2 are random vectors in [0, 1].

To express the hunting procedure mathematically, the alpha, beta and delta wolves are considered. These three wolves have better knowledge about the location of the prey. The dynamics of alpha, beta and delta are presented as Eqs. 5, 6 and 7, respectively [12]. The first three best solutions obtained so far are considered and other agents update their position depending on the position of the target solution using Eqs. 5–11 [12].

$$\vec{D_\alpha} = |\vec{C_1} \cdot \vec{X_\alpha} - \vec{X}| \tag{5}$$

$$\vec{D_\beta} = |\vec{C_2} \cdot \vec{X_\beta} - \vec{X}| \tag{6}$$

$$\vec{D_\delta} = |\vec{C_3} \cdot \vec{X_\delta} - \vec{X}| \tag{7}$$

$$\vec{X_1} = \vec{X_\alpha} - \vec{A_1} \cdot \vec{D_\alpha} \tag{8}$$

$$\vec{X_2} = \vec{X_\beta} - \vec{A_2} \cdot \vec{D_\beta} \tag{9}$$

$$\vec{X_3} = \vec{X_\delta} - \vec{A_3} \cdot \vec{D_\delta} \tag{10}$$

$$\vec{X}(t+1) = \frac{\vec{X_1} + \vec{X_2} + \vec{X_3}}{3} \tag{11}$$

The pseudocodes for grey wolf optimizer are as follows [12]:

- Initialize the grey wolf population X_i (i=1, 2,……, n).
- Initialize a, A, and C.
- Calculate the fitness of each search agent.
- D_{alpha}= the best search agent.
- D_{beta}= the second best search agent.
- D_{delta}= the third best search agent.
- **while** (t<Max number of iteration)
    for each search agent
        Update the position of current search agent by Eq. 11 .
    end for
    Update a, A, and C.
    Calculate the fitness of all search agents.
    Update D_{alpha}, D_{beta}, and D_{delta}.
    t=t+1.
  end while
- return X_{alpha}.

3   Contrast Enhancement Using GWO

3.1   Objective Function Formulation

The idea of using optimization is for retaining the original image characteristics. Among different such characteristics in this paper three major characteristics have been considered, i.e. the brightness distribution, the retention of signal fidelity and the detail retention of the original image. These parameters can be derived from popular metrics, i.e. absolute mean brightness error (AMBE) [13], peak signal-to-noise ratio (PSNR) [14] and structural similarity index measure (SSIM) [14]. The fitness of the solution can therefore be assessed using the function $\phi(\cdot)$ using these three parameters as calculated by Eq. 12.

$$\phi(y) = \frac{\sum_i \sum_j |x(i, j) - y(i, j)|^2}{10 \log(L-1)^2 \times N} \cdot |e(x) - e(y)| \cdot \frac{1}{\theta(x, y)} \qquad (12)$$

where $\phi(y)$ is the fitness of enhanced image y, x is the input original image, L is the total number of available pixel levels, N is the total number of pixels in the image, i and j are the row and column index, respectively, $e(x)$ and $e(y)$ are the statistical means of x and y, respectively and $\theta(x, y)$ is the structural closeness indicator from SSIM.

In case of colour images, a simple colour space conversion method is used due to the inter-channel dynamic properties for colour images. Here the original RGB colour image is first converted to HSV colour mode [15] for perceptual mode realization rather than device-dependent mode like RGB. Maintaining the 'hue' and 'saturation' unchanged to retain the fundamental colour information of the original input image; the 'value' is considered as the input image and applies GWO. Finally, the modified 'value' channel is combined with the unaltered 'hue' and 'saturation' and converted back to RGB for visual representation.

3.2 Pseudocodes for Proposed Methodology

The pseudocodes for proposed methodology are as follows:

- Take low contrast image.
- Compute histogram and unique intensity levels in the image.
- Initialize random solutions related to those levels.
- Initialize the various parameters of GWO algorithm.
- Calculate the fitness values and identified alpha, beta and delta.
- Execute GWO operation.
- Update the final position and generate new solution.
- Stop once the set target is reached.
- Reconstruct the image.

4 Results and Discussions

The proposed methodology has been performed by considering various standard grayscale images and colour images from SIPI image database [16] for grayscale images and TID database [17] for colour images. Some of the test results for grayscale images and colour images are consolidated in Figs. 1 and 2, respectively. In both figures, the proposed results are compared with some standard very much popular techniques for visual comparison purpose. In case of colour images, the V channel has been considered for histogram analysis purpose to enhance the contrast of colour images.

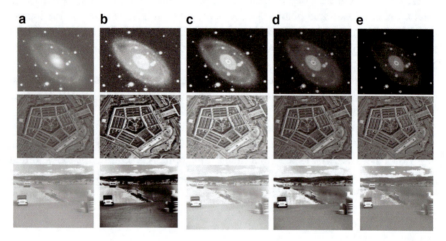

Fig. 1 **a** Grayscale test images and their results using **b** CHE, **c** AGCWD, **d** ABC and **e** GWO

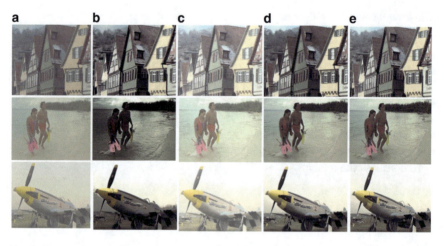

Fig. 2 **a** Colour test images and their results using **b** CHE, **c** AGCWD, **d** ABC and **e** GWO

Table 1 Objective comparisons of gray images

Techniques/Metrics	GHE	AGCWD	ABC	GWO
AMBE	55.74	37.38	30.76	13.52
PSNR	10.58 dB	15.77 dB	18.90 dB	22.69 dB
FSIM	0.4759	0.7982	0.8575	0.8961
BRISQUE	49.70	38.60	32.15	29.26

Table 2 Objective comparisons of colour images

Techniques/Metrics	GHE	AGCWD	ABC	GWO
AMBE	148.73	141.54	4.86	4.89
PSNR	4.44 dB	4.45 dB	17.05 dB	17.54 dB
FSIM	0.6351	0.6275	0.8876	0.8904
BRISQUE	8.66	6.42	6.99	5.95

The results from the grayscale test images and colour images confirm the prospective of GWO for contrast enhancement. In all the cases, the visual appearance is enhanced while preserving the image characteristics that are not well visible in the results of other standard techniques under consideration. The GWO also avoids the tendency of black patches that are visible in the results of other techniques.

The objective evaluations against the standard image quality metrics are also presented here for comparison purpose. In this paper, four important metrics are considered for evaluation, namely, AMBE, PSNR, feature similarity index measure (FSIM) [18] and BRISQUE [19]. AMBE and BRISQUE show improvements with lower values while PSNR and FSIM show improvement with higher values. The metrics used in this paper also indicates image quality in different perspectives, for example, AMBE preserves the mean brightness of the original images, PSNR reflects the information fidelity, while FSIM corresponds to the structural and feature retentively. BRISQUE is different from other above-mentioned metrics and it performs the image quality assessment without any reference image and indicates the genuineness of the image. The objective comparisons are presented in Tables 1 and 2. The average mean values of various standard test images for different methodologies are presented here.

The objective evaluations also confirm the betterment achieved using GWO. The improvements are well visible in case of all the metrics been used in this paper.

5 Conclusions

In this histogram equalization-based grey wolf optimizer algorithm, the parametric values of different image quality metrics are set to achieve the improved contrast images by maintaining the desired image characteristics. Various standard database images are presented here and calculated subjectively as well as objectively to establish the proposed method. In every case, the proposed method showed comparatively better results rather than other techniques. Some of the works, which may be implemented in future, are establishment of more strong objective function, speedy processing time of the method and application of multi-objective optimization techniques for better performance. In general, the presented method showed inspiring potential of GWO in image contrast enhancement.

References

1. S. Jayaraman et al., *Digital Image Processing* (TMH Education 2011)
2. R. Maini, H. Aggarwal, A comprehensive review of image enhancement techniques. J. Comput. **2**(3), 8–13 (2010)
3. T. Arici, S. Dikbas, Y. Altunbasak, A histogram modification framework and its application for image contrast enhancement. IEEE Trans. Image Process. **18**, 1921–1935 (2009)
4. Y.T. Kim, Contrast enhancement using brightness preserving bi-histogram equalization. IEEE Trans. Consum. Electron. **43**(1), 1–8 (1997)
5. A.A. Wadud, M. Kabir, M.H. Dewan et al., A dynamic histogram equalization for image contrast enhancement. IEEE Trans. Consum. Electron. **53**, 593–600 (2007)
6. D. Coltuc, P. Bolon, J.M. Chassery, Exact histogram specification. IEEE Trans. Image Process **15**(5), 1143–1152 (2006)
7. S.C. Huang, F.C. Gneng, Y.S. Chiu, Efficient contrast enhancement using adaptive gamma correction with weighting distribution. IEEE Trans. Image Process. **22**(3), (2013)
8. C.P. Lim, L.C. Jain, S. Dehuri (Eds.), *Innovations in Swarm Intelligence* (Springer, Berlin, 2009)
9. Hashemi et al., An image contrast enhancement method based on genetic algorithm. Pattern Reorg. Lit. **31**, 1816–1824 (2010)
10. J. Kennedy, R.C. Eberhart et al., Particle swarm optimization. IEEE Int. Conf. Neural Netw. **4**, 1942–1948 (1995)
11. A. Draa, A. Bouaziz, An artificial bee colony algorithm for image contrast enhancement. SEC **16**, 69–84 (2014)
12. S. Mirjali, A. Lewis, S.M. Mirjali, Grey wolf optimizer. Adv. Eng. Softw. **69**, 46–61 (2014)
13. S.D. Chen, A new image quality measure for assessment of histogram equalization-based contrast enhancement techniques. Digit. Sign. Proc. **22**, 640–647 (2012)
14. A. Hore, D. Ziou, *Image Quality Metrics: PSNR vs SSIM* (ICPR, Turkey, 2010)
15. K. Kapoor, S. Arora, Colour image enhancement based on histogram equalization. Int. J. Electr. Comput. Eng. **4**(3), 73–82 (2016)
16. http://www.sipi.usc.edu/database. SIPI Image Database
17. http://isp-uv.es. TID database
18. L. Zhang, L. Zhang, X. Mou, D. Zhang, FSIM: a feature similarity index for image quality assessment. IEEE Trans. Image Process **20**(8), 2378–2386 (2011)
19. A. Mittal, A.K. Moorthy, A.C. Bovik, No-reference image quality assessment in the spatial domain. IEEE Trans. Image Process. **21**(2), 4695–4708 (2012)

Analysis of Split Gate Dielectric and Charge Modulated SON FET as Biosensor

Khuraijam Nelson Singh, Amit Jain, and Pranab Kishore Dutta

Abstract Silicon-on-insulator (SOI) MOSFET caters immunity to various short-channel effects (SCEs) but has high fringing electric field, difficulty in controlling the silicon thickness, and high self-heating. The silicon-on-nothing (SON) MOSFET dwindled the said effects which are present in SOI MOSFET. In this work, an analysis of split gate SON MOSFET for biosensing application is presented. The underlapped region in between the split gates is used as biomolecules binding site. Binding of biomolecules in the binding site changes the effective dielectric and charges of the area, which alters the device characteristics. The variation in the characteristics has been used for the sensing purpose. Various electrical parameters of the device such as surface potential, threshold voltage, and sensitivity have been presented, and comparison with SOI and existing structures has also been presented to understand its effectiveness as a biosensor. The study has been performed based on the data obtained from 2-D TCAD simulations.

Keywords Biosensor · Dielectric modulation · MOSFET · Surface potential · SON · Threshold voltage

1 Introduction

The ceaseless miniaturization of the MOSFET dimension has led to the reduction in transistor power consumption, lower threshold voltage, faster switching, and more chips per wafer. However, the miniaturization also gives rise to short-channel effects

K. N. Singh (✉) · P. K. Dutta
Department of Electronics and Communication Engineering, NERIST, Nirjuli, India
e-mail: nelsonkhuraijam16@hotmail.com

P. K. Dutta
e-mail: pkdutta07@gmail.com

A. Jain
Department of Electronics and Communication Engineering, CMRIT, Bangalore, India
e-mail: amit2_8@yahoo.co.in

© Springer Nature Singapore Pte Ltd. 2020
R. Bera et al. (eds.), *Advances in Communication, Devices and Networking*,
Lecture Notes in Electrical Engineering 662,
https://doi.org/10.1007/978-981-15-4932-8_24

215

(SCEs) which affect the device performance. To counter or reduce the effects of SCEs, various variants of MOSFETs like silicon-on-insulator, double-gate MOSFET, junctionless MOSFET, etc. have been introduced [1–3]. These devices not only minimize SCEs but also increase the channel control by the gate. SOI MOSFET is one of the most studied devices due to its immunity to SCEs and simple structure; however, it also has high internal fringing electric field and self-heating due to the embedded oxide layer beneath the silicon channel. The SON MOSFET is one of the best alternative structures to counter the internal fringing electric field and self-heating of the device [4–6]. In this work, SON MOSFET has been analyzed to study its effectiveness in biomolecules detection. The gate is split into two parts with an underlap region in between the two split gates. The underlap region is used as a site for binding the biomolecules.

Biomolecules available in nature can be of two types: (i) chargeless biomolecules, e.g., Protein, APTES, etc. and (ii) charged biomolecules, e.g., DNA. No matter whether they are chargeless or charged biomolecules, all biomolecules possess dielectric values. For example, streptavidin has a dielectric value of 2.1, protein 2.5, biotin 2.63, and APTES 3.57 [7]. On the other hand, charged biomolecules possess both dielectric value and charge attributes, e.g., DNA has a dielectric value of 8 [8] and charge of around $-2 \times 10^{12}C$ cm^{-2} [9]. When the biomolecules get bind in the underlapped region, the effective dielectric and charge of the region change and this in turn changes the electrical characteristics of the device. As biomolecules have some distinct dielectrics and charges, they produce different variations upon their binding [10, 11]. This change is captured to sense the presence of the biomolecules.

The remaining part of the paper is organized as follows. Section 2 presents the structure of the split gate SON FET, Sect. 3 presents the results and discussions, and finally the conclusion of the paper is presented in Sect. 4.

2 Structure of the Device

The schematic structure of the split gate SON FET as a biosensor is presented in Fig. 1. The gate is split into two parts with an underlapped region in between them. Due to the gate splitting, the channel is divided into three regions: region 1 (R-I), region 2 (R-II), and region 3 (R-III). The channel is doped with acceptor material while donor materials are doped both in source and drain. A layer of air is inserted just below the channel regions. Table 1 presents the parameters and their values used in the study.

In sub 65 nm devices, the current leakage through the gate oxide increases due to the thin oxide layer [12]. This can be countered by using high-K oxide instead of the SiO$_2$ as gate oxide [13]. In this study, a high-K oxide (HfO$_2$) has been staked above a thin layer of SiO$_2$. This not only decreases the leakage of current through the gate oxide but also provides a space large enough for the biomolecules to be bonded. The thin oxide layer of SiO$_2$ plays as an interfacial coat amid the device and the biomolecules.

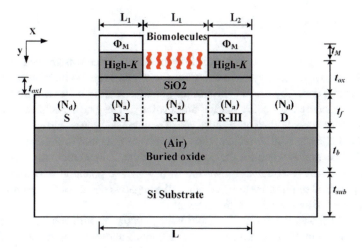

Fig. 1 Structure of split gate SON FET

Table 1 Structural parameters of split gate SON FET

Structural parameter	Symbol	Value
Channel length	L	40 nm
R-I and R-III length	L_1, L_3	10 nm
R-II length	L_2	20 nm
SiO$_2$ thickness	t_{ox1}	1 nm
High-K thickness	$t_{ox} - t_{ox1}$	10 nm
Channel thickness	t_f	10 nm
Channel doping	N_a	1×10^{17} cm^{-3}

The presence of chargeless biomolecules in this study has been simulated by inserting an oxide with a dielectric value greater than 1 ($K > 1$) with no charge; $K = 1$ means no biomolecule is present in the binding site. Various studies have been made by varying the K ranging from 3 to 10. The presence of charge biomolecules has been simulated by fixing the charge to $K = 5$ and changing its charge from $0 \times 10^{11} C$ cm^{-2} to $-10 \times 10^{11} C$ cm^{-2} [14]. The simulation of the device has been obtained by using ATLAS (SILVACO). Models used in the simulations are CONMOB, FLDMOB, SRH, AUGER, and BOLTZMANN.

3 Results and Discussions

Figure 2 presents the surface potential variation of the device with the changes of the biomolecule properties. Figure 2a shows that when there are no biomolecules in the channel (represented by $K = 1$), the surface potential minimum lies toward R-I as R-III is influenced by the electric field of drain voltage. The gate capacitance increases when the biomolecule dielectric increases. The increasing gate capacitance causes the vertical electric field to reduce; thus, the surface potential in R-II gets lowered. Figure 2b shows that when the dielectric value of the charged biomolecules is fixed ($K = 5$) and charge increases, lowering in the surface potential is observed. The surface potential minima are lowered due to the negative charge of the biomolecules, which prevents the inversion in the channel.

Figure 3a shows the device threshold voltage variance with change in dielectric values of the chargeless biomolecules. As mentioned above, the increment in dielectric values of the biomolecules lowers the vertical electric field, which produces an increment in the device threshold voltage. Figure 3b shows the variation in device threshold voltage with changing biomolecule charges. Increase in biomolecule negative charges repels the minority charge carriers responsible for inversion of charge in the channel; thus, the device threshold voltage again increases. The figure also shows the comparison of split gate SON and split gate SOI FET threshold voltages. It shows that SOI has larger threshold voltage than SON owing to the larger internal fringing electric field due to the buried SiO_2 in SOI.

The sensitivity of the device is defined as [15]

$$\Delta V_{th} = V_{th}(\text{with biomolecule}) - V_{th}(\text{with air}) \tag{1}$$

where V_{th} is the threshold voltage of the device.

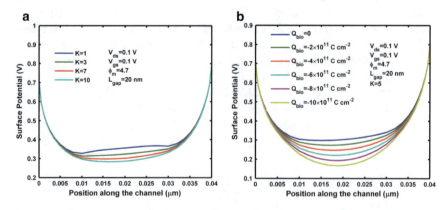

Fig. 2 a Surface potential variation of the device with change of chargeless biomolecule dielectric values. **b** Surface potential variation of the device with change of charged biomolecule charges

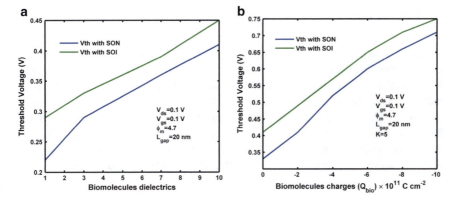

Fig. 3 **a** Threshold voltage variance of the device with change in dielectric values of charge-less biomolecules. **b** Threshold voltage variance of the device with change in charge of charged biomolecules

Figure 4a and 4b shows that the sensitivity of the device increases when dielectric and charges of the biomolecules increase. It also shows that sensitivity of the split gate SON FET is better than that of split gate SOI FET as threshold voltage variation of split gate SON FET is more significant than that of the later.

When $K = 5$ and $Q_{\text{bio}} = -10 \times 10^{11} C$ cm^{-2}, the sensitivity of the proposed device is found to be 380 mV while for the same parameters, split gate SOI FET has a sensitivity of 340 mV, and the device proposed by Singh et al. (2018) has a sensitivity of 350 mV. Thus, the proposed device is a good choice for sensing biomolecules.

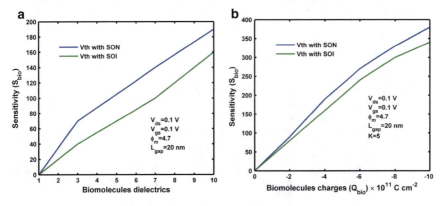

Fig. 4 **a** Variation in sensitivity of the device with chargeless biomolecule dielectric values. **b** Variation in sensitivity of the device with charged biomolecule charges

4 Conclusion

To understand the applicability of split gate SON FET as a biosensor, an extensive study of the device has been presented in this study. The consequence of binding biomolecules on the characteristics of the device has been studied by varying the dielectric and charge properties of the biomolecules and examining its effect on the surface potential and the threshold voltage of the device. The study shows that the proposed device has higher sensitivity than split gate SOI FET even though its threshold voltage is lower than the later. The lower threshold voltage of the proposed device also ensures greater miniaturization ability than the SOI-based structure.

References

1. A. Chaudhry, M.J. Kumar, Controlling short-channel effects in deep-submicron SOI MOSFETs for improved reliability: a review. IEEE Trans. Device Mater. Reliab. **4**(1), 99–109 (2004). https://doi.org/10.1109/TDMR.2004.824359
2. P.K. Dutta, N. Bagga, K. Naskar, S.K. Sarkar, Analysis and simulation of dual metal double gate son MOSFET using hafnium dioxide for better performance, in *Michael Faraday IET International Summit 2015*, Kolkata, India (2015), pp. 399–402. https://doi.org/10.1049/cp.2015.1665
3. C. Liu, J. Xu, L. Liu, H. Lu, Y. Huang, A threshold-voltage model for small-scaled GaAs nMOSFET with stacked high-k gate dielectric. J. Semiconductors **37**(2), 024004 (2016). https://doi.org/10.1088/1674-4926/37/2/024004
4. J. Pretet, S. Monfray, S. Cristoloveanu, T. Skotnicki, Silicon-on-nothing MOSFETs: performance, short-channel effects, and backgate coupling. IEEE Trans. Electron. Dev. **51**(2), 240–245 (2004). https://doi.org/10.1109/TED.2003.822226
5. S. Dubey, P.K. Tiwari, S. Jit, A two-dimensional model for the potential distribution and threshold voltage of short-channel double-gate metal-oxide-semiconductor field-effect transistors with a vertical Gaussian like doping profile. J. Appl. Phys. **108**(3), 034518 (2010). https://doi.org/10.1063/1.3460796
6. P. Banerjee, S.K. Sarkar, 3-D analytical modeling of dual-material triple-gate silicon-on-nothing MOSFET. IEEE Trans. Electron. Dev. **64**(2), 368–375 (2017). https://doi.org/10.1109/TED.2016.2643688
7. A. Chakraborty, A. Sarkar, Analytical modeling and sensitivity analysis of dielectric-modulated junctionless gate stack surrounding gate MOSFET (JLGSSRG) for application as biosensor. J. Comput. Electron. **16**(3), 556–567 (2017). https://doi.org/10.1007/s10825-017-0999-2
8. A. Cuervo, P.D. Dans, J.L. Carrascosa, M. Orozco, G. Gomila, L. Fumagalli, Direct measurement of the dielectric polarization properties of DNA. Proc. Natl. Acad. Sci. **111**(35), E3624–E3630 (2014). https://doi.org/10.1073/pnas.1405702111
9. R. Narang, M. Saxena, R.S. Gupta, M. Gupta, Dielectric modulated tunnel field-effect transistor—a biomolecule sensor. IEEE Electron. Dev. Lett. **33**(2), 266–268 (2012). https://doi.org/10.1109/LED.2011.2174024
10. B. Buvaneswari, N.B. Balamurugan, 2D analytical modeling and simulation of dual material DG MOSFET for biosensing application. AEU Int. J. Electron. Commun. **99**, 193–200 (2019). https://doi.org/10.1016/j.aeue.2018.11.039
11. P. Venkatesh, K. Nigam, S. Pandey, D. Sharma, P.N. Kondekar, A dielectrically modulated electrically doped tunnel FET for application of label free biosensor. Superlattices Microstruct. **109**, 470–479 (2017). https://doi.org/10.1016/j.spmi.2017.05.035

12. A. Chaudhry, Nanoscale effects: gate oxide leakage currents, in A. Chaudhry (ed.) *Fundamentals of Nanoscaled Field Effect Transistors*, pp. 25–36 (2013). https://doi.org/10.1007/978-1-4614-6822-6_2
13. R. Basak, B. Maiti, A. Mallik, Effect of the presence of trap states in oxides in modeling gate leakage current in advanced MOSFET with multi-oxide stack. Superlattices Microstruct. **129**, 193–201 (2019). https://doi.org/10.1016/j.spmi.2019.03.023
14. S. Singh, B. Raj, S.K. Vishvakarma, Analytical modeling of split-gate junction-less transistor for a biosensor application. Sens. Bio-Sens. Res. **18**, 31–36 (2018). https://doi.org/10.1016/j.sbsr.2018.02.001
15. S. Khuraijam Nelson, P.K. Dutta, Comparative analysis of underlapped silicon on insulator and underlapped silicon on nothing dielectric and charge modulated FET based biosensors, in *2019 Devices for Integrated Circuit (DevIC)*, Kolkata, India (2019), pp. 231–235. https://doi.org/10.1109/DEVIC.2019.8783714

Algorithmic Improvisation of Music Rhythm

Sudipta Chakrabarty, Anushka Bhattacharya, Md Ruhul Islam, and Hiren Kumar Deva Sarma

Abstract Music has two fundamental elements like the rhythm and melody. Rhythm is a perfect arrangement of notes that is the most inseparable component of music containing the length of note structures in a musical composition. Knowledge sharing on rhythmic patterns and their applications for creating new Music is a difficult task for learners of musicology. The memetic algorithm has been shown to be very effective in finding near-optimum solutions to the modeling of rhythmic patterns. The tournament selection strategy of evolutionary algorithms has been used to construct the fruitful rhythmic patterns for music composition. The work has been introduced to identify the parent rhythms and then create the versatile offspring rhythms using memetic algorithm. The tournament selection mechanism has been initiated for parent rhythms selection process for generating offspring rhythmic structures.

Keywords Music rhythm · Memetic algorithm · Tournament selection · Evolutionary algorithm · Local search · Mean fitness

1 Introduction

Automatic music composition and music information retrieval have been a challenging, fascinating, and not very explored area as it is very critical to distinguish the rhythmic structures which are qualified significantly through automatic composition. Regardless of complex nature, automatic composition of music is fully controlled

S. Chakrabarty (✉)
Department of MCA, Techno India, Salt Lake, Kolkata, West Bengal, India
e-mail: chakrabarty.sudipta@gmail.com

A. Bhattacharya
Department of CSE, Bengal Institute of Technology, Kolkata, West Bengal, India

M. R. Islam
Department of CSE, SMIT, Majhitar, Rangpo, East Sikkim, India

H. K. D. Sarma
Department of IT, SMIT, Majhitar, Rangpo, East Sikkim, India

© Springer Nature Singapore Pte Ltd. 2020
R. Bera et al. (eds.), *Advances in Communication, Devices and Networking*,
Lecture Notes in Electrical Engineering 662,
https://doi.org/10.1007/978-981-15-4932-8_25

by the compositional intelligent systems. In a few cases, listeners might to create the versatile music without involving a composer or expert, and achieving this kind of intelligence has become one of the computer researchers' thoughts. The variations of rhythmic structures over a specific vocal performance have been implemented to mechanize a composition. Hence, at some point, it is essential completely to exploit some structural models to facilitate decision-making from a list of possible alternate outcomes dependent on the restraint.

The rhythmic structures have been modeled as 16 beat rhythms, based on Indian Music and its listeners' concept. Memetic algorithm (MA) has been used for the process of parent rhythms selection using tournament selection. The local search criteria are the primary advantage of MA which might be effective to model problems and the improvisation processes are already familiar. How to improvise a rhythmic structure by applying evolutionary algorithm is principally aware to illustrate. After that a few rules has been extracted from Indian Music Concept.

2 Related Works

Various researchers have worked in music rhythm exploration though unexplored areas are there. Two papers [1, 5] present musical patterns of perceptual applications using machine recognition and measurement of rhythm complexity. Two papers [2, 6] propose the musical pattern recognition. Some papers introduce the study of music improvisation by genetic algorithm using different GA operators [3–5]. Papers [7, 11] propose the concept of genetic algorithm for generating offspring rhythm automatically. Paper [10] proposes the music pattern generation using median filter. Two important papers present music modeling based on polymorphism and inheritance [8, 9]. Two papers [12, 13] propose modeling music through Petri Nets. One work is to generate a context-aware teaching-learning model based on pervasive computing system to create offspring rhythms by genetic algorithm [14].

Paper [15] represents the concept of automatic raga recognition of a song using song pitch contour [15]. The song music origin can be represented by object-oriented modeling [16]. One paper introduces time-based music recommendation system [17]. A paper [19] presents neural-network-based raga recommender. Two excellent papers [18, 22] introduce to identify the music pattern similarity through coefficient of variance and correlation of coefficient, respectively. Two very good papers propose to determine the music rhythm density and complexity, respectively [20, 23]. The paper represents the implications on different computational music research papers [21]. The paper [24] proposes a web-based application that applies for a system which recommends songs based on users' choice.

3   Method of Analysis

3.1   An Outline of Memetic Algorithm

Memetic Algorithm is the amalgamation of local and global search techniques prepared by each of the individuals [25]. Memetic algorithms are only the specially designed genetic algorithm with great use of local search.

3.2   The Algorithm

```
1.    START
2.    Initilization of population:
Population P = initialize();
3.    Local_Search (P);
4.    Fitness_Value  (F)=  Equivalent  Decimal  value  of  each  binary
population(P1 ,........., P8);
Binary_string(Pi) = number;
while (Condition_number) {
Reminder = number%10;
F= F+ reminder*base;
number=number/10;
base= base*2; }
Print Decimal= Fitness (F);
5.    Selection: Two parents from Population (P) randomly.
random (int_array (value));
iteration (condition) {
random = array[random(F)]; }
print(random population Pi & Pj);
6.    Decimal to Binary Population Pi and Pj
while (decimal>0) {
decimal [array]= decimal%2;
array++;
decimal= decimal/2; }
        for (j=i-1; j>=0; j++) {
        print ("%d", decimal[j]); }
```

7.   Crossover (P) on Multipoints in Binary Population Pi and Pj
Offspring (O) = Crossover (P);
8.   Pi and Pj fitness values(F)
Step 4 repeats
9.   Mutation (O);
10.   Evaluate (P, O);
11.   P= regenerate (P,O);
12.   Iteration from step 3 to step 11
        while (! Stop_Condition) {
Search_Local(P);
offspring (O) = crossover(P);
mutate_combination(O);
evaluate_combination(P,O);
P = regenerate_result(P,O);
}
13.   STOP

4   Result Analysis and Discussion

In the tournament method, Table 1 depicts the different Rhythm Chromosomes and
their fitness value and Table 2 depicts the selections of chromosomes using Tournament Selection Mechanism. Chromosome 5 and Chromosome 7 have not participated
in the selection mechanism.

Hence, the two offsprings are string 1(1111111100000111) and string 4
(1101100100000110) (Fig. 1).

Table 1 Sixteen-bit rhythm population and their corresponding fitness

Population or string no.	Chromosome type	Population	Fitness
1	Rhythm1	1111111100000111	65287
2	Rhythm2	1001100110000011	39299
3	Rhythm3	1111100000000000	63488
4	Rhythm4	1101100100000110	55558
5	Rhythm5	1001001111001101	37837
6	Rhythm6	1100110000000001	52225
7	Rhythm7	1000001110010000	33680
8	Rhythm8	1110111000010010	60946

Table 2 Selection using tournament selection mechanism randomly

Tournament number	Randomly selected two individuals		Winner chromosome number	Winner chromosome	Times of winning
	Chromosome 1	Chromosome 2			
01	2	4	4	1101100100000110	2
02	3	8	3	1111100000000000	1
03	1	3	1	1111111100000111	2
04	4	5	4	1101100100000110	2
05	1	6	1	1111111100000111	2
06	2	5	2	1001100110000011	1
07	4	8	8	1110111000010010	1
08	2	6	6	1100110000000001	1

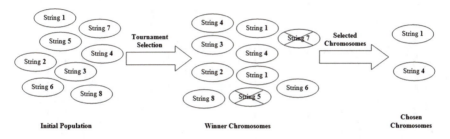

Fig. 1 Diagram of demonstration of winner chromosome selection

In Table 3, the population selection procedure has been described. The eight rhythm chromosomes have been initiated based on the rhythmic variation structures. Two randomly selected chromosomes out of eight have been selected as parent chromosomes based on their maximum occurrences as winners for participating in crossover operation.

From Table 3, the chromosome fitness mean value has been calculated to find the optimum selection of parents for further crossover and mutation operations. The fitness mean after tenth iteration has been found as shown below:

$$\text{Mean Fitness after 10th Iteration} = \left[\left(\sum \text{Offspring1}\right)/10 + \left(\sum \text{Offspring2}\right)/10\right]/2$$
$$= [539176/10 + 555822/10]/2$$
$$= [53917.6 + 55582.2]/2 = 109499.8/2 = 54749.9 \quad (1)$$

After first selection of parent chromosomes (after first iteration), the fitness mean has been calculated as follows:

Table 3 Selection of population with offspring creation

Iteration	Tournament number	Randomly selected two individuals		Winner chromosome number	Selected parent chromosomes 1 and 2	Offspring 1 and 2 after crossover	New fitness
		Chromosome 1	Chromosome 2				
01	01	2	4	4			
	02	3	8	3			
	03	1	3	1			
	04	4	5	4	1	1111100100000110	63750
	05	1	6	1	4	1101111100000111	57095
	06	2	5	2			
	07	4	8	8			
	08	2	6	6			
02	01	1	4	1			
	02	5	8	8			
	03	3	8	3			
	04	5	7	5	3	1111110000000001	64513
	05	2	7	2	8	1100111100000111	52999
	06	4	6	4			
	07	6	8	8			
	08	2	3	3			
03	01	1	5	1			
	02	2	6	6			
	03	1	4	1			
	04	1	7	1	1	1111110000000001	64513

(continued)

Table 3 (continued)

Iteration	Tournament number	Randomly selected two individuals		Winner chromosome number	Selected parent chromosomes 1 and 2	Offspring 1 and 2 after crossover	New fitness
		Chromosome 1	Chromosome 2				
	05	5	6	6	6	1100100100000110	51462
	06	2	8	8			
	07	3	7	3			
	08	4	5	4			
04	01	2	7	2			
	02	3	4	3			
	03	2	4	4			
	04	3	8	3	2	100111001000001	40065
	05	2	5	2	3	111110010000011	63747
	06	1	8	1			
	07	3	6	3			
	08	3	4	3			
05	01	5	7	5			
	02	4	5	4			
	03	4	8	4			
	04	2	5	2	2	100111110000111	40839
	05	2	7	2	4	110111000000001	56321
	06	1	5	1			
	07	6	7	6			

(continued)

Table 3 (continued)

Iteration	Tournament number	Randomly selected two individuals		Winner chromosome number	Selected parent chromosomes 1 and 2	Offspring 1 and 2 after crossover	New fitness
		Chromosome 1	Chromosome 2				
06	08	7	8	8			
	01	1	3	1			
	02	2	8	8			
	03	1	8	1			
	04	5	8	8	1	111111100000111	65287
	05	3	4	3	8	110011000000000001	52225
	06	2	7	2			
	07	5	7	5			
	08	6	8	8			
07	01	2	5	2			
	02	2	7	2			
	03	5	6	6			
	04	6	7	6	2	100110011000110	39302
	05	2	8	8	6	110011100000111	52999
	06	5	7	5			
	07	4	5	4			
	08	1	4	1			
08	01	4	2	4			
	02	4	8	4			

(continued)

Table 3 (continued)

Iteration	Tournament number	Randomly selected two individuals		Winner chromosome number	Selected parent chromosomes 1 and 2	Offspring 1 and 2 after crossover	New fitness
		Chromosome 1	Chromosome 2				
	03	2	6	6			
	04	6	7	6	4	11011111100000111	57095
	05	1	3	1	6	11001100000000001	52225
	06	4	5	4			
	07	3	7	3			
	08	2	7	2			
09	01	1	6	1			
	02	3	2	3			
	03	3	4	3			
	04	2	8	8	1	11111001000000011	63747
	05	2	7	2	3	11111111100000111	65287
	06	3	5	3			
	07	1	6	1			
	08	1	3	1			
10	01	2	7	2			
	02	5	8	8			
	03	3	4	3			
	04	2	3	3	2	10011001000000001	40065
	05	1	8	1	8	11001001000000110	51462

(continued)

Table 3 (continued)

Iteration	Tournament number	Randomly selected two individuals		Winner chromosome number	Selected parent chromosomes 1 and 2	Offspring 1 and 2 after crossover	New fitness
		Chromosome 1	Chromosome 2				
	06	2	8	8			
	07	2	5	2			
	08	7	8	8			

Fig. 2 Graphical representation of iteration versus offspring fitness

$$\text{Mean Fitness after 1st Iteration} = \left[\sum \text{Offspring1} + \sum \text{Offspring2}\right]/2$$
$$= [63750 + 57095]/2 = 120845/2 = 60422.5 \tag{2}$$

Equation (1) describes that the mean fitness after tenth iteration and Eq. (2) describes the mean fitness after first iteration of parent selection and creating offspring. Hence, initial parent rhythms produce better results for child rhythms and gradually it degraded. Chromosome 1 and chromosome 4 are the optimum parents for performing better to create rhythmic offspring in this specific approach. Figure 2 depicts the graphical representation of offspring fitness and mean fitness.

5 Conclusion

The foremost intention of the contribution is to generate variations of a specific rhythmic structure for the real-life music composition. From the outcomes of experimentation of eight initial rhythmic patterns, the quality of rhythm structures over a vocal performance has been improved with tournament selection mechanism that is evidently concluded. The capability of reproduction, local search, crossover, and mutation, that is, all the evolutionary operators of Memetic Algorithm to generate better outcomes which are the prime explanation that has been developed into one of the most improvised mechanisms for generating new musical rhythmic structures. In this tiny endeavor, a forward solving system developing tool has been implemented to assist and estimate the quality of every fundamental rhythmic pattern.

References

1. I. Shmulevich, O. Yli-Harja, E.J. Coyle, D. Povel, K. Lemstrm, *Perceptual Issues in Music Pattern Recognition Complexity of Rhythm and Key Finding. Computers and the Humanities* (Kluwer Academic Publishers, 2001), pp. 23–35
2. S. Chakraborty, D. De, Pattern classification of indian classical ragas based on object oriented concepts. Int. J. Adv. Comput. Eng. Archit. **2**, 285–294 (2012)

3. A. Gartland-Jones, P. Copley, The suitability of genetic algorithms for musical composition. Contemp. Music Rev. **22**(3), 43–55 (2003)
4. D. Matic, A genetic algorithm for composing music. Proc. Yugoslav J. Oper. Res. **20**, 157–177 (2010)
5. M. Dostal, Genetic algorithms as a model of musical creativity—on generating of a human-like rhythimic accompainment. Comput. Inform. **22**, 321–340 (2005)
6. S. Chakraborty, D. De, Object oriented classification and pattern recognition of indian classical ragas, in *Proceedings of the 1st International Conference on Recent Advances in Information Technology (RAIT)* (IEEE, 2012), pp. 505–510
7. M. Alfonceca, M. Cebrian, A. Ortega, A fitness function for computer-generated music using genetic algorithms. WSEAS Trans. Inf. Sci. Appl. **3**(3), 518–525 (2006)
8. D. De, S. Roy, Polymorphism in Indian classical music: a pattern recognition approach, in *Proceedings of International Conference on Communications, Devices and Intelligence System (CODIS)* (IEEE, 2012), pp. 632–635
9. D. De, S. Roy, Inheritance in Indian classical music: an object-oriented analysis and pattern recognition approach, in *Proceedings of International Conference on RADAR, Communications and Computing (ICRCC)* (IEEE, 2012), pp. 296–301
10. M. Bhattacharyya, D. De, An approach to identify that of Indian classical music, in *Proceedings of International Conference of Communications, Devices and Intelligence System (CODIS)* (IEEE, 2012), pp. 592–595
11. S. Chakrabarty, D. De, Quality measure model of music rhythm using genetic algorithm, in *Proceedings of the International Conference on RADAR, Communications and Computing (ICRCC)* (IEEE, 2012), pp. 203–208
12. S. Roy, S. Chakrabarty, P. Bhakta, D. De, Modelling high performing music computing using petri nets, in *Proceedings of the International Conference on Control, Instrumentation, Energy and Communication* (IEEE, 2013), pp. 757–761
13. S. Roy, S. Chakrabarty, D. De, A framework of musical pattern recognition using petri nets, in *Emerging Trends in Computing and Communication 2014* (Springer-Link Digital Library, 2013), pp. 242–252
14. S. Chakrabarty, S. Roy, D. De, Pervasive diary in music rhythm education: a context-aware learning tool using genetic algorithm, in *Advanced Computing, Networking and Informatics*, vol. 1 (Springer International Publishing, 2014), pp. 669–677
15. S. Chakrabarty, S. Roy, D. De, Automatic raga recognition using fundamental frequency range of extracted musical notes, in *Proceedings of the Eight International Multiconference on Image and Signal Processing (ICISP 2014)* (Elsevier, 2013), pp. 337–345
16. S. Chakrabarty, P. Gupta, D. De, Behavioural modelling of ragas of Indian classical music using unified modelling language, in *Proceedings of the Second International Conference on Perception and Machine Intelligence (PerMIn'15)* (ACM Digital Library, 2015), pp. 151–160
17. S. Chakrabarty, S. Roy, D. De, Time-slot based intelligent music recommender in Indian music, *International Book* Chapter, ISBN13: 9781522504986, ISBN10: 1522504982, (IGI Global, USA, 2016)
18. S. Chakrabarty, M.R. Islam, D. De, Modelling of song pattern similarity using coefficient of variance, in *International Journal of Computer Science and Information Sequrity* (ISSN 1947–5500) (2017), pp. 388–394
19. S. Roy, S. Chakrabarty, D. De, Time-based raga recommendation and information retrieval of musical patterns in Indian classical music using neural network, in *IAES International Journal of Artificial Intelligence (IJ-AI)* (ISSN: 2252-8938) (2017), pp. 33–48
20. S. Chakrabarty, M.R. Islam, D. De, Reckoning of music rhythm density and complexity through mathematical measures, in *Proceedings of the International Conference on Advanced Computational and Communication Paradigms* (Springer, 2018), pp. 387–394
21. S. Chakrabarty, S. Roy, D. De, A foremost survey on state-of-the-art computational music research, in *Proceedings of the Recent Trends in Computations and Mathematical Analysis in Engineering and Sciences (CRCMAS)* (International Science Congress Association, 2015), pp. 16–25

22. S. Chakrabarty, M.R. Islam, H.K.D. Sarma, An approach towards the modeling of pattern similarity in music using statistical measures, in *Proceedings of the Fifth International Conference on Parallel, Distributed and Grid Computing (PDGC)* (IEEE, 2018), pp. 436–441

23. S. Chakrabarty, S. Banik, M.R. Islam, H.K.D. Sarma, Measuring song complexity by statistical techniques, in *Proceedings of the 2nd International Conference on Communications, Device and Computing (ICCDC 2019)* (HIT, Haldia, West Bengal, 2019)

24. S. Chakrabarty, S. Banik, M.R. Islam, H.K.D. Sarma, Context aware song recommendation system, in *Proceedings of the Conference on Communication, Cloud, and Big Data (CCB)* (SMIT, Majhitar, East Sikkim, 2018)

25. R. Dawkins, *The Selfish Gene* (Oxford University Press, Oxford, 1976)

Multipath Mitigation Techniques for Range Error Reduction in PN Ranging Systems

Sathish Nayak, K. Krishna Naik, and Odelu Ojjela

Abstract In this paper, we investigate the Psuedo-Noise (PN)-sequence-based ranging technique for tracking an object by range measurements. PN-sequence-based signals for range measurements provide better range accuracies with secured data transmission. For range calculations, the delay parameter plays a major role. In real-time applications, the Line-of-Sight (LOS) link is not always guaranteed. The signal received at the receiver may contain reflected signals along with the direct path signals. Similar kind of range measurements are performed in Global Navigation Satellite System (GNSS) receivers, and the same can be applied for PN ranging sequence based object tracking, like UAVs, missiles, etc. The basic difference between GNSS and PN system is the ranging signal. In GNSS, for example, GPS and IRNSS, the Standard Positioning Service (SPS) comprises coarse acquisition code as the ranging sequence, which is of length 1023 chips transmitted at 1.023 Mbits/s. As per the Consultative Committee for Space Data Systems (CCSDS), two types of PN ranging sequences are proposed, weighted-voting-balanced Tausworthe codes abbreviated as T2B and T4B. Both the codes have the same lengths of 1,009,470 chips with different properties. Here the acquisition of the signal will be different in comparison with GNSS. Multipath signal propagation error can distort the actual signal, which leads to range measurement errors. In GNSS to reduce multipath error the conventional methods are based on Delay Locked Loop (DLL) feedback techniques. In order to reduce the multipath error, different methods which can be used to improve the range accuracy are discussed.

S. Nayak (✉) · K. K. Naik
Department of Electronics Engineering, Defence Institute
of Advanced Technology, Pune, India
e-mail: sathishnayak52@gmail.com

K. K. Naik
e-mail: krishnakaramtoti@gmail.com

O. Ojjela
Department of Applied Mathematics, Defence Institute
of Advanced Technology, Pune, India
e-mail: odelu@diat.ac.in

© Springer Nature Singapore Pte Ltd. 2020   237
R. Bera et al. (eds.), *Advances in Communication, Devices and Networking*,
Lecture Notes in Electrical Engineering 662,
https://doi.org/10.1007/978-981-15-4932-8_26

Keywords Multipath mitigation · Range measurements · DLL · Object tracking

1 Introduction

In object tracking applications, Psuedo-Noise (PN)-sequence-based ranging system can be used to get accurate and secured range measurements compared to conventional methods [1]. In case of GNSS for positioning applications, there are many possible errors that will affect the range values as well as the position of the object, such as ionospheric delay, tropospheric delay, ephemeris errors, clock errors, multipath, etc [2]. In Consultative Committee for Space Data Systems (CCSDS) [3] document, the range measurement of the object can be calculated by a known ranging signal modulated on a carrier and transmitted to transceiver mounted onboard. The received signal is retransmitted by object (onboard) and received at the ground station receiver. The range of an object is given by the round trip delay multiplied by speed of light [3]. The process of onboard reception and retransmission can be of two types: non-regenerative (transparent) and regenerative, with different Signal-to-Noise Ratio (SNR) specifications [4]. The CCSDS blue book [5] proposed two PN ranging sequences, which are Weighted-Voting Tausworthe (TVB) ($V = 2$ or 4) ranging sequences.

The multipath signal propagation error is mainly dependent on the receiver antenna position. The secondary signals consist of different path lengths, hence different delays, which distort the magnitude and phase of the actual signal. To reduce this error, earlier methods used antenna-based multipath mitigation, where special type of antennas are used to eliminate the reflected signals based on signal direction of arrival, signal polarization, etc. In order to find the range with PN-sequence-based measurement, the delay is measured with the help of the correlation technique, i.e., correlation peak magnitude denotes the delay with respect to the actual transmitted signal. To reduce this error, a suitable method can be chosen based on the scenario, from GNSS multipath mitigation techniques. Advanced multipath mitigation techniques which use estimation of LOS signal parameters based on measurement metrics are explained in [6, 7].

2 PN Ranging Codes

Ranging sequence is a periodic unit bipolar sequence which is modulated over a carrier. The ranging sequence should have certain desirable properties like acquisition time and spectral characteristics. The CCSDS standards proposed two PN code sequences with similar structure and from the same polynomial differ in some properties. Both sequences are generated by using six component sequences C1–C6, and the combination logic used is same, only the voting factor (V) is different as shown in Fig. 1. The length of the code $L = 2 \cdot 7 \cdot 11 \cdot 15 \cdot 19 \cdot 23 = 1,009,470$ chips.

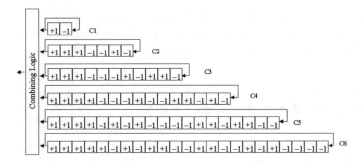

Fig. 1 TVB PN code generation $V = 2$ for T2B and $V = 4$ for T4B

Table 1 Ranging sequence properties [3]

Code type	Sequence length	Number of 1s	Number of −1s	Longest run of 1s	Longest run of −1s
T4B	1,009,470	504,583	504,887	7	5
T2B	1,009,470	504,033	505,437	9	9

$$C = \text{sign}(VC_1 + C_2 - C_3 - C_4 + C_5 - C_6) \tag{1}$$

a. T2B PN code: This PN sequence is called weighted-voting-balanced ($V = 2$) Tausworthe code. This code has faster acquisition time (5.2 s for $P_r/N_0 = 30$ dBHz [5]) compromising with jitter in the range measurement. It has a weaker ranging clock, which does not provide good ranging accuracy as compared to T4B code. The combinational logic to form T2B sequence is shown in Fig. 1 with Eq. 1.

b. T4B PN code: The PN sequence is called weighted-voting-balanced ($V = 4$) Tausworthe code. This code has a strong ranging clock component: hence it will provide good ranging accuracy compared to T2B code, but it takes slightly longer acquisition time (85.7 s for $P_r/N_0 = 30$ dBHz [5]). T4B code is preferred where the ranging accuracy is the primary concern [4]. The combinational logic to form T4B sequence is shown in Fig. 1 with Eq. 1.

T2B and T4B codes are generated in MATLAB using shift registers, and verified using characteristics given in the CCSDS green book [3] like sequence length, number of 1s and −1s, and consecutive longest run of 1s and −1s as given in Table 1. The difference in number of 1s and −1s results in a DC component in PN code spectrum. This code imbalance is reduced in T4B code since energy in the DC component will not be used for ranging. The unambiguous range can be calculated as given in [3].

$$U = 1/2 \, c \cdot L \cdot T_c = \frac{c \cdot L}{4 f_{RC}} \tag{2}$$

where L is the sequence length , f_{RC} is the range-clock frequency, and c is the speed of the light. For T2B and T4B sequence code length L is 1,009,470 and if range-clock frequency f_{RC} is 1 MHz then the unambiguous range is 75,710 km.

3 Range Measurements of an Object

Tracking an object, for example, tracking an Unmanned Aerial Vehicle (UAV) or testing a missile trajectory involves range measurement. PN-sequence-based tracking can be used with various advantages in terms of accuracy and security. PN sequence is used as a ranging sequence, which is modulated over a carrier and used for range measurement. As per the CCSDS standards, the ranging system consists of following stages [5] as shown in Fig. 2.

a. Ground Station Uplink: Ground station will generate the transmitting signal, which consists of a ranging code sequence as discussed in Sect. 2, modulated on a carrier. The modulation scheme can be a linear phase modulation. The detailed uplink chip rates and frequency bands are explained in [5]. The generated signal is transmitted from the ground station uplink transmitter [8].

b. Onboard Processing: Onboard transponder is responsible for receiving the signal transmitted from the ground station and retransmitting it to the ground station with some processing. Two kinds of onboard processing techniques can be implemented: non-regenerative and regenerative ranging techniques.

Non-regenerative ranging technique is a transparent (turnaround) technique in which the received signal is retransmitted to the ground station by an onboard transponder in a turnaround manner. The received signal will be phase demodulated and the remodulation of the carrier only: acquisition of ranging sequence will not be performed [4]. In this method, the uplink noise also gets modulated on the downlink carrier. This technique can be implemented where the acquisition time is

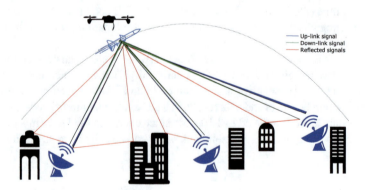

Fig. 2 Object tracking in multipath scenario

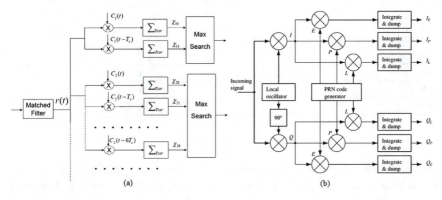

Fig. 3 **a** Parallel acquisition [3], **b** code tracking with DLL [9]

the primary concern compared to ranging accuracy. The advantage of this method is lower complexity compared to regenerative ranging method.

In regenerative ranging, the signal received at the onboard consists of ranging signal from which ranging clock is acquired, and then the ranging code position is searched, acquired, and tracked [4]. Onboard transponder then generates the ranging code coherently with the uplink code and phase modulates on the downlink carrier with regenerated ranging code. The generated signal is retransmitted to the ground station. In this regenerative process, the uplink noise gets eliminated from the downlink signal, thereby increasing the Signal-to-Noise Ratio (SNR), up to 30 dB substantial power gain compared to non-regenerative ranging. It can be used where the ranging accuracy is the main concern.

c. Ground Station Downlink: The ground station downlink processing performs similar to the onboard regenerative ranging. Received signal ranging clock component is acquired, and then ranging code position is searched, acquired, and tracked [8]. In the code phase acquisition, the length of the code is large resulting in long serial acquisition time. Instead, parallel acquisition can be performed with each cyclic shift of each probing sequence as discussed in Sect. 2. The total number of correlators required is 76 (for C1 probing sequence, single correlator is sufficient). Based on maximum search algorithm, the code position is calculated. The ground station acquisition processing is shown in Fig. 3a. The time taken for the signal transmitted from ground station (uplink) to the signal received at the ground station (downlink) is measured. The range of an object can be calculated from the delay measurement.

4 Multipath Mitigation

In the process of signal transmitted from ground station to onboard receiver (uplink) and onboard transmitter to ground station receiver (downlink), the received signal may not be the same as the transmitted signal. When the radio wave interacts with an

Table 2 Multipath components

Parameter	Code phase	Amplitude
LOS signal	0	1
MP1	0.08	0.791
MP2	0.28	0.588
MP3	0.35	0.416

obstruction or reflector, the signal properties like amplitude, phase, and direction of propagation get altered [9] depending on the type of the reflector object. The signal received at the receiver is usually the sum of the direct path signal along with its reflected or diffracted version replicas. The reflected paths will have longer traveling path than the direct path, and hence the extra delay relative to direct path [2]. This multipath error distorts the correlation between received and locally generated signals at the receiver, introducing errors in pseudorange values and carrier phase measurements. The normalized correlation function for a scenario where three multipath signals are with different delays and amplitudes is represented in (4). Code phase in terms of chips and amplitude values for the same scenario is shown in Table 2. In PN ranging system, ground station receiver receives the signal from onboard. For range measurements, the first step is to know the starting position of the PN sequence. This process is known as acquisition, which is correlation between received signal and locally generated signal by generating the local carrier with all the possible phases. When the sequence is lengthy then it becomes complicated. In this PN ranging technique, ranging sequence used is of length 1,009,470 chips. Correlation process requires locally generated sequence with all possible phase-shifted versions, i.e, 1,009,470 sequences. To avoid this problem, we use the PN code generating logic as a key to search based on probing sequence position. The sequence is generated at the transmitter by using six probing sequences C1–C6 in a particular manner to reduce the complexity of acquisition [10] as shown in Fig. 3a.

The locally generated sequences are only six probing sequences and its possible phase-shifted versions. Hence, all six combinations of correlations require 76 correlators. Based on correlator outputs, maximum search logic gives the combination of phases for which received signal gets correlated. The block diagram of acquisition is shown in Fig. 3a. Once the acquisition is done, then continuous measure of code delay can be done by tracking, i.e., code tracking. In GNSS as shown in Fig. 3b, the similar code tracking is performed and the same process can be adopted in the PN tracking. The basic purpose of tracking operation is to perfectly align the locally generated sequence with the received signal. The code tracking is done by DLL. The incoming signal is first multiplied by a perfectly aligned local carrier. After that the signal is multiplied by locally generated code replica. Three replica codes are, namely, Prompt, Early, and Late. Prompt code at which received code must align, Early code is 0.5 chip advanced version and Late code is 0.5 chip delayed version of Prompt code. These three versions of codes are multiplied with received sequence

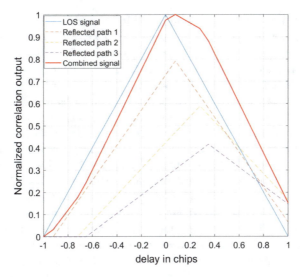

Fig. 4 Correlation function shape in multipath scenario

and integrated for a particular time duration [9]. The output from the integrators provides numerical values representing how much code replica is correlated with the received sequence, which is used to generate feedback signal to the local code generator to adjust the delay for perfect alignment as shown in Fig. 4. There are two types of discriminator functions: coherent and non-coherent. Coherent discriminator is simple Early minus Late value calculation. But in case of carrier phase misalignment, non-coherent method will be useful and the discriminator function is given in [2] as

$$\text{DLL}_{\text{Discriminator}} = \frac{(I_E^2 + Q_E^2) - (I_L^2 + Q_L^2)}{(I_E^2 + Q_E^2) + (I_L^2 + Q_L^2)} \tag{3}$$

where I_E, Q_E, I_L, and Q_L are correlation magnitudes of in-phase, quadrature-phase early and in-phase, quadrature-phase late code components, respectively.

The space between Early, Prompt, and Late represents the noise bandwidth in the DLL, typically 0.5 chip. If the space is larger than 0.5 chip, DLL is able to handle wider noise range: if lesser than 0.5 chip then it is more precise [1]. Ma et al. [11] explain methods for optimal correlator spacing for code tracking discriminator.

Multipath error due to the reflected signals from nearby objects is difficult to eliminate. For distant object reflections, the delay is more than one chip, which can be easily eliminated by correlation-based techniques. To mitigate these errors in GNSS, there are several methods which can be used in PN ranging system. The conventional methods are based on EML DLL correlation technique [2], in which three correlators (EML) are arranged in such a way that the received signal correlation peak is adjusted at prompt correlation with feedback delay estimator. The zero crossing of the discriminator function gives the path delay of the received signal [12]. To get precise correlation peak, the correlation spacing should be less. Narrow correlator technique follows this method of reducing correlator spacing. The space between early and late correlators is less than half chip but requires large bandwidth of the received signal. There are different methods following EML techniques such as double delta, multipath elimination methods, where the number of correlators is more and the correlator spacing is less. These multi-correlator-based techniques are used for estimating the channel properties for precise code delay measurements [12–14]. In case of closely spaced paths, these correlation-based techniques do not perform effectively. Maximum Likelihood (ML)-based methods perform better than the conventional methods in GNSS. Multipath Eliminating Delay Lock Loop (MEDLL) is one of the best methods to reduce multipath, where each channel uses multiple correlators to determine the shape of multipath-affected correlation function. ML-based estimation is used for determining the best combination of LOS and Non-LOS signal components by comparing a reference correlation function. The reference correlation function is formed by averaging the measured correlation function for a significant time duration. Estimation of number of multipath signals is based on SNR of received signal and it requires accurate SNR value. Reduced Search Space Maximum Likelihood (RSSML) DLL is also based on ML approach. At the input correlation function, a nonlinear curve fitting is performed by matching the amplitude, phase, and delay with the ideal correlation functions. RSSML delay estimator is proposed in [12] for GNSS. The same technique can be adopted in PN ranging system for object tracking in multipath scenario.

5 Conclusion

In this paper, we have discussed the PN ranging sequence based object tracking technique. Two types of ranging codes, namely, T2B and T4B and their applicability in different requirements are discussed. T2B and T4B codes of length 1,009,470 are generated and validated with the characteristics given as per CCSDS standards. Acquisition method is used at the ground station, and tracking technique based on EML DLL with non-coherent discriminator function can be used. In case of multipath error, the suitable value for EML correlator spacing may be used to mitigate small errors. For dense multipath scenarios, advanced multipath mitigation methods may be useful.

References

1. X. Jin, N. Zhang, K. Yang, X. Shen, X. Zhaobin, C. Zhang, Z. Jin, PN ranging based on noncommensurate sampling: zero-bias mitigation methods. IEEE Trans. Aerosp. Electron. Syst. **53**(2), 926–940 (2017)
2. E. Kaplan, C. Hegarty, *Understanding GPS: principles and applications* (Artech house, 2005)
3. Pseudo Noise (PN) ranging systems Green book standard CCSDS 414.0-G2 (2014)

4. K.S. Angkasa, J.S. Border, P.W. Kinman, C.B. Duncan, M.M. Kobayashi, Z.J. Towfic, T.J. Voss, Regenerative ranging for JPL software-defined radios. IEEE Aerosp. Electron. Syst. Mag. **34**(9), 46–55 (2019)
5. Pseudo Noise (PN) ranging systems Blue book. Standard CCSDS 414.1-B2 (2009)
6. A. Pirsiavash, A. Broumandan, G. Lachapelle, K. OKeefe, GNSS code multipath mitigation by cascading measurement monitoring techniques. Sensors **18**(6), 1967 (2018)
7. L. Cheng, K. Wang, M.F Ren, X.Y. Xu, Comprehensive analysis of multipath estimation algorithms in the framework of information theoretic learning. IEEE Access **6**, 5521–5530 (2018)
8. G. Boscagli, P. Holsters, E. Vassallo, M. Visintin, PN regenerative ranging and its compatibility with telecommand and telemetry signals. Proc. IEEE **95**(11), 2224–2234 (2007)
9. K. Borre, D.M. Akos, N. Bertelsen, P. Rinder, S.H. Jensen, *A Software-Defined GPS and Galileo Receiver: A Single-Frequency Approach* (Springer Science & Business Media 2007)
10. M. Maffei, L. Simone, G. Boscagli, On-board PN ranging acquisition based on threshold comparison with soft-quantized correlators. IEEE Trans. Aerosp. Electron. Syst. **48**(1), 869–890 (2012)
11. C. Ma, Z. Lv, X. Tang, Z. Xiao, G. Sun, Zero-bias mitigation method based on optimal correlation interval for digital code phase discriminator. Electron. Lett. **55**(11), 667–669 (2019)
12. Bhuiyan et al., Advanced multipath mitigation techniques for satellite-based positioning applications. Int. J. Navig. Obs. (2010)
13. N. Sokhandan, J.T. Curran, A. Broumandan, G. Lachapelle, An advanced GNSS code multipath detection and estimation algorithm. GPS Solutions **20**(4), 627–640 (2016)
14. N. Blanco-Delgado, F.D. Nunes, Multipath estimation in multicorrelator GNSS receivers using the maximum likelihood principle. IEEE Trans. Aerosp. Electron. Syst. **48**(4), 3222–3233 (2012)

Raspberry Pi Based Smart and Automated Irrigation System

Madhushree Dere, Naiwrita Dey, Subhasish Roy, Sagnik Roy, Dibyarup Das, and Debabrata Bhattacharya

Abstract The development of a smart and automated irrigation system has been reported in this paper. The system is developed using Raspberry Pi 3b+model and it integrates multiple sensors such as resistive soil moisture sensor, temperature and humidity sensor along with a water sprinkler unit. Daily life application is the motivation behind the present work. So a smart and automated irrigation system is designed here which is capable of automatically turning on or off the motor depending on certain parameters such as soil moisture content, environmental temperature, and humidity. To make it a low cost system, an Arduino board is used here in serial mode to interface the soil moisture sensor instead of A/D converter unit. An observation table is presented for different soil condition testing.

Keywords Smart irrigation · Raspberry pi · Serial connection · Automate

1 Introduction

Agriculture is the spine of the world's economy as well as the economy of a developing country like India. In this very moment of twenty-first century, the technology

M. Dere · N. Dey (✉) · S. Roy · S. Roy · D. Das · D. Bhattacharya
Department of AEIE, RCCIIT, Kolkata, West Bengal, India
e-mail: naiwritadey@gmail.com

M. Dere
e-mail: dere.madhushree@yahoo.in

S. Roy
e-mail: subham.sr26@gmail.com

S. Roy
e-mail: roniroy067@gmail.com

D. Das
e-mail: dibyarup7@gmail.com

D. Bhattacharya
e-mail: dxb.cin@gmail.com

© Springer Nature Singapore Pte Ltd. 2020
R. Bera et al. (eds.), *Advances in Communication, Devices and Networking*,
Lecture Notes in Electrical Engineering 662,
https://doi.org/10.1007/978-981-15-4932-8_27

and its advancement have become so fluent in every sector of life that it is the duty of technology people to take it forward through the infrastructure for development of agriculture as well [1]. Agriculture also contributes a significant role to gross domestic product (GDP) and it is well known that irrigation plays the paramount role in agriculture. In the existing system, a farmer has to work physically to control the irrigation system, and traditional instrumentation based on discrete and wired solutions presents difficulties in large geographical areas in our country; this also goes against the development of long-term agricultural production and sustainable utilization of water resources. Conventional irrigation system leads to wastage of time and wastage of water. Whereas, automation helps reducing consumption of electricity, decreases the wastage of water, uses less manpower, and helps in saving energy. The increase in urbanization and rapid industrialization causes a large void in agricultural activities [2]. Throughout the history of human civilization, urbanization patterns have been the strongest near large bodies of water. And conventional irrigation has led to deforestation and what not! To overcome these problems an automated smart irrigation system was required. The agricultural sector is the biggest user of water, followed by the domestic sector, and the industrial sector. There have been a lot of works carried out to automate the agricultural sector or to make irrigation smart, like uses of wireless networks, uses of RF module, and using Arduino to process the conventional irrigation into an automated smart one [3, 4]. India is the largest freshwater user in the world, and the country's total water use is greater than any other continent [5]. The system may prove to be a substitute to traditional farming method. And adopting an optimized irrigation is a necessity nowadays due to the lack of world water resource [6, 7]. Raspberry Pi based irrigation system has been implemented. Star Zigbee topology serves as backbone for the communication between raspberry pi and end devices. Soil moisture sensor and temperature sensors are placed in the root zone of plant and the gateway unit handles the sensor information and transmits data to a web application. The Internet or WiFi module is interfaced with the system to provide data inspection. Raspberry pi is the heart of the system [8]. Capacitive soil moisture sensor is connected to raspberry pi board through a node along with temperature and humidity sensor [9]. The raspberry pi board is connected as a main component and the relay directs the water pump to work along with the change in temperature, humidity, moisture, and such environmental parameters. One ultrasonic sensor is used to connect the pump with the tank [10]. Distributed wireless network of temperature sensor and soil moisture sensor are also reported in literature for smart irrigation system. IoT based has also implemented for the same [11, 12].

This paper presents a raspberry pi based smart and automated irrigation system to control the water supply to each plant automatically depending on values of temperature, humidity, and soil moisture sensors. An Arduino Uno board is used for serial communication with the Raspberry pi board to interface the analog soil moisture sensor instead of A/D converter. The paper is organized as follows: after a brief introduction, methodology explains the overall working of the system followed by hardware prototyping, experimental result, and conclusion.

2 Methodology

The block diagram of the system has been shown in Fig. 1 and the method of working of the proposed scheme is described below stepwise.

Step 1:
In this proposal two different types of sensors—Temperature and humidity, soil moisture sensor are used to detect the required scarcity of water in the soil by measuring temperature, humidity, and moisture quantity of soil for proper undergoing of the process, that is, for proper irrigation.

Step 2:
Required action with the help of controller (i.e., Raspberry pi 3 model b+) is to be taken for the deficit of the physical parameters of the soil for a healthy process.

Step 3:
An autonomous system is implemented here for taking the action by supplying adequate amount of raw water through the pump. The overall system is indeed controlled by the Raspberry Pi, the controller itself.

For the exact amount of water in the soil to be measured one has to read the analog output of the Moisture Sensor. But since, Raspberry Pi cannot read analog outputs, an external open source microcontroller, Arduino Uno is used. This part is shown in hardware prototyping of the proposed system.

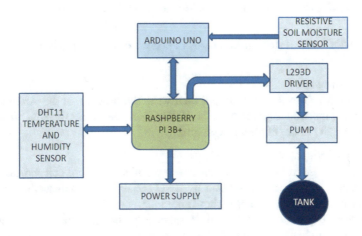

Fig. 1 Block-diagram of the proposed system

Fig. 2 Flowchart for setting
up Raspberry Pi to read data
from Arduino

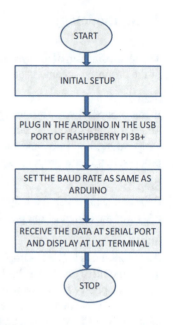

3 Hardware Prototyping of the Proposed System

3.1 Serial Connection of Raspberry Pi with Arduino

It is to be done in two steps given below. Flowchart of the process is given below in
Fig. 2.

Arduino Uno setup:
Connect the UNO to the PC first and then write the program in the Arduino IDE
software and upload the program to the UNO. Then disconnect the UNO from PC.
Attach the UNO to the PI after programming.

Raspberry Pi setup:
After that a program for Pi is written to receive this data from soil moisture sensor connected to Arduino, being sent by UNO serially. Serial connection between
raspberry pi and Arduino board is shown in Fig. 3.

Real-time experimental setup of Raspberry pi based smart and automated
irrigation system has been shown in Fig. 4a, b, c, d.

4 Experimental Result

Soil moisture content is checked for different conditions and shown in Table 1.

Fig. 3 Raspberry and Arduino Interfacing

Fig. 4 a Real-time setup of the Raspberry pi. **b** Automatic water sprinkler connected immersed pump. **c** Real-time temperature and humidity data acquired in Pi. **d** Test set up

Table 1 Observation table of the soil moisture sensor output data for different types of soil

Condition	Voltage (mV)	Statement
Open air	5.2	Resistance between the probes is least
Sand	5.15	Resistance is first observed between probes in this condition
Dry soil	4.74	Resistance rises and voltage drop is noted
Medium wet soil	2.91	Resistance is duly noted as voltage drop is low compared to the previous one
Wet sand	2.22	Voltage drop still occurs as there is high amount of water present in this condition
Wet soil (Moist)	1.63	Maximum Water a soil sample can absorb, so resistance between the probes is very high, as seen by the voltage reading
Water	1.275	Voltage reading noted in this condition is the least, which implies that resistance is highest

5 Conclusion

The implemented system presented in this paper is integrated with multi-sensors such as soil moisture sensors, humidity, and temperature sensors. The entire system is integrated using microcomputer Raspberry Pi 3B+model. It is capable of automatic watering of plants depending upon monitored parameters such as soil moisture content, temperature, and humidity of the surrounding environment.

The future of this work can be comprehended with an addition of camera module to monitor plant health along with the present system.

References

1. Y.G. Gawali, D.S. Chaudhuri, H.C. Chaudhuri, Review paper on automated irrigation system using WSN. IJARECE **5**(6) (2016)
2. A. Joshi, L. Ali, Survey on auto irrigation system. IEEE Catalog Number CFP17D81-POD (2017)
3. T.J. Kazmierski, SSRG Int. J. Electron. Commun. Eng. (2017)
4. D. Rane, P.R. Indurkar, Review paper on automatic irrigation system using RF module. IJAICT **1**(9) (2015)
5. B.K. Chate, J.G. Rana Smart irrigation system using Raspberry pi. IRJET **3**(5) (2016)
6. B.H. Fouchal, O. Zytoune, D. Aboutajdine, Drip irrigation system using wireless sensor networks **44**(4) (2018)
7. J. Gutiérrez, J. Francisco Villa-Medina et al., Automated irrigation system using a wireless sensor network and GPRS module. IEEE **63**(1), 166–176 (2014)
8. N. Agrawal, S. Singhal, Smart drip irrigation system using Raspberry pi and Arduino. IJIRCCE **5** (Special Issue 4) (2017)
9. G. Ashok, G. Rajasekar, Smart drip irrigation system using Raspberry Pi and Arduino. IJSETR **1742**(3) (2018)

10. P.S. Shwetha, Survey on automated irrigation system. TROINDIA (2) (2017)
11. K.J. Vanaja, A. Suresh, S. Srilatha, K.V. Kumar, M. Bharath, IOT based agriculture system using Node MCU. IRJET **5**(3) (2018)
12. J. Gutiérrez, J. Francisco Villa-Medina et al., Automated irrigation system using a wireless sensor network and GPRS module. IEEE **63**(1), 166–176

A Rectangular Annular Slotted Frequency Reconfigurable Patch Antenna

Abha Sharma, Rahul Suvalka, Amit Kumar Singh, Santosh Agrahari, and Amit Rathi

Abstract A low-profile annular slotted microstrip antenna having frequency reconfigurable characteristics along with circular polarization has been described here. The antenna consists of FR-4 substrates of 48 mm × 44 mm with dielectric constant 4.4. A rectangular patch of 26 mm × 20 mm acts as a radiator. The proposed structure gives multi bands at 5.5, 6.5, 8.8, 9.7, 12.1, and 14 GHz. The proposed reconfigurable antenna shows switching involving multiple working bands at 2.1–2.2, 5.4–5.6, 10.8, 11.7, and 12.1 GHz with different diode configurations. Microstrip line feed technique is applied for excitation. This type of device can be used in spectrum management cognitive radio application.

Keywords Annular · Reconfigurable antenna · Polarization · Cognitive radio

1 Introduction

For the comprehensive employment of wireless communication systems, existing technologies is not plenty enough to meet up the challenges due to the spectrum crisis. To deal with spectrum scarceness, current research has recommended reconfigurable antenna to overcome the complexity of wireless devices [1–3].

The reconfigurable antennas offer diverse characteristics of polarization, resonant frequency, radiation pattern, and any combination of two in order to contend with the changing system requirement [4]. The evolution of reconfigurable antennas has made immense advancement in modern existence. As compared to traditional antennas, reconfigurable antennas are smarter with respect to their size, weight, and cost [5–16].

A. Sharma · A. K. Singh · A. Rathi (✉)
Department of Electronics & Communication Engineering, Manipal University Jaipur, Jaipur 303007, Rajasthan, India
e-mail: amitrathi1978@gmail.com

A. Sharma
e-mail: sharmaabha0307@gmail.com

R. Suvalka · S. Agrahari
Department of Electronics & Communication, Poornima University, Jaipur, India

© Springer Nature Singapore Pte Ltd. 2020
R. Bera et al. (eds.), *Advances in Communication, Devices and Networking*,
Lecture Notes in Electrical Engineering 662,
https://doi.org/10.1007/978-981-15-4932-8_28

Frequency reconfigurable antennas are broadly investigating in the ancient times together with patch antennas, monopole antennas, and slot antennas [17–19]. In common, there are generally two methods to realize this class of reconfigurable antennas. One system is to utilize PIN diodes and conventional bandpass filters to switch among many states for various services. The further is to implement reconfigurable filters along with varactor diodes to constantly tune the operating states. Using bandpass filters in the feed line structure is critical to design filtering function or reconfigurability but the system complexity and cost of the device increases.

1.1 Design of Proposed Antenna

This paper described a reconfigurable antenna having a substrate of FR-4 with rectangular-shaped patch of 26 mm × 20 mm. A circular ring is cut out into the patch along with two rectangular-shaped apertures at opposite end of the radiator. These two slots are providing circular polarization (Fig. 1).

Figure 2 defines the return loss (S_{11}) plot for the proposed design which shows for frequency bands at 5.5 GHz, 6.5 GHz, 8.8 GHz, and 9.7 GHz with the return loss of -15 dB, -24 dB, -14 dB, and -24 dB correspondingly.

2 Implementation of Reconfigurable Antenna

In the next structure, circular ring size has been changed with two diodes for switching which show the resonance at 6.5, 12.1, and 14 GHz (Figs. 3 and 4).

Further, we have cut two rectangular slots into the patch and then embedded two diodes in the rectangular slots. The different configuration of diode switching has shown different frequency bands (Fig. 5).

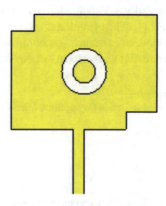

Fig. 1 Front sight of the proposed designed antenna

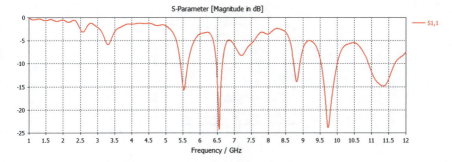

Fig. 2 Return loss (S_{11}) plot for proposed designed antenna

Fig. 3 Front sight of proposed designed reconfigurable antenna

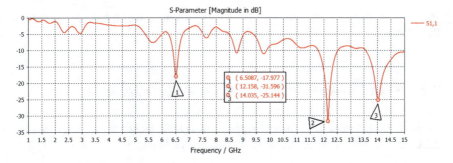

Fig. 4 Return loss (S_{11}) plot for proposed designed antenna

2.1 While Both Diodes Are Turned ON

While both the diodes D1 and D2 are switched ON, the anticipated design resonates at three frequencies of 2.1 GHz, 10.8 GHz, and 12.1 GHz with the return

Fig. 5 Front sight of proposed designed reconfigurable aerial

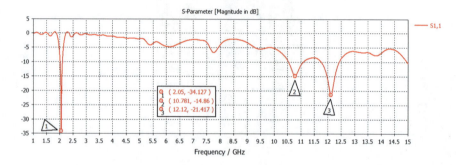

Fig. 6 Return loss plot for D1 ON and D2 OFF

loss of −34 dB, −15 dB, and −21 dB, correspondingly, which has been shown in Figs. 6 and 7.

2.2 When Diode D1 Is Turned ON and D2 Is Turned OFF

As diode D1 is turned ON, and D2 is switched to OFF, there are four resonances at 2.2 GHz, 5.4 GHz, 10.8 GHz, and 12.1 GHz as seen in Fig. 8.

2.3 When Diode D1 Is Turned OFF and D2 Is Turned ON

Once diode D1 is OFF, and D2 is ON, there are four resonances at 2.2 GHz, 5.6 GHz, 10.8 GHz, and 11.7 GHz as seen in Fig. 9.

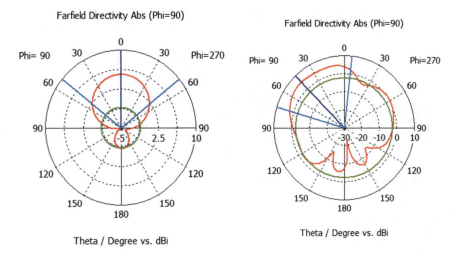

Fig. 7 Radiation pattern for D1 ON and D2 OFF

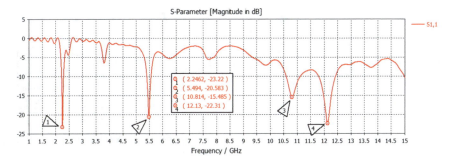

Fig. 8 Return loss plot for D1 ON and D2 OFF

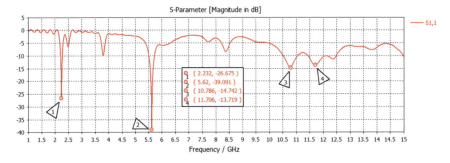

Fig. 9 Returns loss plot for D1 OFF and D2 ON

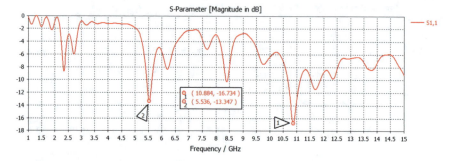

Fig. 10 Returns loss plot for D1 OFF and D2 OFF

Fig. 11 Comparative
analysis for different
switching conditions

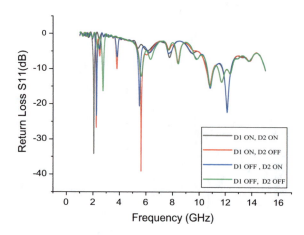

2.4 When Diode D1 Is OFF and D2 Is OFF

When both diodes D1 and D2 are OFF, two frequency band resonances at 5.6 GHz, 10.8 GHz as shown in Fig. 10.

Figure 11 shows the comparative analysis of the different configurations of switching conditions of diode.

3 Conclusion

Different structures have been designed which show different operating frequency range. The proposed antenna designs can switch at diverse frequencies via changing the switch configuration and can be used for cognitive radio application.

References

1. C. Borda-Fortuny, K.-F. Tong, K. Chetty, Low-cost mechanism to reconfigure the operating frequency band of a Vivaldi antenna for cognitive radio and spectrum monitoring applications. IET Microw. Antennas Propag. **12**(5), 779–782 (2018)
2. S. Kingsly, D. Thangarasu, M. Kanagasabai, M. Gulam Nabi Alsath, T. Rama Rao, P. Sandeep Kumar, Y. Panneer Selvam, S. Subbaraj, P. Sambandam, Multiband reconfigurable filtering monopole antenna for cognitive radio application. IEEE Antennas Wirel. Propag. Lett. **17**(8) (2018)
3. FCC, First report and order, revision of part 15 of the commission's rules regarding "Ultra-Wideband Transmission Systems FCC," Washington, DC (2002)
4. A. Mansoul, F. Ghanem, M.R. Hamid, M. Trabelsi, A selective frequency-reconfigurable antenna for cognitive radio applications. IEEE Antennas Wirel. Propag. Lett. **13** (2014)
5. A. Khidre, F. Yang, A.Z. Elsherbeni, Patch antenna with a varactor-loaded slot for reconfigurable dual-band operation. IEEE Trans. Antennas Propag. **63**(2), 755–760 (2015)
6. P.-Y. Qui, F. Wei, Y. Jay Guo, Wideband-to-narrow tunable antenna using a reconfigurable filter. IEEE Trans. Antennas
7. A. Sulaiman Al-Zayed, M. Abdelazim Kourah, S. Famir Mahmoud, Frequency-reconfigurable single and dual-band designs of a multi-mode microstrip antenna. IET Microw. Antennas Propag. **8**(13), 1105–1112 (2014)
8. A. Tariq, H. Ghafouri-Shiraz, Frequency-reconfigurable monopole antennas. IEEE Trans. Antennas Propag. **60**(1), 44–50 (2012)
9. Valizade, C. Ghobadi, J. Nourinia, M. Ojaroudi, A novel design of reconfigurable slot antenna with switchable band notch and multiresonance functions for UWB applications. IEEE Antennas Wirel. Propag. Lett. **11**, 1166–1169 (2012)
10. S. Nikolaou, N.D. Kingsley, G.E. Ponchak, J. Papapolymerou, M.M. Tentzeris, UWB elliptical monopoles with a reconfigurable band notch using MEMS switches actuated without bias lines. IEEE Trans. Antennas Propag. **57**(8), 2242–2251 (2009)
11. A. Rathi, R. Vijay, Optimization of MSA with swift particle swarm optimization. Int. J. Comput. Appl. (IJCA) **8**, 28–33 (2010). Article, ISBN-978-93-80747-23-7
12. A. Rathi, R. Vijay, Expedite particle swarm optimization algorithm (EPSO) for optimization of MSA, in *Swarm Evolutionary and Memetic Computing, Lecture Notes in Computer Science (LNCS)* (Springer, 9th Sep 2010), ISBN 978-3-642-175626, pp. 163–170
13. R. Suvalka, S. Agrahari, A. Rathi, Design and analysis of compact planner microstrip antenna for UWB and wireless applications using DGS, in *Second IEEE International Conference on Recent Advances and Innovations in Engineering (ICRAIE-2016)* (IEEE RECORD #39140) (23–25 December 2016)
14. S. Loizeau, A. Sibille, Reconfigurable ultra-wideband monopole antenna with a continuously tunable band notch. IET Microw. Antennas Propag. **8**(5), 346–350 (2013)
15. B. Badamchi, J. Nourinia, C. Ghobadi, A. Valizade Shahmirzadi, Design of compact reconfigurable ultra-wideband slot antenna with switchable single/dual band notch functions. IET Microw. Antennas Propag. **8**(8), 541–548 (Nov. 2013)
16. J. Tan, W. Jiang, S. Gong, T. Cheng, J. Ren, K. Zhang, Design of a dual-beam cavity-backed patch antenna for future fifth generation wireless networks. IET Microw. Antennas Propag. **12**(10) (2018)
17. R. Hussain, M.S. Sharawi, Planar meandered-F-shaped 4-element reconfigurable multiple-input multiple output antenna system with isolation enhancement for cognitive radio platforms. IET Microw. Antennas Propag. **10**(1), 45–52 (2016)
18. G. Srivastava, A. Mohan, A. Chakrabarty, Compact reconfigurable UWB slot antenna for cognitive radio applications. IEEE Antennas Wirel. Propag. Lett. **16** (2017)
19. Y.-M. Cai, K. Li, Y. Yi, S. Gao, A low-profile frequency reconfigurable grid-slotted patch antenna. IEEE J. Mag. **6**, 36305–36312 (2018)

Fractional Order Control of Unstable and Time-Delayed Linear Time-Invariant (LTI) Plants

Reetam Mondal and Jayati Dey

Abstract The present work describes the formulation of the fractional order (FO) proportional–integral–derivative $PI^\lambda D^\mu$ controllers entailing FO integrator and FO differentiator, a procedure which allows defining the numerical terms of the FO controller. These are inferences to be the general form of PIDs, the output of which is a linear combination of the input, a fractional derivative of the input, and a fractional integral of the input. They have more tuning freedom and a wider region of parameters that may stabilize the plant under control surpassing the performance of the integer-order (IO) counterparts, specifically in servo control applications. A consequential issue in control engineering is presented by plants and processes that are unstable. Therefore, in this work unstable as well as time-delayed plants have been considered which are frequently encountered in the field of control.

Keywords Fractional order controller · Fractional derivative · Fractional integral · Servo system control · Unstable and time-delayed plants

1 Introduction

Application of FO models is more apposite for the representation and interpretation of these real dynamical systems than its IO versions. FO-$PI^\lambda D^\mu$ controller is therefore naturally suitable for these FO plants. They are being largely employed by many researchers in order to pull off and accomplish the most robust conduct of the plants under control [1–3]. Integration and derivatives with non-integer orders come into sight if the controller or the system is described by differential equations with generalized orders. The comparative study [4, 5] calls attention to the supremacy of availing these FO-$PI^\lambda D^\mu$ controllers over traditional PID controllers with settling time and maximum peak overshoot diminished with content. To a great extent, the

R. Mondal (✉) · J. Dey
National Institute of Technology (NIT) Durgapur, Durgapur, West Bengal, India
e-mail: reetammondal2008@gmail.com

J. Dey
e-mail: deybiswasjayati@gmail.com

© Springer Nature Singapore Pte Ltd. 2020
R. Bera et al. (eds.), *Advances in Communication, Devices and Networking*,
Lecture Notes in Electrical Engineering 662,
https://doi.org/10.1007/978-981-15-4932-8_29

adjusting and regulating statutes of the numerical terms of the controllers proposed in the literature are at most applicable to self-regulating asymptotically stable processes, whereas integral processes and unstable processes have been disregarded and are not heeded to. Besides, investigation of stability of the system counteracted with the FO controller is dispensed with Riemann surface [6]. In this work, the controller parameters are selected by root locus technique with the non-integer orders graphically tuned to achieve satisfactory stability margins and sensitivity peaks. Numerical examples illustrated and explained elaborately to validate the design approach made. The results fed here entitle to express for a given process, the performance advancement, and upgrade that can be attained by bringing into utilization the FO controllers in lieu of the IO one.

2 Generalized Non-integer Order $PI^\lambda D^\mu$ Controller

A PID controller is a comprehensive feedback control composition extensively and broadly exercised on industrial applications for several decades [1]. In the perpetuation of the derivative and integrator orders from integer to fractional numerals yields a better adaptable adjusting scheme of the $PI^\lambda D^\mu$ controllers and consequently an elementary approach to achieve control requisites in contrast to its integer equivalent [2]. The control action affects the system behavior by increasing the pace of the dynamic reaction by slashing error in steady state. The command signal from the controller $u(t)$ adapts effortlessly with the rate of change of error signal $e(t)$ by the combination of the three continuous controller modes as [3],

$$u(t) = K_P e(t) + K_I \int_0^t e(t)dt + K_D \frac{d}{dt}e(t) \tag{1}$$

The three adjustable control parameters here are proportional control (K_P), integral control (K_I), and derivative control (K_D) gains. The control law is thus sustained in the form as [4],

$$C_{PID}(s) = K_P + \frac{K_I}{s} + K_D s \tag{2}$$

In complement to the typically common PID controllers, with the two ancillary variable quantities of $PI^\lambda D^\mu$ (FOPID) controllers, it is essential to study the additional design statements that can be fulfilled as far as performance with plant uncertainties and high-frequency noise is concerned. The continuous-time transfer function representation of the generalized structure of the controller is contemplated as [5],

$$C_{PI^\lambda D^\mu}(s) = K_P + \frac{K_I}{s^\lambda} + K_D s^\mu \tag{3}$$

with the non-integer orders λ and μ specified between the range 0 and 2 [1]. These are barely sensitive to plant uncertainties, generating vigorous performance against gain changes and noise [6].

3   Fundamental Design Philosophy

The design philosophy conceived in the available literature mostly deals with the choice of gain cross-frequency (ω_{cg}) and Phase Margin (PM). This has been feasible due to the structure of the controllers considered. However, FO-$PI^\lambda D^\mu$ controllers, on the contrary, do not provide such opportunity to the designer. Therefore, the design philosophy adopted here according to the scheme in Fig. 1 is not the same and is quite different.

The following section describes a procedure which allows defining the parameters of a FO-$PI^\lambda D^\mu$ controller. It is obvious from the formation of the PID controller in Eq. (2) that it has two zeros and one pole at the origin in the s-plane as

$$C_{PID}(s) = \frac{K_D s^2 + K_P s + K_I}{s} = \frac{K(s + z_1)(s + z_2)}{s} \tag{4}$$

Hence, one may initiate the design with placing of these zeros simply in the LHP of the complex s-plane. Root locus analysis seems to be advantageous in this direction. Now, FO-$PI^\lambda D^\mu$ controller does not clearly have these zeros. The zeros of this non-integer order controller does not clearly have these zeros. The zeros of the FO-$PI^\lambda D^\mu$ controllers have the zeros shifted from this positions which lead to the change in the properties of stability, stability margins, and gain cross-over frequency. Therefore, the next step is to probe and investigate the stability margins of the compensated system with the variation of λ and μ in the ranges between 0 and 2. The designer may have stability margin specifications well defined beforehand. The graphical solution of these non-integer orders λ and μ of the controller with the zeros previously located completes the process of controller design and synthesis. In a design problem, one of the objectives is to keep the steady-state error to a minimum while at the same time the transient response must satisfy a certain set of performance specifications. It is therefore necessary to formulate some corrective system to drive the plant under

Fig. 1 Overview of fractional order (FO) control scheme

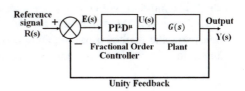

control to meet the desired specifications. Given a plant, a set of attributes is defined to formulate a suitable controller so that the overall system converges to reach them. The design specifications to be met by the overall system merely include minimum deviation of the output from the reference input in steady state, % overshoot which depends on the damping ratio, with settling time and rise time as small as possible with stability margins in frequency domain within specified desired limits to achieve the design goals. This is basically a graphical iterative method. It tries to select the parameters of the controller so that the closed-loop poles are placed suitably in the stable half of the complex s-plane.

4 Proposed Design Algorithm

With the dynamic model of the plant known, the modulating scheme of the FO-$PI^\lambda D^\mu$ controller can be analyzed by root locus technique. The following steps are pursued to realize the generalized compensation with two independent fractional orders of the FO controller for the system as

Step 1: Plotting the Root locus of the system along with an integrator $\left[K \cdot \frac{G(s)}{s} \right]$ for the gain K in the range, $0 < K < \infty$.

Step 2: Selection of the two finite zeros at $-z_1$ and $-z_2$ to obtain the PID controller parameters leading to increase in damping with minimum percentage overshoot and settling time.

Step 3: The non-integer parameters λ and μ of the FO-$PI^\lambda D^\mu$ controller are varied and adjusted so that it is competent to reach and fulfill preferable positive Gain Margin (GM), Phase Margin (PM), and peak sensitivity margin.

Step 4: Now, on fulfilling the desired time and frequency domain criteria, the required FO-$PI^\lambda D^\mu$ controller is obtained. Otherwise, repeating the entire algorithm described from **step 1** with the new values of $-z_1$ and $-z_2$ until desired stability margins and sensitivity peak magnitudes are accomplished.

Step 5: Here, the three controller gains of the controller are determined as $K = K_D$, $K_P = K_D(z_1 + z_2)$ and $K_I = K_D(z_1 z_2)$.

This leads to the determination of the controller gains utilizing where the non-integer orders of the fractional form are varied to attain the required stability margins. The parameters of the PID controllers are optimal in terms of frequency domain objectives adopted with desirable stability margins and sensitivity peak magnitudes to achieve satisfactory robust stabilization, with fast speed of response along with the achievement of control inputs within restricted limits. Thus, the numerical terms of the PID controller will be employed here as a part of the retuning procedure of the FO-$PI^\lambda D^\mu$ controller illustrated through examples. The rationale behind taking up this particular method is its simplicity with satisfying more design specifications resulting in remarkable up-gradation in control quality and execution.

5 Numerical Examples with Results

Example 1 The following unstable plant with a pair of RHP pole at ± 46.67 has been considered [7] as

$$G(s) = \frac{-3518.85}{s^2 - 2177.8} \tag{5}$$

to verify the proposed root locus method of formulating the FO-$PI^\lambda D^\mu$ controller for this case. The design objectives to be met to determine the controller parameters are defined to be (a) Damping ratio,$\xi \leq 0.9$, (b) Settling Time within 1 s, and (c) Maximum percentage Peak Overshoot within 10%. Two zeros of Eq. (4) is selected at -1 and -14 to reorient the root locus path from the unstable region of the complex s-plane due to the unstable location of the RHP poles to the left half of the stable region as shown in Fig. 2a. The non-integer orders of the proposed FO-$PI^\lambda D^\mu$ controller in the form of Eq. (3) is then varied and adjusted between 0 and 2 [4] to achieve the desirable positive GM, PM, and maximal magnitude of sensitivity tuned graphically as exhibited in Fig. 3a, b, and c. The FO-$PI^\lambda D^\mu$ controller thus takes the form as

$$C_{PI^\lambda D^\mu}(s) = 15 + \frac{15}{s^{1.2}} + s^{0.98} \tag{6}$$

The variation of the magnitude and phase of the loop transfer function with the plant in Eq. (5) and the FO controller in Eq. (6) reveals PM $= 90°$ at ω_{cg} of 3100 rad/s in Fig. 2b. in juxtapose to the IO controller which has a PM of only 85° at ω_{cg} of 3000 rad/s with same controller gains. The corresponding step response is also compared in Fig. 3d. The sensitivity plot with the FO-$PI^\lambda D^\mu$ controller confirmed to be maintained, $|S|_{max} < 2$. The time domain interpretation of the transient response of the unstable plant with the integer and non-integer controllers has been summarized in Table 1.

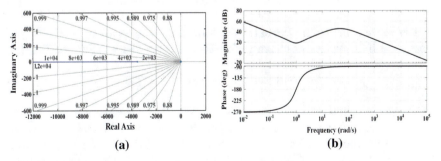

(a)   (b)

Fig. 2 **a** Root locus plot of the compensated plant and **b** Variation of magnitude and phase with frequency for the FO controller

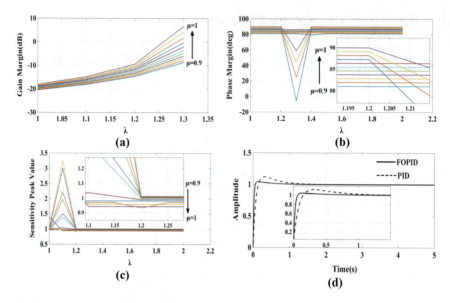

Fig. 3 Variation of **a** GM (dB). **b** PM(deg). **c** Maximum Sensitivity with the fractional orders λ
and μ. **d** Response of the plant to unit step input

Table 1 Comparative attributes of the FO controller with IO controller	Sl. No.	Controllers	Transient performance specifications achieved	
	1.	$PI^\lambda D^\mu$	Settling time (t_s) (s)	% Overshoot
			0.5	4.44
	2.	PID	0.75	12

The closed-loop characteristic equation obtained by implementing the FO-$PI^\lambda D^\mu$
controller in Eq. (6) is

$$s^{3.2} + 3518.8s^{2.18} + 50605s^{1.2} + 52783 = 0 \qquad (7)$$

The above equation in (7) can be transformed to the w-plane by $w = s^{1/m}$ [6].
Here, $m = 100$ indicates the quantity of sheets within the Riemann Surface (Fig. 4a).

$$w^{320} + 3518.8w^{218} + 50605w^{120} + 52783 = 0 \qquad (8)$$

The stable region in the s-plane, $\phi_s > \frac{\pi}{2}$ transforms to the sector in $\phi_w > \frac{\pi}{2m}$ in
w-plane [8]. Stability is confirmed if the principle sheet poles in w-plane lie within
this segment. The closed-loop poles of the above Eq. (8) in the principle Riemann
Sheet are $-1.0828 \pm 0.0334i$ and $-1.0005 \pm 0.0258i$ which has arguments $|\varphi_{w1}|$
$= 0.0308$ and $|\varphi_{w2}| = 0.0258$, both of which are $> \frac{\pi}{2m}$ [6, 8]. It is clearly noticed
from diagram in Fig. 4b that absence of poles lying within the unstable segment or

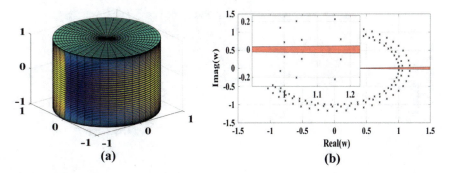

Fig. 4 a Riemann surface of function $w = s^{1/100}$ with $m = 100$ Riemann sheets. **b** Closed-loop pole locations in complex w-plane

sector $-\frac{\pi}{m} < \varphi_w < \frac{\pi}{m}$ confirms that the arguments have $|\varphi_w| > \frac{\pi}{2m}$ which reasserts stability.

Example 2 A Type-0 plant of a DC motor servo system with delay time is considered as [2]

$$G(s) = \frac{166.1038}{0.75507s + 1} e^{-0.1s} \qquad (9)$$

Considering the undelayed dynamics of the same plant as [2]

$$G(s) = \frac{166.3714}{0.83907s + 1} \qquad (10)$$

the FO-$PI^\lambda D^\mu$ controller is designed to satisfy performance specifications of (a) Damping ratio,$\xi \leq 0.9$, (b) Settling Time within 5 s, and (c) Maximum percentage Peak Overshoot within 10% following the same design method put forward in Sect. 4. To satisfy these objectives, a fractional order (FO) controller of the form in Eq. (3) has been formulated in a similar way as

$$C_{PI^\lambda D^\mu}(s) = 0.005 + \frac{0.02}{s^{0.8}} + 0.0014s^{0.5} \qquad (11)$$

The root locus of the compensated plant using the integer equivalent form of PID controller is as depicted in Fig. 5a.

Utilizing the same values of the controller gains for the generalized form in Eq. (3) the non-integer orders λ and μ are varied and tuned graphically in between the range 0 and 2 [4] to achieve desirable satisfactory stability margin as in Fig. 5b. Here, to select the values of the fractional orders it is observed from the plot in Fig. 6 that sensitivity peak relatively remains constant till $\lambda = 0.9$ with the variation of μ in the

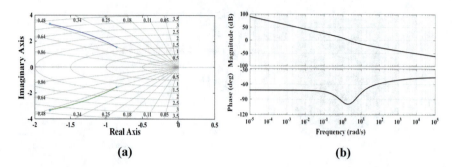

Fig. 5 **a** Root locus plot of the compensated plant and **b** Variation of magnitude and phase with frequency for the FO controller

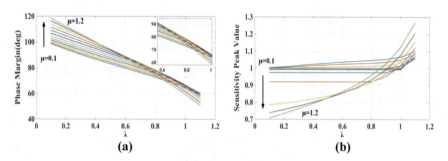

Fig. 6 Variations of **a** Phase margin (PM) and **b** Sensitivity peak magnitude with the fractional orders λ and μ of the FO controller

range (0.1, 1.2) beyond which it increases. It is also observed here that till this value λ, the Phase Margin (PM) decreases drastically beyond the range μ = 0.6. With these values, the system step response has been plotted in Fig. 7a. Now, to ensure minimum % overshoot with reduced settling time the non-integer orders have been fixed up at λ = 0.8 and μ = 0.6 to perceive the controller as in Eq. (11).

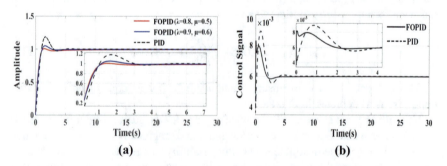

Fig. 7 **a** Response of the plant to unit step input and **b** Control signal

Table 2 Comparative attributes of the FO controller with IO controller

Sl. No.	Controllers	Transient performance specifications achieved	
1.	$PI^\lambda D^\mu$	Settling time (t_s) (s)	% Overshoot
		1.3	0
2.	PID	2.65	18.14

Fig. 8 Step response with and without time delay

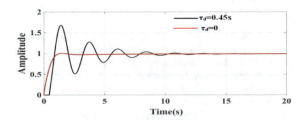

The frequency response plots in Fig. 5b makes it visible that the frequency domain specification is well satisfied achieving GM = *InfdB* with PM = 76.9° at ω_{cg} = 2.26 rad/s in contrast to the PID controller which could ascertain PM of only 63.2° at ω_{cg} = 1.71 rad/s employing the same controller gains. The bandwidth is 2 rad/s for PID controller whereas it is spotted at 3 rad/s in the case of the FO $PI^\lambda D^\mu$ controller. Practically, to have an acceptable robust design, the peak sensitivity value attained in the mid-frequency region is anticipated to be less than 2 to restrict and fulfill the GM and PM specifications satisfactorily [9, 10]. The peak magnitude of sensitivity is < 1 in the mid-frequency fragment with a very low gain (<< 1) at lower frequencies and gain reaching to unity at higher frequencies. The quantitative analysis of the system step response of the proposed controller compared with its classical counterpart is tabulated in Table 2. It is noted here from Fig. 7b that the amplitude of the control signal is diminished in case of the $PI^\lambda D^\mu$ controller with reduced settling time and % overshoot in collation with the IO-PID controller. The stability analysis can be confirmed in a similar way as in Example 1.

Now, since, it has been verified that the loop transfer function yields a delay margin τ_d = 0.595*s* calculated mathematically by $\tau = (\pi \times PM)/(180 \times \omega_{cg})$ which has led to determination of the delay margin through simulation to be 0.45 s much greater than the predefined value of 0.1 s in Eq. (9). The nominal step response and the reaction of the plant with dead-time are displayed through Fig. 8.

6   Concluding Remarks

FO-$PI^\lambda D^\mu$ controller is conveyed here for unstable and time-delayed plants. The controller parameters are selected by root locus technique with the non-integer orders graphically tuned to achieve satisfactory stability margins and sensitivity peaks. The advancement in the behavior of the system shown by availing the FO controller has been enumerated by numerical examples, stability of which is ratified by Riemann analysis. An adequate delay margin compensation for time-delayed plants is shown to accommodate the dead-time keeping hold of stability.

References

1. R.S. Barbosa, J.A.T. Machado, I.S. Jesus, On the fractional PID control of a laboratory servo system, in *Proceedings of the 17th World Congress the International Federation of Automatic Control (IFAC)* (Seoul, Korea, 2008), pp. 15273–15278
2. A. Tepljakov, E.A. Gonzalez, E. Petlenkov, J. Belikov, C.A. Monje, I. Petras, Incorporation of fractional-order dynamics into an existing PI/PID DC motor control loop. ISA Trans. Elsevier **60**, 262–273 (2016)
3. K. Bettou, A. Charef, Control quality enhancement using fractional $PI^\lambda D^\mu$ controller. Int. J. Syst. Sci. Taylor Francis **40**(8), 875–888 (2009)
4. P. Shah, S. Agashe, Review of fractional PID controller. Mechatron. Sci. Direct, Elsevier **38**, 29–41 (2016)
5. C. Yeroglu, N. Tan, Note on fractional-order proportional-integral-differential controller design. IET Control Theory Appl. **5**(17), 1978–1989 (2011)
6. C.A. Monje, Y.Q. Chen, B.M. Vinagre, D. Xue, V. Feliu, *Fractional Order Systems and Controls—Fundamentals and Applications, Advances in Industrial Control (AIC)* (Springer, 2010)
7. S. Pandey, P. Dwivedi, A.S. Junghare, A novel 2-DOF fractional order $PI^\lambda D^\mu$ controller with inherent anti-wind up capability for magnetic levitation system. Int. J. Electron. Commun. (AEU), Elsevier **79**, 158–171 (2017)
8. I. Petras, *Fractional-Order Nonlinear Systems-Modeling, Analysis and Simulation* (Non-linear Physical Science, Springer, 2010)
9. W.A. Wolovich, *Automatic Control Systems* (Saunders College Publishing, 1994)
10. J.C. Doyle, B.A. Francis, A.R. Tannenbaum, *Feedback Control Theory* (Macmillan, New York, 1993)

Applications and Approaches for Texture Analysis and Their Modern Evolution

Ayushman Ramola, Amit Kumar Shakya, and Anurag Vidyarthi

Abstract In this research work, we have presented a detailed study of various methods used for the texture analysis. These methods are classified as model-based methods and statistical-based methods. These methods are widely used in the field of remote sensing, document processing, automated inspection, and biomedical imaging applications. Texture plays a very prominent role in various applications related to image processing. These methods are discussed in brief in this research work.

Keywords Texture · Model-based methods · Statistical-based methods · Remote sensing · Biomedical imaging

1 Introduction

The texture is an important feature of the image surface [1]. The texture is the fundamental descriptor of an image and is also important for human visual perception [2]. The texture is the characteristic property of all surfaces like the pattern of crops in fields, weave of the fabric tree bark, etc., therefore it is an important factor for object recognition and classification. Human evaluates the texture qualitatively as fine, coarse, smooth, rippled, irregular, regular, etc., but quantitative measurement of texture is a difficult process. A quantitative assessment of texture is possible by structural and statistical methods. The paper aims to identify important model-based descriptors of texture that are suitable for texture analysis and have wide applications in the real world. Figure 1 shows different textured surfaces, illustrated in the Brodatz album [3].

The paper is coordinated in the following sections. Section 2 presents a brief review of important applications of texture analysis. Section 3 overviews the different

A. Ramola · A. K. Shakya (✉)
Department of Electronics and Communication Engineering, Sant Longowal Institute of Engineering and Technology, Sangrur, Punjab, India
e-mail: xlamitshakya.gate2014@ieee.org

A. Vidyarthi
Graphic Era University, Dehradun, Uttarakhand, India

© Springer Nature Singapore Pte Ltd. 2020
R. Bera et al. (eds.), *Advances in Communication, Devices and Networking*,
Lecture Notes in Electrical Engineering 662,
https://doi.org/10.1007/978-981-15-4932-8_30

273

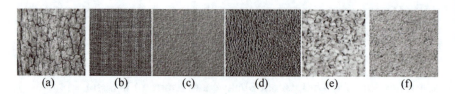

Fig. 1 Examples of common texture surfaces. **a** Bark. **b** Herringbone weave. **c** Raffia. **d** Pressed calf leather. **e** Plastic bubble. **f** Beach Sand (Courtesy of Brodatz texture photographic album)

approaches for the texture analysis and Sects. 4 and 5 give discussion with concluding remarks, respectively.

2 Applications of Texture Analysis

The texture is an active and intense research field until this date. Several areas like remote sensing, medical imaging, petrography, metallography, lumber processing, industries, etc., make extensive use of texture features. Texture is an important part of pattern recognition, human visual perception, computer vision, and digital image processing [4]. Due to discriminating property of texture it is used by computer vision for object classification and recognition. Major fields of applications of texture are remote sensing, biomedical imaging, automated inspection, document processing.

2.1 Remote Sensing

Remotely sensed images use texture to classify land use categories (agriculture, forest, water bodies, urban regions, etc.) [5–7]. The first attempt to classify different land uses by texture analysis method was done by Lenders and Stanley [8]. In [9] remotely sensed images were analyzed using texture features to obtain accurate classification results. Synthetic Aperture Radar (SAR) images are high-resolution imagery that provides finer details of texture information. SAR images shown in Fig. 2 categorize various land uses by observing the texture of the image. Leen-Kiat Soh states that texture features can be used successfully for analyzing and classifying SAR images of sea ice [10]. In [11] texture descriptions of the usefulness and use of various texture features to analyze SAR images are presented. The process of segmentation is performed to subdivide an image into meaningful regions that are used for further analysis. In [12] authors showed successfully texture segmentation of the SAR images in order to achieve the classification of water into different categories like newly formed ice, older ice, hard ice, soft ice, multi-year ice. Texture descriptors are suitable to obtain land use information from satellite images of high resolution

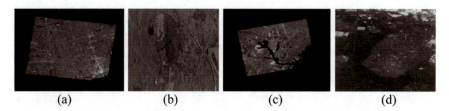

Fig. 2 Example of SAR images for texture analysis; **a** Canada, Toronto; **b** Russia, Monino; **c** India, Vishakhapatnam Port; **d** Strip map, Kuwait (Courtesy of Airbus Defence & Space)

[13]. Over the years texture analysis has become very important in the classification and analysis of remote sensing data.

2.2 Biomedical Imaging

The texture in medical imaging is mainly used for the classification of various medical conditions such as the classification and categorization of normal and abnormal tissues. Today texture classification of the medical image is playing an important role in the early detection of fatal diseases. Here we are discussing X-ray and ultrasound cases of patients. X-Ray image shown in Fig. 3a is diagnosed by texture analysis. Sutton and Hall [14] use of texture analysis methods in the diagnosis of pulmonary disease is related to human bodies. X-ray images of lungs infected by interstitial fibrosis show textural changes which help in the classification of normal and abnormal conditions, therefore texture features can be used for identifying healthy lungs from infected lungs. Textural features combined with color information were used in the diagnosis of the Leukemic malignancy in the sample of stained blood cells [15]. This combination of the texture and color features significantly shows improved results in comparison with the classification done using color features alone. Different types of cells can be classified by the use of texture analysis. Texture features can be used

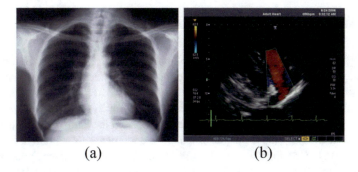

(a)   (b)

Fig. 3 **a** X-ray of lungs. **b** Ultrasound of the heart

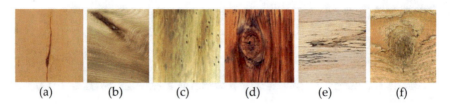

   (a)   (b)   (c)   (d)   (e)   (f)

Fig. 4 **a** Mineral rot. **b** Elm bark pocket. **c** Grub hole. **d** Knot. **e** Incipient rot. **f** Decayed knot source

to calculate tissue scattering parameters from ultrasound images [16]. Ultrasound images of livers can be used for diagnoses with the help of texture analysis [17]. Ultrasound images of the heart shown in Fig. 3b can also diagnose with the help of texture features [18].

2.3 Automated Inspection

Industries use texture analysis for automated inspection of different surfaces, the application includes defect detection in textile industries, a surface inspection of carpet-ware, wood defects, and automobile paints. In [19] a structural approach of texture analysis is used to identify defects in textured wood images. Figure 4 shows some common defects generated in wood. Corner et al. [20] used texture features, to analyze defects in the wood, lumber, such as note decay, mineral streak, insect bites on wood surface, etc. The texture is an important feature in quality control, as the appearance of the texture is the basis of inspection of the various surfaces of materials. In [21] authors proposed a texture analysis method for assessing the quality of carpet-ware. They found a statistical technique for texture analysis, which is capable of characterizing carpet-ware. The assessment of the quality of the painted metal surface was done by using texture analysis in [22].

2.4 Document Processing

The "paperless" office concept gives rise to the conversion of papers in electronic form so as to increase faster additions, searches, modifications, and the most important prolonged life of the records. Machine vision uses texture features for document processing in different ways. The texture is used to identify and retrieve documents, including newspapers, faxes, business letters, etc., from the large database [23]. The problem of identifying the script and language of documents is common in document processing. In [24] authors showed the effectiveness of texture to identify scripts containing multiple fonts and style. The characterization of the old document's images without any prior knowledge can be performed by texture features as described in

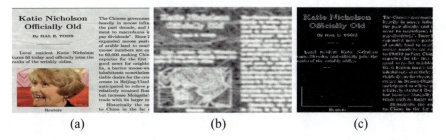

(a)   (b)   (c)

Fig. 5 **a** Scan copy of newspaper i.e., digital copy. **b** Texture segmentation is performed in Matlab 2013 to extract a region of interest. **c** Text region is identified

[25]. Texture segmentation methods can be used to isolate regions of interest in document images for pre-processing of images of documents [26]. Figure 5 shows the use of texture segmentation to identify the regions of interest. Hence the application of texture analysis in document processing is vast.

3   Different Approaches for Texture Analysis

Approaches to describe, analyze, and identify the texture are structural and statistical-based methods. Other approaches include model-based methods which are as Markov Random Field (MRF) model, fractal model, Autoregressive (AR) model, and synthetic model. They are briefly described as follows.

3.1   Structural Method

Structural approaches are based on texture primitives [1]. Primitive is a group of pixels with the same attributes. The structural textural analysis is a two-step process. In first step, one primitive is extracted and in second step placement rules are defined. The choice of the primitives and placement rules can be functions of the location, where chosen primitive have to be placed in neighboring primitive's location [2]. A better symbolic description of the images can be obtained by a structural approach, therefore making it more useful for texture synthesis than texture analysis. The use of complex primitives is one of the disadvantages of a structural approach limiting its application in texture analysis.

3.2 Model-Based Methods

Numbers of model-based methods are introduced to analyze texture [27]. In model-based methods, mathematical operations are used to model texture. Estimation of the coefficients for these models is a key problem. Model-based methods are used to synthesize texture.

3.2.1 Fractal Model

Mandelbrot [28] was the first one to purpose a fractal model. The fractal model is characterized by self-similarity, looking the same at any magnification. The self-similarity is quantized by its fractal dimension. The self-similarity concept to define fractal is presented in [2]. In a Euclidean space, a bounded set X is given. If X is a union of N number of copies and each copy is scaled down by a ratio of r, then set X is said to be similar. Fractal Dimension (FD) is defined as follows:

$$FD = \frac{\log N}{\log(\frac{1}{r})} \tag{1}$$

The FD estimates the roughness of the surface, larger the dimension rougher the texture will be. Various numbers of methods are proposed for the calculation of FD. Jennie and Harba [29] characterize the fractal dimension of signals by a fractional Brownian motion *(fBm)*.

$$FD = E + 1 - H \tag{2}$$

E is the Euclidean distance/dimension and (*E is* 2 for a curve and 3 for surface), $0 < H < 1$ is Hurst parameter.

3.2.2 Markov Random Field Model

Markov Random Field (MRF) is a probabilistic process. The MRF model assumes each pixel intensity is dependent on the intensity of its neighboring pixel [30]. The inter-relationship between two pixels is defined in MRF. Gibbs Random Field (GRF) is equivalent to the conditional probability density function of MRF, used by the MRF model. The Probability Mass Function (PMF) used along with MRF is calculated from GRF.

$$P(Z = z) = \frac{1}{Z}e^{-u(z)} \forall z \in \Omega \tag{3}$$

where $u(z)$ is an energy function, Z is normalization constant. The energy function is expressed in terms of the potential function $p_f(.)$ over cliques C_Q.

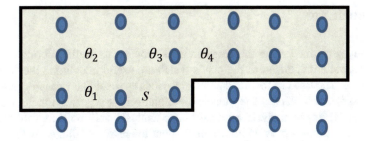

Fig. 6 Local neighborhood of the image I

$$u(z) = \sum_{f \in C_Q} P_f(z) \qquad (4)$$

There exists a unique Markov field for every Gibbs random field and vice versa [31]. The above theorem concludes that the total energy can be used to model texture globally.

3.2.3   Autoregressive (AR) Model

The AR model considers the local neighboring relationship between gray levels of the image. Gray values are weighted sum of the pixel on interest and the pixels surrounding the pixel of interest. AR model describes various generated shapes and features within an image by exploring the relationships between various pixels and their pixel groups. Figure 6 presents an example of a local neighborhood of image I.

4   Discussion

In this research work, we have investigated the various model-based methods based on texture analysis for various categories of images. The textured surface of DICOM images, remote sensing images, and document text is analyzed. The model-based approach provides useful information regarding pixel arrangement through which texture is analyzed significantly. Fractal Model, Markov Random Field Model, and Autoregressive (AR) model provide global information regarding the structural arrangement of image pixels.

5 Conclusion

The study performed under this research work provides us useful information about the structural arrangement of the pixels which are useful in defining the texture of the surface. The Brodatz texture database provides us spectral information about the various types of surfaces. The models are effectively used for Synthetic Aperture Radar (SAR) image analysis, medical image analysis, soil moisture retrieval purposes. These models provide information about the region of diagnosis thus play a prominent role in DICOM image analysis. Various kinds of defects developing in the wooden structure can be easily analyzed through these models. Finally, document image analysis is also performed with the assistance of these models.

References

1. R.M. Haralick, Statistical and structural approaches to texture. IEEE **67**(5) (1979)
2. M. Turceryan, A. Jain, *Handbook of Pattern Recognition and Computer Vision* (World Scientific Publishing, 1993), pp. 235–276
3. P. Brodatz, *Textures-Photographic Album for Artists and Designers* (Dover, New York, 1966)
4. A.K. Shakya, A. Ramola, A. Kandwal, P. Mittal, R. Prakash, Morphological change detection in terror camps of area 1 and 2 by pre- and post-strike through MOAB: a, in *Lecture Notes in Electrical Engineering* (Sikkim, Springer, Singapore, 2019), pp. 253–263
5. A.K. Shakya, A. Ramola, A. Kandwal, Estimating change percentage in texture developed by the water turndown of Bolivia's Lake Poopo, in *IEEE, International Conference on Automation and Computational Engineering (ICACE)* (Noida, 2018)
6. A.K. Shakya, R. Prakash, A. Ramola, D.C. Pandey, Change detection from pre and post urbanisation LANDSAT 5TM multispectral images, in *IEEE, International Conference on Innovations in Control, Communication and Information Systems (ICICCI)* (Nodia, 2017)
7. A.K. Shakya, A. Ramola, A. Kandwal, R. Prakash, Comparison of supervised classification techniques with ALOS PALSAR sensor for Roorkee region of Uttarakhand India. Int. Arch. Photogramm. Remote Sens. Spat. Inf. Sci. **XLII**(5), 693–701 (2018)
8. G.G. Lendaris, G. Stanley, Diffraction-pattern sampling for automatic pattern recognition. Proc. IEEE **58**(2), 198–216 (1978)
9. Haralick, Texture features for image classification. IEEE Trans. Syst. Men Cybern. (1973)
10. L.-k. Soh, Texture analysis of SAR sea ice imagery using gray level co occurence matrices. IEEE Trans. Geo Sci. Remote Sens. (1999)
11. E. Rignot, R. Kwokl, Extraction of textural features in SAR images: statistical model and sensitivity, in *Proceedings of International Geoscience and Remote Sensing Symposium*, pp. 1979–1982 (1990)
12. L.J. Du, Texture segmentation of SAR images using localized spatial filtering, in *Proceedings of International Geoscience and Remote Sensing Symposium*, pp. 1983–1986 (1990)
13. A.K. Shakya, A. Ramola, A. Kandwal, P. Mittal, R. Prakash, Morphological change detection in terror camps of area 3 and 4 by pre- and post-strike through MOAB: b, in *Lecture Notes in Electrical Engineering* (Sikkim, Springer, Singapore, 2019), pp. 265–275
14. R. Sutton, E.L. Hall, Texture measures for automatic classification of pulmonary disease. IEEE Trans. Comput. **c-21**, pp. 667–676 (1972)
15. Harms, U.H. Gunzer, H.M. Aus, Combined local color and texture analysis of stained cells. Comput. Vis. Graph. Image Process. **33**, 364–376 (1986)
16. M.F. Insana, R.F. Wagner, B.S. Garra, D.G. Brown, T.H. Shawker, Analysis of ultrasound image texture via generalized Rician statistics. Opt. Eng. **25**, 743–748 (1986)

17. C.C. Chen, J.S. DaPonte, M.D. Fox, Fractal feature analysis and classification in medical imaging. IEEE Trans. Med. Imaging **8**, 133–142 (1989)
18. A. Lundervold, Ultrasonic tissue characterization—a pattern recognition approach (Oslo, Norway, 1992)
19. J. Chen, A.K. Jain, A structural approach to identify defects in textured images, in *Proceedings of IEEE International Conference on Systems, Man and Cybernetics* (Beijing, 1988)
20. R.W. Conners, C.W. McMillin, K. Lin, R.E. Vasquez-Espinosa, Identifying and locating surface defects in wood: part of an automated lumber processing system, IEEE Trans. Pattern Anal. Mach. Intell. **PAMI-5**, pp. 573–583 (1983)
21. L.H. Siew, R.M. Hodgson, E.J. Wood, Texture measures for carpet wear assessment. IEEE Trans. Pattern Anal. Mach. Intell. **PAMI-10**, 92–105 (1988)
22. A.K. Jain, F. Farrokhnia, D.H. Alman, Texture analysis of automotive finishes, in *Proceedings of SME Machine Vision Applications Conference* (Detroit, MI, 1990)
23. J.F. Cullen, J.J. Hull, P.E. Hart, Document image database retrieval and browsing using texture analysis. IEEE (1997)
24. A. Busch, W.W. Boles, S. Sridharan, Texture for script identification. IEEE Trans. Pattern Anal. Mach. Intell. **27**(11), 1720–1732 (2005)
25. N. Journet, J.Y. Ramel, R. Mullot, V. Eglin, Document image characterization using a multiresolution analysis of the texture: application to old documents (Springer, 2008)
26. A.K. Jain, S.K. Bhattacharjee, Address block location on envelopes using gabor filters, in *Proceedings of 11th International Conference on Pattern Recognition* (1992)
27. I. Majumdar, B. Chatterji, Texture feature matching methods for content based image retrieval. Tech. Rev. **24**(4), 257–269 (2007)
28. B.B. Mandelbrot, *The Fractal Geometry of Nature* (Freeman, 1983)
29. R. Jennane, R. Harba, Fractional Brownian motion: a model for image texture. Signal Process. VII: Theor. Appl. (1994)
30. A. Zisserman, A. Blake, *Visual Reconstruction* (The MIT Press, 1987)
31. J. Besag, Spatial interaction and the statistical analysis of lattice systems. R Stat. Soc. **B-36**, 344–348 (1974)

Landcover Pattern Recognition Through Texture Classification Using LANDSAT Data of Dallas

Amit Kumar Shakya, Ayushman Ramola, and Anurag Vidyarthi

Abstract In this investigation we have taken pre- and post-multispectral images of Dallas before and after the urbanization. Both the images are classified through a Grey Level Co-occurrence Matrix (GLCM) texture classification technique and on the basis of this, we have identified a novel pattern in the changing texture of the landcover. This similarity in the changing texture suggests that this can be assumed as a standard pattern for the case of urbanization.

Keywords Multispectral · Texture · GLCM · Pattern · Landcover

1 Introduction

The texture is a very important property to the point of human visual perception and computer vision [1]. The texture of the satellite images plays a dominant role in the field of remote sensing and visual image interpretation. Through texture images can be classified in many categories, i.e., fine, coarse, regular, rough, smooth, etc., every category of the texture have its own property and they can be interpreted according to the need of the researcher [2]. Urban LC classification is a current area of investigation [3, 4]. Yan [5] used LiDAR data of Canadian Geobase to monitor the changes that got developed in urban areas of the region. Homer [6] used time-series Landsat data to obtain the changes in the LC that got developed in 2011 with the assistance of the National LC database of the United States. Mohan and Kandya [7] monitored the changes that got developed due to urbanization in tropical India using remote sensing data. Poursanidis et al. [8] used and compared the data acquired from Landsat 5 TM and Landsat 8 Operational Land Imager (OLI) for the urban and peri-urban mapping of East Attica Greece. Remote monitoring of the LC is quite a useful technique

A. K. Shakya (✉) · A. Ramola
Department of Electronics and Communication Engineering, Sant Longowal Institute of Engineering and Technology, Sangrur, Punjab, India
e-mail: xlamitshakya.gate2014@ieee.org

A. Vidyarthi
Graphic Era University, Dehradun, Uttarakhand, India

© Springer Nature Singapore Pte Ltd. 2020
R. Bera et al. (eds.), *Advances in Communication, Devices and Networking*,
Lecture Notes in Electrical Engineering 662,
https://doi.org/10.1007/978-981-15-4932-8_31

because through this we can gather information about various types of statistical changes in limited time. United States Geological Survey (USGS) and the National Aeronautics and Space Administration (NASA) have successfully developed the Landsat Program [9]. The changes developed in the LC can be estimated using different available techniques. Change Detection (CD) techniques are classified into two types (a) pre-classification and (b) post-classification [10]. In pre-classification CD technique message about the changes is in yes or no only, whereas in the post-classification we get complete information about these changes that got established in the LC [10]. Pre-classification procedures which provides only binary information about the changes are image differencing [11–13], image ratio [14], image regression [15], Change Vector Analysis (CVA) [16], Principal Component Analysis (PCA) [17], Kauth-Thomas Transformation (KCT) [18]. Post-classification CD techniques include machine learning, Geographical Information System (GIS) based, composite classification or direct multi-date classification, texture analysis-based CD [19], deep learning CD approach. The Landsat satellite program was initially started to explore Mother Earth from outer space [20]. The only objective of this mission was observation and imaging of the Earth but as the program becomes successful the idea of extending the range of the satellite's imaging also came into the picture and now beside Earth-imaging the satellite began to provide information regarding weather forecast, agriculture sector, deep Earth resources, water resources, etc. Since then USGS and NASA are continuously working to gather vital information on the Earth surface [20]. Till today NASA and USGS have successfully completed seven Landsat missions, Landsat 6 was an unsuccessful mission as it was unable to reach its orbit [21]. NASA is planning to launch its latest Landsat 9 by 2020 [22]. In this research work, we have used Landsat 5 dataset. This satellite was also awarded Guinness World Record for the longest-operating Earth observation satellite [23]. Landsat 5 was designed for the work period of just 3 years, but due to the loss of Landsat 6 additional work was assigned to Landsat 5. It works effectively in the outer space for the period of 25 years and was decommissioned officially on June 5, 2013 [24]. TM sensor installed on the Landsat 5 contains seven bands. The bands were effectively operating over a wide range of wavelengths. The screen size captured by the Landsat 5 was of dimension 170 km $\times$ 185 km (106 mi $\times$ 115 mi).

a

i	1	2	3	4	5
1	5	4	2	5	5
2	2	2	4	4	2
3	1	1	5	1	4
4	4	3	3	2	1
5	2	1	2	3	3

b

i	1	2	3	4	5
1	1	1	0	1	0
2	2	1	1	1	1
3	0	1	2	0	0
4	0	2	1	1	0
5	1	0	0	1	1

Fig. 1 **a** Test image having dimension 5 × 5. **b** GLCM image of dimension 5 × 5

a

i	1	2	3	4	5
1	1/25	1/25	0	1/25	0
2	2/25	1/25	1/25	1/25	1/25
3	0	1/25	2/25	0/25	0/25
4	0/25	2/25	1/25	1/25	0
5	1/25	0	0	1/25	1/25

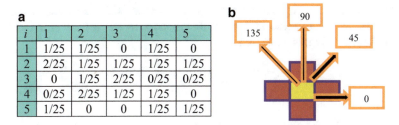

Fig. 2 **a** Normalized GLCM image of dimension 5 × 5. **b** Angular orientation representation of the pixel of interest

2 Procedure to Create GLCM from an Imput Image

The GLCM was developed by Haralick et al. [25]. He proposed "14" different TF as "angular second moment, contrast, correlation, energy, homogeneity, entropy, variance, inverse difference moment, sum average, sum entropy, sum variance, difference variance, difference entropy, information measures of correlation". Gotlieb and Kreyszig [26] categorized the TF in four different categories based on correlation measures, entropy measures, statistical measures, and remote sensing texture visual measures. Texture visual measures include contrast, correlation, energy, and homogeneity. This suggests that surfaces that are visually smooth and coarse mostly have some sought pattern. Let us assume an input image "i" having a dimension of 5 × 5 (Fig. 1).

The GLCM image of the input image is obtained following the principle "one pixel toward the right", i.e., in the input image DN 2 is followed by 1 in two occasions, therefore in the GLCM image 2 by 1 is 2, similarly the frequency of occurrence of 3 followed by 3 in the input image is 2, therefore in the GLCM image 3 followed by 3 is 2, and so on [27]. Now the normalization of the GLCM values is done, and a new matrix is created. It is the ratio of DN of the GLCM matrix to that of the total number of pixels in the image.

GLCM of the image is calculated along four directions as shown in Fig. 2, i.e., 0°, 45°, 90°, and 135°. These directions are East, North East, North, and northwest, respectively. Texture visual features obtained from GLCM are contrast, correlation, energy, and homogeneity, their mathematical notation along with their normalized range are listed as follows.

a. **Contrast**: It is the discrepancy developed between the highest (most) and the lowest (least) values of a contiguous set of pixels, the range of variation is (*range of contast is* > 0). Mathematically it is expressed as

$$\sum_{i,j} \left|i - j^2\right| \times p(i, j). \tag{1}$$

b. **Correlation**: It is the outcome in the grey value measurement for the first element *I* of the displacement vector, whereas *j* is associated with the second element of the displacement vector. The range of correlation is $-1 \leq (correlation) \geq 1$. Mathematically it is expressed as

$$\sum_{i,j} \frac{\left|i - \mu \times i\right|\left|j - \mu \times j\right| p(i, j)}{\sigma_i \sigma_j}. \tag{2}$$

Here, μ is the mean of the image pixels, σ_i and σ_j represents the standard deviation of the ith and jth pixels, respectively.

c. **Energy**: It is the measure of the pixel pair's repetition when pixel under consideration is homogeneous under similar grey levels,i.e., pixel has attained the same grey levels. The range of variation is ($0 < energy < 1$). Mathematically it is expressed as

$$\sum_{i,j} \left|p(i, j)\right|^2. \tag{3}$$

d. **Homogeneity**: Assumes large values for the same grey tone differences in the paired element. The range of variation of homogeneity is ($0 < homogeniety < 1$). Mathematically it is expressed as

$$\sum_{i,j} \frac{p(i, j)}{1 + \left|i - j\right|}. \tag{4}$$

3 Methodology and Pattern Identification

Here Brodatz images are used as source images. This dataset contains combinations of several textural surfaces like Grass, Bark, Straw, Herringbone weave, Beach sand, Water, Wood Grain, Raffia, and so on. Now we evaluate TF of bark and sand which are rough surface and smooth surfaces (Fig. 3).

Fig. 3 a Bark (Brodatz No: 1.2.02). **b** Sand (Brodatz No: 1.4.12)

(a)   (b)

Here we have calculated the TF for both the images and on the basis of the observations we have observed a systematic pattern in the change of TF (Table 1).

Now the average values of the texture of Bark and Sand are compared and thus we have obtained a arrangement in the changing behavior of texture features (Table 2).

From the above conclusions, we have identified a arrangement of the changing texture from "coarse" to "smooth" and vice versa. Now we are applying this technique on the pre- and post-Landsat images of Dallas which have grown at a very rapid rate in the time period of the 25 years. The pre- and post-images are captured by Landsat 5 with the assistance of TM sensor (Fig. 4).

We have obtained the grey level data for the pre-urbanization and post-urbanization through GLCM. These images represented the spread of urbanization in Dallas.

Now the TF of the pre-urbanization of Dallas is calculated, all the four features are computed for the various distances varying from d = 1 to d = 4, later the average of all the four features will be computed and make the GLCM direction independent (Table 3).

Now to compute the difference developed in the LC we have used absolute image differencing technique, through which we have subtracted pre- and post-urbanization images and obtained the developed urban settlements which are can be visually identified also. Here from the image derived from the absolute image differencing, shows the development of the new urban colonies in Dallas shown in Fig. 5). This certainly concludes that the change in the visual TF has occurred. Now compare the average values of the TF of pre- and post-urbanization images of Dallas. The bar plot is shown in the Fig. 6 suggests that the TF contrast, have developed an 18.93 % positive change, whereas TF energy and homogeneity have developed 30.40 and 4.72 % negative change.

Table 1 GLCM classification of (Brodatz No: 1.2.02) and SAND (Brodatz No: 1.4.12)

S.No	D*	T.F	BARK (Brodatz No: 1.2.02)				SAND (Brodatz No: 1.4.12)			
			0 D	45 D	90 D	135 D	0 D	45 D	90 D	135 D
	d = 1	Con*	1.8428	2.5187	1.4989	2.8483	0.4742	0.6729	0.4607	0.6775
1		Corr*	0.8258	0.7616	0.8581	0.7304	0.9047	0.8643	0.9071	0.8633
2		E*	0.0353	0.0306	0.0394	0.0288	0.1160	0.1010	0.1184	0.1001
3		H*	0.6471	0.6103	0.6803	0.5921	0.8069	0.7642	0.8123	0.7611
4	d = 2	Con*	4.1313	4.9918	3.3395	5.5423	0.9601	1.2171	0.8949	1.2318
5		Corr*	0.6095	0.5272	0.6836	0.4751	0.8071	0.7535	0.8189	0.7506
6		E*	0.0238	0.0213	0.0272	0.0201	0.0874	0.0809	0.0911	0.0803
7		H*	0.5424	0.5166	0.5798	0.4975	0.7148	0.6823	0.7295	0.6792
8	d = 3	Con*	5.8808	6.6061	4.5477	7.1776	0.12677	1.4494	1.2340	1.4565
9		Corr*	0.4444	0.3746	0.5691	0.3205	0.7452	0.7054	0.7494	0.7040
10		E*	0.0196	0.0183	0.0229	0.0175	0.0792	0.0767	0.0806	0.0765
11		H*	0.4910	0.4730	0.5367	0.4567	0.6736	0.6567	0.6810	0.6553
12	d = 4	Con*	7.1945	7.7237	5.3704	8.2399	1.3864	1.5179	1.4292	1.5208
13		Corr*	0.3206	0.2692	0.4912	0.2204	0.7213	0.6903	0.7087	0.6898
14		E*	0.0176	0.0169	0.0208	0.0165	0.0770	0.0760	0.0770	0.0759
15		H*	0.4597	0.4474	0.5121	0.4335	0.6594	0.6515	0.6599	0.6504

Con* = Contrast, Corr* = Correlation, E* = Energy, H* = Homogeneity, D = Degree, D* = Distance

Table 2 Pattern identification from texture feature "SAND" and "BARK"

S.No	Texture features	Coarse (Bark)	Smooth (Sand)	Coarse to smooth	Smooth to coarse
1.	Contrast	4.9658	1.1156	▬	✚
2.	Energy	0.0235	0.0871	✚	▬
3.	Homogeneity	0.5297	0.7023	✚	▬

Fig. 4 **a** Pre-urbanization (August 31, 1984). **b** Post-urbanization (August 4, 2009)

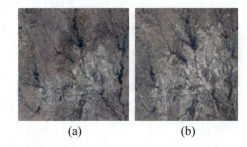

(a)   (b)

4   Conclusion

The proposed technique of texture classification through GLCM is based on extensive simulation of the texture features. The pattern obtained in the texture visual parameters suggests that due to large time gap and rapid urbanization the changing texture of the surface will follow the same pattern.

Table 3 Texture features of pre-urbanization image and post-urbanization image

S.No	D*	T.F	Pre-urbanization image				Post-urbanization image			
			0 D	45 D	90 D	135 D	0 D	45 D	90 D	135 D
1	d = 1	Con*	0.3394	0.5088	0.3510	0.5036	0.4400	0.6968	0.4731	0.6823
2		Corr*	0.8694	0.8043	0.8650	0.8063	0.8719	0.7971	0.8623	0.8014
3		E*	0.1474	0.1258	0.1470	0.1364	0.1151	0.0958	0.1135	0.0966
4		H*	0.8559	0.8124	0.8546	0.8138	0.8279	0.7758	0.8229	0.7781
5	d = 2	Con*	0.6573	0.8705	0.6533	0.8669	0.8912	1.1990	0.9005	1.1900
6		Corr*	0.7471	0.6652	0.7487	0.6665	0.7405	0.6510	0.7379	0.6536
7		E*	0.1148	0.1013	0.1161	0.1016	0.0879	0.0770	0.0886	0.0773
8		H*	0.7869	0.7505	0.7899	0.7515	0.7525	0.7128	0.7539	0.7138
9	d = 3	Con*	0.8597	1.0728	0.8435	1.0671	1.1606	1.4565	1.1485	1.4485
10		Corr*	0.6692	0.5874	0.6756	0.5896	0.6621	0.5761	0.5785	0.5785
11		E*	0.1024	0.0919	0.1042	0.0923	0.0786	0.0704	0.0707	0.0707
12		H*	0.7539	0.7210	0.7589	0.7222	0.7202	0.6858	0.6868	0.6868
13	d = 4	Con*	1.0023	1.2184	0.9845	1.2126	1.3483	1.6460	1.6375	1.6375
14		Corr*	0.6144	0.5315	0.6214	0.5337	0.6075	0.5211	0.5235	0.5235
15		E*	0.0953	0.0864	0.0967	0.0867	0.0733	0.0663	0.0665	0.0665
16		H*	0.7323	0.7007	0.7367	0.7019	0.6991	0.6664	0.6673	0.6673

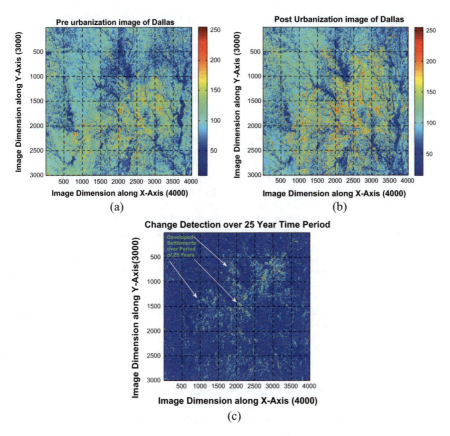

Fig. 5 **a** False color pre-urbanization image. **b** False color post-urbanization image. **c** Difference image through image differencing

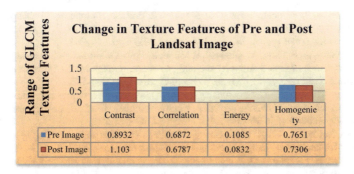

Fig. 6 Arrangement obtained in the texture features by comparing pre- and post-landsat data

References

1. J. Yuan, D.L. Wang, R. Li, Remote sensing image segmentation by combining spectral and texture features. Trans. Geosci. Remote Sens. **52**(1), 16–24 (2014)
2. B.M. Aïssoun, S.-D. Hwang, K.H. Khayat, Influence of aggregate characteristics on the workability of super workable concrete. Mater. Struct. (2015)
3. A.K. Shakya, A. Ramola, A. Kandwal, P. Mittal, R. Prakash, Morphological change detection in terror camps of area 1 and 2 by pre- and post-strike through MOAB: a, in *Lecture Notes in Electrical Engineering* (Springer, Singapore, Sikkim, India, 2019), pp. 253–263
4. A.K. Shakya, A. Ramola, A. Kandwal, P. Mittal, R. Prakash, Morphological change detection in terror camps of area 3 and 4 by pre- and post-strike through MOAB: b, in *Lecture Notes in Electrical Engineering* (Springer, Singapore, Sikkim, India, 2019), pp. 265–275
5. W.Y. Yan, A. Shaker, N. El-Ashmawy, Urban land cover classification using airborne LiDAR data: a review. Remote Sens. Environ. 3–35 (2014)
6. B.C. Homer et al., Completion of the 2011 national land cover database for the conterminous United States—representing a decade of land cover change information. Photogram. Eng. Remote Sens. 346–354 (2015)
7. M. Mohan, A. Kandya, Impact of urbanization and land-use/land-cover change on diurnal temperature range: a case study of tropical urban airshed of India using remote sensing data. Sci. Total Environ. **506–507**, 453–465 (2015)
8. D. Poursanidis, N. Chrysoulakis, Z. Mitrakaa, Landsat 8 vs. Landsat 5: a comparison based on urban and peri-urban land cover mapping. Int. J. Appl. Earth Obs. Geoinf. **35**, 259–269 (2015)
9. U.S. Department of the Interior and U.S. Geological Survey. (2018, February) USGSs [Online]. https://landsat.usgs.gov/
10. M. Hussain, D. Chen, A. Cheng, H. Wei, D. Stanley, Change detection from remotely sensed images: from pixel-based to object-based approaches. ISPRS J. Photogram. Remote Sens. **80**, 91–106 (2013)
11. A.K. Shakya, R. Prakash, A. Ramola, D.C. Pandey, Change detection from pre and post urbanisation LANDSAT 5TM multispectral images, in *IEEE, International Conference on Innovations in Control, Communication and Information Systems (ICICCI)* (Noida, 2017), pp. 1–6
12. A.K. Shakya, A. Ramola, A. Kandwal, Estimating change percentage in texture developed by the water turndown of Bolivia's Lake Poopo, in *IEEE, International Conference on Automation and Computational Engineering (ICACE)* (Noida, 2018), pp. 133–138
13. A.K. Shakya, A. Ramola, A. Kandwal, P. Mittal, R. Prakash, Morphological change detection in terror camps of area 1 and 2 by Pre- and post-strike through MOAB: a, *Lecture Notes in Electrical Engineering*, vol. 537, no. 1 (February 2019), pp. 253–263
14. A.K. Shakya, A. Ramola, A. Kandwal, P. Mittal, R. Prakash, Morphological change detection in terror camps of area 3 and 4 by pre-and post-strike through MOAB: b, *Lecture Notes in Electrical Engineering*, vol. 537, no. 1 (February 2019), pp. 265–275
15. J. Aguirre-Gutiérrez, A.C. Seijmonsbergen, J.F. Duivenvoorden, Optimizing land cover classification accuracy for change detection, a combined pixel-based and object-based approach in a mountainous area in Mexico. Appl. Geogr. **34**, 29–37 (2012)
16. J. Chen, X. Chen, X. Cui, J. Chen, Change vector analysis in posterior probability space: a new method for land cover change detection. Geosci. Remote Sens. Lett. **8**(2), 319–321 (2011)
17. A. Guisan, B. Petitpierre, O. Broennimann, C. Daehler, Unifying niche shift studies: insights from biological invasions. Trends Ecol. Evol. **29**(5), 260–268 (2014)
18. Rodríguez-Galiano, Abarca-Hernández, Ghimire, and Chica-Olmo, Incorporating spatial variability measures in land-cover classification using random forest. Procedia Environ. Sci. **3**, 44–49 (2011)
19. A.K. Shakya, A. Ramola, A. Kandwal, Estimating change percentage in texture developed by the water turndown of Bolivia's Lake Poopo, in *IEEE, International Conference on Automation and Computational Engineering (ICACE)* (Nodia, 2018), pp. 133–138

20. U.S. Department of the Interior and U.S. Geological Survey. (2018, February) USGS Science of changing the World [Online]. https://landsat.usgs.gov
21. L. Rocchio, J.R. Irons, M.P. Taylor, Landsat Science [Online] (2018, March). https://landsat.gsfc.nasa.gov/landsat-6/
22. L. Rocchio, J.R. Irons, M.P. Taylor, Landsat 9 [Online] (2018, March). https://landsat.gsfc.nasa.gov/landsat-9/
23. L. Betz, Landsat 5 Sets Guinness World Record For 'Longest Operating Earth Observation Satellite' [Online] (2013, February). https://www.nasa.gov/mission_pages/landsat/news/landsat5-guinness.htm
24. U.S. Department of the Interior and U.S. Geological Survey, (2018, February) USGS Science for a Changing World [Online]. https://landsat.usgs.gov
25. R.M. Haralick, K. Shanmugam, I. Dinstein, Texture features for image classification. Trans. Syst. Man Cybern. SMC(3), 610–621 (1973)
26. C.C. Gotlieb, H.E. Kreyszig, Texture descriptors based on co-occurrence matrices. Comput. Vis. Graph. Image Process. **51**(1), 70–86 (1990)
27. A. Ramola, A.K. Shakya, D. Van Pham, Study of statistical methods for texture analysis and their modern evolutions. Eng. Reports 2, e12149 (2020). https://doi.org/10.1002/eng2.12149

Computational Design of Multilayer High-Speed MTJ MRAM by Using Quantum-Cellular-Automata Technique

Rupsa Roy and Swarup Sarkar

Abstract An "MTJ MRAM" is a non-volatile memory design which is more acceptable memory design to reduce the power leakage than other random-access memories. The bit-wise word line operation of this memory-circuit can be represented through "CMOS" technique. But, to get better scalability word line delay and power-leakage reduction the previous technique should be replaced by using 3-D quantum-cell technology word line reversibility. This paper presents a multilayer structure of a "MTJ MRAM" by using "Fredkin" gate and reversible "NAND". This design is compared with "45 nm CMOS technology" and also the synthesized-outcome of "Xilinx-Software" using "VHDL" code. The synthesized schematic-figures of the proposed RAM circuit are also presented here.

Keywords Fredkin Gate · Multiple-layers · MTJ MRAM · Quantum-cell · RTL-Schematic

1 Introduction

Magnetoresistive RAM has the superiority of non-volatility [1]. It can work by "Electron-Tunneling-Phenomena" where electrons cross a thin-layer of insulator through a tunnel because of "Tunneling-Magneto-Resistance". So, this structure is known as "Magnetic-Tunnel-Junction" or MTJ [2] which is introduced here in Fig. 1. A bit-word line operation of "MTJ" using "reversible Fredkin Gate" is formed in this paperwork by "QCA" which is used to reduce the delay by Nano-technique and its superiority to give better scalability with high-frequency-range is a reason of "CMOS" technology replacement [3–5]. This paper also shows the basic quantum-cell diagrams word line logic "0" and "1" in Fig. 2a, b, and c, respectively.

R. Roy (✉)
Jakir Hossain Institute of Polytechnic (JHIP), Murshidabad, India
e-mail: onerupsa@gmail.com

S. Sarkar (✉)
Sikkim Manipal University, Gangtok, Sikkim, India
e-mail: swarup.s@smit.smu.edu.in

© Springer Nature Singapore Pte Ltd. 2020
R. Bera et al. (eds.), *Advances in Communication, Devices and Networking*,
Lecture Notes in Electrical Engineering 662,
https://doi.org/10.1007/978-981-15-4932-8_32

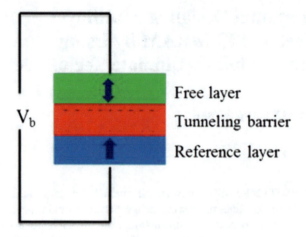

Fig. 1 Figure of "MTJ MRAM"

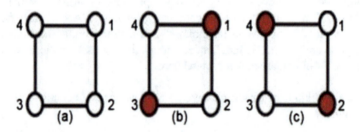

Fig. 2 **a** Figure of basic QCA-cell. **b** Figure of logic "1" **c** Figure of logic "0" [5]

Fig. 3 Representation of clock-zones [6]

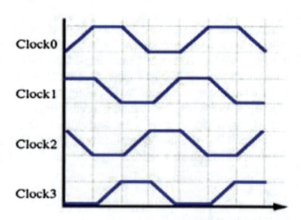

In "QCA" there are four different clock-schemes such as clock-zone 1, 2, 3, and 4, which are known as switch, hold, release, and relax [6–8]. The four phases of clock-scheme are presented in Fig. 3. In this work, we choose multilayer design to reduce the formation-size and also to reduce the required clocking-zone in "QCA".

2 Proposed Design

This research-work selects "Fredkin" reversible-gate to form the proposed memory formation to ignore the "KTln2" energy-loss/single-bit (same as irreversible-logic) and to save the information [9]. This paper presents the "Fredkin" gate by making the 3rd input "0" and ignoring the garbage-value in Fig. 4 and also presents the truth-table in Table 1. In this design the 3rd input is made "0" to use it properly in the "MTJ MRAM" functions and this figure is formed without any garbage output which reduces the size of the formation. Here the "AND" value of the two input is changed and it presents "OR" operation between input "A" and complement of input "B" for the structure-size reduction.

This paper shows a "MTJ MRAM" structure using "CMOS" technology which was presented in paper [10] (in Fig. 5). This paper also presents the bit-word lines operative-phenomena of this magnetic RAM. This is a "45 nm" technology which is used to design "MRAM" functions by "CMOS" configuration.

In our work we take the pulse as a select-line or Sl. If word line is one, bit-line is one and select-line is zero then it stores "1" and if the Wl is one, Bl is zero, and Sl is one or zero it presents "0". This phenomenon is presented here in Table 2 and we form this design in "QCA" presented in Fig. 6.

In the above figure, the bit-wise functionality of the MTJ magnetoresistive RAM is presented by a quantum-cell formation. It uses reversible "Fredkin" gate with reversible "AND" gate to reduce the information-loss of the memory. This is a multilayer or "3-D" design which uses only two clocking-zones (clock-zone "0" and

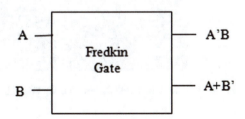

Fig. 4 Representation of "Fredkin" reversible-gate

Table 1 Truth-table of presented "Fredkin" gate	A as input	B as input	A′B as output	A + B′ as output
	0	0	0	1
	0	1	1	0
	1	0	0	1
	1	1	0	1

Fig. 5 Representation of "Fredkin" reversible-gate using "CMOS" technology [10]

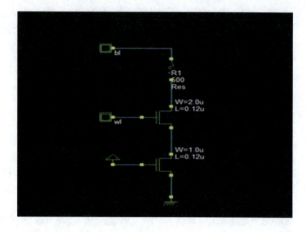

Table 2 Truth-table of presented "MTJ MRAM"

Wl (word line)	Sl (select-line)	Bl (bit-line)	Output
1	0	0	0
1	0	1	1
1	1	0	0
1	1	1	0

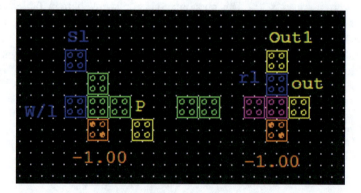

Fig. 6 3-D "MTJ MRAM" formation using "Fredkin" reversible-gate and reversible "AND" gate

"1") and the delay of the suggested formation is also reduced by applying multilayer design (three layers). The first layer shows the proposed "Fredkin" gate with two outputs and without any area-loss for garbage output. The 2nd layer is the transmission line between two reversible gates and the 3rd layer is used to present the reversible "AND" structure where "out" is the representation of the "MTJ MRAM" function and "out 1" is the representation of bit-line. Here the word line is presented as "W/l", select-line is presented as "S1" and bit-line is presented as "rl". The 2nd input of the presented "Fredkin" gate is always "1" in our work-done because this input presents the word line.

3 Results of Simulated "MTJ"

This structure reduces the power up to 37 % and delay up to 75 % from the previous technique and the result of the suggested magnetic memory formation is attached here in Fig. 7. This proposed structure is also synthesized in "Xilinx" and the outputs are compared with the "QCA" output and enlisted here through Table 3.

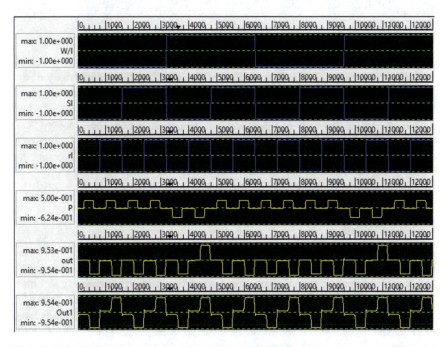

Fig. 7 Simulated outcomes of presented multi-layer "MTJ MRAM" formation using "Fredkin" reversible-gate and reversible "AND" gate

Table 3 Comparison-table between "QCA" and "Xilinx" outputs

Designs	Area	Delay	Leakage power
"Xilinx"	6 IOBs	6.236 ns	203 mW
"QCA Designer"	0.007 μm2	0.5 clock-cycles (Frequency in THz)	0.008 μW for 0.7 Ek

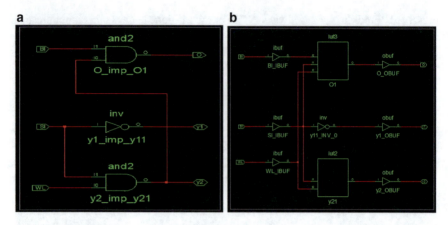

Fig. 8 **a** "RTL" schematic of "MTJ MRAM". **b** "Technology" schematic of "MTJ MRAM"

This above design in Fig. 7 presents the output of the proposed memory design ("out") which only shows logic "1" when word line and bit-line are "1" but the select-line is zero and it also represents the output of "Fredkin" gate by output "P" which shows the "AND" value of word line and complement of select-line. Here the "out" value presents the bit-line value. Figure 8 presents the "RTL" and "Technology" schematic of the proposed magnetic memory structure. The "logic-diagram" and k-map of the used circuits can be shown from this simulation by "Xilinx".

4 Conclusion

The advantages of quantum-cell-automata as a proper and effective Nano-scale design is higher compared to the other technologies (which are discussed here) is again proved by our proposed "MTJ MRAM" design and also the "MRAM" is an advance non-volatile memory-circuit. But, there is a major challenge is faced by "MRAM" which is the reduction of "read to write current-ratio" due to high-scalability. In future, this problem can be solved by using the phenomena of quantum-dot-cellular automata. In future, the proposed design of the "MTJ" magnetoresistive random-access memory can be fabricated by using "FPGA" board for real-world application.

References

1. L. Liu, C.F. Pai, Y. Li, Spin-torque switching with the giant spin hall effect of tantalum. Science **336**, 555–558 (2012)

2. L. Liu, O.J. Lee et al., Current induced switching of perpendicularly magnetized magnetic layers using spin torque from the spin hall effect. APS Phys. Rev. Lett. **109** (2012)
3. M. Chakraverty et al., First principle simulations of various magnetic tunnel junctions for applications in magnetoresistive random access memories. IEEE Trans. Nanotechnol. **12**(6), 971–977 (2013)
4. D. Abedi, G. Jaberipur, M. Sangsefidi, Coplanar full adder quantum-dot cellular automata via clock-zone based crossover. IEEE Trans. Nanotechnol. **99** (2015)
5. G. Sing, B. Line Raj, R.K. Sarin, Design and performance analysis of a new efficient coplanar quantum-dot cellular automata adder. Indian J. Pure Appl. Phys. **55**, 97–103 (2017)
6. J. Chaharlang, M. Mosleh, An overview on RAM memories in QCA technology. MJEE **11**(2) (2017)
7. T. Endoh, H. Hanzo, A recent progress of spintronics devices for integrated circuit applications. JLPEA **8**, 44 (2018)
8. P.R. Yelekar, S.S. Chiwande, Introductions to reversible logic gates and its application. NCICT (2011)
9. A. Purna Ramesh, Implementation of low power carry skip adder using reversible logic. IJRTE **8**(3). ISSN. 2277-3878 (2019)
10. S. Hamsa, N. Thangadurai, A.G. Ananth, Magnetic tunnel junction design in magnetoresistive random-access memory. IJEAT **8**(5). ISSN. 2249-8958 (2019)

Discrete Decoupling and Control of Time-Delayed MIMO Systems

Sumit Kumar Pandey, Jayati Dey, and Subrata Banerjee

Abstract In this paper, a method of discrete decoupling and control of time-delayed MIMO systems s proposed. As to control the time-delayed MIMO plants in continuous domain, there is a requirement of the Padé approximation to rationalize the time-delayed MIMO plants. The Padé approximation technique may degrade the time domain performance of the MIMO systems. To avoid the Padé approximation, discrete-time approach is applied to decouple and control the time-delayed MIMO systems in which there will be no approximation required. Discrete-time PID controller is designed using bacterial foregoing optimization (BFO) technique. The proposed method is implemented to both square and non-square plants.

Keywords Time-delay systems · Decoupling · MIMO · Discrete · PID

1 Introduction

The decoupling control problem for discrete-time MIMO system is studied in this work. The motivation towards this work is because the decoupled output of the system can be controlled independently which will increase the efficacy of the results. To this end, some work is found in the literature based on state feedback method of decoupling for the discrete MIMO system which is difficult since the states are not always available for the measurement [1, 2]. Therefore, in the present work, the transfer matrix approach is employed for the sake of decoupling which is comparatively easier than the reported method.

S. K. Pandey (✉) · J. Dey · S. Banerjee
Department of Electrical Engineering, National Institute of Technology (NIT), Durgapur, West Bengal, India
e-mail: skpdmk@gmail.com

J. Dey
e-mail: deybiswasjayati@gmail.com

S. Banerjee
e-mail: bansub2004@yahoo.com

© Springer Nature Singapore Pte Ltd. 2020
R. Bera et al. (eds.), *Advances in Communication, Devices and Networking*,
Lecture Notes in Electrical Engineering 662,
https://doi.org/10.1007/978-981-15-4932-8_33

In the recent years, researchers showed the interest towards the designing of decoupler to minimize the coupling effect. One approach, by [3], designed a decoupler for the discrete MIMO plants on the basis of polynomial theory and pole placement. The other decoupling controller is described in [4] in which adjoint decoupling method is implemented to achieve the decoupling. The stability analysis of the decoupling is also examined by the researchers [5] in which the internal stability problem is analysed. The Padé approximation method is used to approximate the delay-free transfer matrix for time-delayed system [6]. The stability of the approximated system may deteriorate; therefore, there is a need to develop a technique that directly decouples the time-delayed MIMO system.

The remaining part of the paper is arranged in such a way that Sect. 2 describes the decoupling of the MIMO systems. In Sect. 3, the procedure of designing the PID controllers is described. Different examples are taken to test the proposed method in Sect. 4 and last section concludes the paper.

2   Decoupling of MIMO Systems

This section describes the novel methodology of decoupling for discrete MIMO plants.

2.1   Case I: Square MIMO Plant

Consider a discrete-time square plant $G(z)$ as given below,

$$G(z) = \begin{bmatrix} g_{11}(z) & g_{12}(z) & \cdots & g_{1p}(z) \\ g_{21}(z) & g_{22}(z) & \cdots & g_{2p}(z) \\ \vdots & \vdots & \ddots & \vdots \\ g_{p1}(z) & g_{p2}(z) & \cdots & g_{pp}(z) \end{bmatrix} \tag{1}$$

The present work proposes a decoupler as shown in Fig. 1 which is obtained by the multiplication of the inverse of the static gain value of the system with the diagonal matrix of the system. Hence, the decoupling matrix is written as,

Fig. 1 Square plant with decoupler in open loop

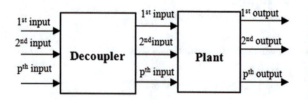

$$G_D(z) = G^{-1}(z)\big|_{z=1} \times G_I(z) \tag{2}$$

For the plant (1), the diagonal form is given by Eq. (3).

$$G_I(z) = \begin{bmatrix} g_{11}(z) & 0 & 0 & 0 \\ 0 & g_{22}(z) & 0 & 0 \\ 0 & 0 & \cdots & 0 \\ 0 & 0 & 0 & g_{pp}(z) \end{bmatrix} \tag{3}$$

The decoupled plant $G_N(z)$ is found out in the form as given by

$$G_N(z) = G(z) \times G_D(z) \tag{4}$$

Case II: Non-square MIMO plant:

Consider a discrete-time square plant $G(z)$ as given below with q number of inputs and p number of outputs, where $p \neq q$.

$$G(z) = \begin{bmatrix} g_{11}(z) & g_{12}(z) & \cdots & g_{1q}(z) \\ \cdots & \cdots & \cdots & \cdots \\ g_{p1}(z) & g_{p2}(z) & \cdots & g_{pq}(z) \end{bmatrix} \tag{5}$$

In this, a decoupler as shown in Fig. 2 is developed by multiplication of the pseudo-inverse of the static gain matrix of the plant with the diagonal matrix of the plant as described in Eq. 2. The inverse is based on the formula of right inverse as given below.

$$G1^{-1}(z) = G^T(z)(G(z)G^T(z))^{-1} \tag{6}$$

The diagonal matrix is obtained by considering the first diagonal element of each row as below.

$$G1_I(z) = \begin{bmatrix} g_{11}(z) & 0 & 0 \\ 0 & \cdots & 0 \\ 0 & 0 & g_{pp}(z) \end{bmatrix} \tag{7}$$

In this, the decoupling matrix is obtained as given below.

Fig. 2 Plant with decoupler in open loop

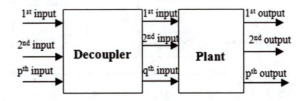

$$G1_D(z) = G1^{-1}(z)\big|_{z=1} \times G1_I(z) \tag{8}$$

The decoupled plant is again obtained by using the Eq. (4) as described in section A.

3 PID Controller Design

The discrete PID control structure is depicted in Fig. 3 of transfer function represented as below for an ensuing decoupled system.

$$G_c(z) = K_p + \frac{T}{2}\frac{z+1}{z-1}K_i + \frac{z-1}{Tz}K_d \tag{9}$$

where, K_p = proportional gain, K_i = integral gain, K_d = derivative gain and T = sampling time. Let nth input–output pair of a MIMO system be considered and each input–output can be treated as a subsystem. Hence, the discrete-time transfer function of the nth subsystem of the MIMO system is given as follows.

$$G_{pn}(z) = \frac{c(z)}{r(z)} = \frac{b_0 + b_1 z + b_2 z^2 + b_3 z^3 + \cdots + b_m z^m}{a_0 + a_1 z + a_2 z^2 + a_3 z^3 + \cdots + a_n z^n}, m \le n \tag{10}$$

Let the loop transfer function of the MIMO system be written as below.

$$G_{Ln}(z) = \left(\frac{2T K_p(z^2 - z) + T^2 K_i(z^2 + z) + 2K_d(z^2 + 1 - 2z)}{2T(z^2 - z)} \right).$$
$$\left(\frac{b_0 + b_{1z} + b_2 z^2 + \cdots + b_m z^m}{a_0 + a_{1z} + a_2 z^2 + \cdots + a_n z^n} \right) \tag{11}$$

The characteristic equation is written as,

$$1 + G_{Ln}(z) = 0 \tag{12}$$

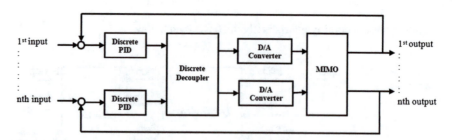

Fig. 3 Design of discrete PID controller for MIMO system

The aim is to get the optimum value of controller gains so that the system shows best performance. It is seen that there are infinite possible combinations between the three gains of PID controller and it is really a difficult task to find best possible combination to achieve robust performance by using classical approach. Hence, bacterial foraging optimization (BFO) technique [7] is used to get best performance of PID controller. The objective function is considered as w for determining the controller gains represented through Eq. (13) [8].

$$W = \left(1 - e^{-\beta}\right).\left(M_p + E_{ss}\right) + e^{-\beta}(t_s - t_r) \tag{13}$$

The pictorial display of the design steps of the BFO method is portrayed in Fig. 4 and the parameters are taken from [9].

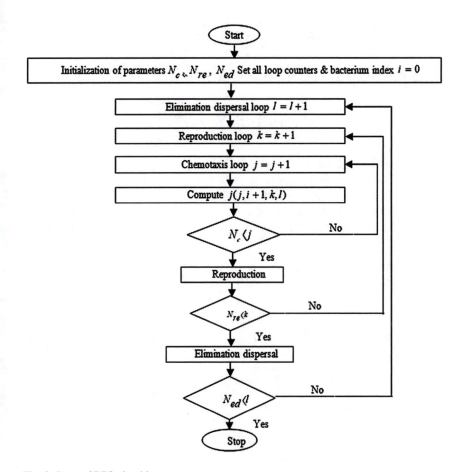

Fig. 4 Steps of BFO algorithm

4 Numerical Examples

A. *Square time-delayed Plant*:

The wood-berry binary distribution plant [10] is considered as a time-delayed plant which transfer matrix is written as below.

$$G(s) = \begin{bmatrix} \frac{12.8e^{-s}}{16.7s+1} & \frac{-18.9e^{-3s}}{21s+1} \\ \frac{6.6e^{-7s}}{10.9s+1} & \frac{-19.4e^{-3s}}{14.4s+1} \end{bmatrix} \tag{14}$$

The corresponding discrete-time transfer matrix of above equation is obtained [11] as below with sampling time 1 s.

$$G(z) = \begin{bmatrix} \frac{z^{-1}(0.744)}{z-0.9419} & \frac{-z^{-3}(0.8789)}{z-0.9535} \\ \frac{z^{-7}(0.5786)}{z-0.9123} & \frac{-z^{-3}(1.302)}{z-0.9329} \end{bmatrix} \tag{15}$$

Now the decoupler matrix is obtained following Eq. (8) as below,

$$G_D(z) = \begin{bmatrix} \frac{0.1168z^{-1}}{z-0.9419} & \frac{0.199z^{-3}}{z-0.9324} \\ \frac{0.0397z^{-1}}{z-0.9419} & \frac{0.1348z^{-3}}{z-0.9324} \end{bmatrix} \tag{16}$$

The decoupler is connected as portrayed in Fig. 1. A step input is introduce on first channel, while zero input is applied to second channel and it is seen in simulation results that only first output is obtained, whereas the second output remains zero as displayed in Fig. 5a. Again step input is fed on second channel, while zero input is applied to first channel and it is found through simulation results that only second

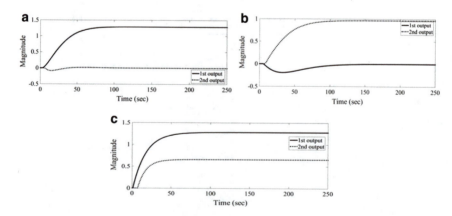

Fig. 5 **a** Output response of decoupled system when step response is applied to 1st channel & zero input is applied to 2nd channel. **b** Output response of Decoupled System when zero input is applied to 1st channel & step response is applied to 2nd channel. **c** Output response of the plant without decoupler when step response is applied to 1st channel& zero input is applied to 2nd channel

output is obtained, whereas the first output remains zero as shown in Fig. 5b. It is also verified that without a decoupler, if only the step input is applied to first channel, while zero input is applied to second channel then both the outputs are the step response which show the coupling effect of the plant as shown in Fig. 5c.

B. *Non-square time-delayed MIMO plant*:

The proposed discrete decoupling method is also applied on non-square 2×3 time-delayed MIMO plant considered by Ogunnaike and Ray (1994) [12]. The system considered is given as follows.

$$G(s) = \begin{bmatrix} \frac{0.5}{3s+1}e^{-0.2s} & \frac{0.07}{2.5s+1}e^{-0.3s} & \frac{0.04}{2.8s+1}e^{-0.03s} \\ \frac{0.004}{1.5s+1}e^{-0.4s} & \frac{-0.003}{s+1}e^{-0.2s} & \frac{-0.001}{1.6s+1}e^{-0.4s} \end{bmatrix} \tag{17}$$

The corresponding discrete-time transfer matrix of above equation is obtained [11] as below with sampling time 1 s.

$$G(z) = \begin{bmatrix} \frac{0.024z^{-1}+0.117}{z-0.716} & \frac{0.005z^{-1}+0.017}{z-0.671} & \frac{0.0003z^{-1}+0.011}{z-0.699} \\ \frac{0.0006z^{-1}+0.001}{z-0.514} & \frac{0.024z^{-1}-0.001}{z-0.367} & \frac{0.0001z^{-1}-0.0003}{z-0.535} \end{bmatrix} \tag{18}$$

The inverse of $G(z)|_{z=1}$ is given below,

$$G^{-1}(z)\big|_{z=1} = \begin{bmatrix} 1.66 & 41.87 \\ 1.96 & -247.3 \\ 0.776 & -90.53 \end{bmatrix} \tag{19}$$

Following Eq. (10), the resultant decoupled matrix of $G(z)$ is found out as below.

$$G_D(z) = \begin{bmatrix} \frac{0.042z^{-1}+0.20}{z-0.72} & \frac{-0.0712-0.010z^{-1}}{z-0.3679} \\ \frac{0.04z^{-1}+2.2}{z-0.72} & \frac{0.0618z^{-1}+0.42}{z-0.3679} \\ \frac{0.02z^{-1}+0.09}{z-0.72} & \frac{0.02z^{-1}+0.1539}{z-0.3679} \end{bmatrix} \tag{20}$$

Next, the decoupled plant $G_N(z)$ is obtained at the steady-state condition as $G_N(z)|_{z=1}$ for the computation of relative gain array with

$$G_N(z) = \begin{bmatrix} \frac{0.24z^3-0.26z^2+0.02z+0.03}{z^4-2.08z^3+1.45z^2-0.33z} & \frac{0.002z^3-0.001z^2-0.001z}{z^4-2.08z^3+1.45z^2-0.33z} \\ \frac{-0.01z^3+0.01z^2}{z^4-1.41z^3+0.66z^2-0.1z} & \frac{-0.001z^3+0.001z^2}{z^4-1.41z^3+0.66z^2-0.1z} \end{bmatrix} \tag{21}$$

The controller is designed only for the diagonal transfer function $G_{N11}(z)$ & $G_{N22}(z)$ of the decoupled MIMO system described by Eq. (26). The closed-loop characteristics equation of $G_{N11}(z)$ with PID controller is obtained as defined by Eq. (14) is given below.

Table 1 Controller gains obtained by BFO method

$G_{11}(z)$			$G_{22}(z)$		
K_p	K_i	K_d	K_p	K_i	K_d
0.45	0.16	0.55	−18.05	−18.18	−3.95

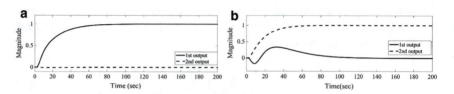

Fig. 6 **a** Step response of y_1 and interaction in y_2 due to step input in r_1. **b** Step response of y_2 and interaction in y_1 due to step input in r_2

Fig. 7 Step response of y_1 and y_2 due to unit step inputs in both r_1 and r_2

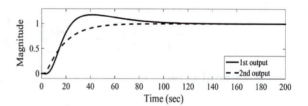

The controller gains are determined on the basis of performance in time domain of the plant and depicted in Table 1.

Figure 6a shows the output feature of y_1 and cross-coupling effect in y_2 due to step input in r_1. Figure 6b shows output feature of y_2 and cross-coupling effect in y_1 due to step input in r_2. It is worth to mention that the cross-coupling effect is negligibly small with the proposed decoupler. Next, step input command is given to both inputs r_1 and r_2. With these controller gains as indicated in Table 1, the decoupling control scheme is used to track the reference input successfully. Fig. 7 displayed the output performance of y_1 and y_2 with respect to r_1 and r_2, respectively.

5 Conclusion

This paper has addressed the problem of decoupling control of time-delayed MIMO systems with strong cross-coupling between inputs and outputs. In this direction, the discrete decoupling control scheme is proposed to eradicate the interaction effects of the time-delayed MIMO systems and to achieve the desired output responses. The proposed approach has been successfully implemented to the time-delayed MIMO systems in simulation to show the efficacy of method.

References

1. A. Kaldmae, U. Kotta, Input–output decoupling of discrete-time non-linear systems by dynamic measurement feedback. Eur. J. Control **34**, 31–38 (2017)
2. H. Nijmeijer, W. Respondek, Dynamic input–output decoupling of nonlinear control systems. IEEE Trans. Autom. Control **33**(11), 1065–1070 (1988)
3. M. Kubalcik, V. Bobal, Continuous-time and discrete multivariable decoupling controllers. IEEE Trans. Autom. Control **9**, 327–335 (2014)
4. L.B. Xiea, C.Y. Wub, L.S. Shiehc, J.S.H. Tsaib, Digital decoupling controller design for multiple time-delay continuous-time transfer function matrices. Int. J. Syst. Sci. **46**(4), 577–589 (2015)
5. B. Srinivasan, P. Mullhaupt, T. Baumann, D. Bonvin, A discrete-time decoupling scheme for a differentially cross coupled system, in *13th IFAC Triennial World Congress* (San Fmncisco, 1996), pp. 301–306
6. M. Morari, E. Zafiriou, *Robust process control, englewood cliffs* (Prentice Hall, NJ, 1989)
7. J. Li, J. Dang, F. Bu, J. Wang, Analysis and improvement of the bacterial foraging optimization algorithm. J. Comput. Sci. Eng. **8**(1), 1–10 (2014)
8. Z.L. Gaing, A particle swarm optimization approach for optimum design of PID controller in AVR system. IEEE Trans. Energy Convers. **19**(2), 384–391 (2004)
9. S.K. Pandey, J. Dey, S. Banerjee, Design of robust proportional–integral-derivative controller for generalized decoupled twin rotor multi-input-multi-output system with actuator non-linearity. J. Syst. Control Eng. **232**(8), 971–982 (2018)
10. K.J. Astrom, K.H. Johansson, Q.G. Wang, Design of decoupled PID controllers for MIMO system, in *Proceedings of the 2001 American Control Conference* (2001). https://ieeexplore.ieee.org/xpl/conhome/7520/proceeding
11. K. Oggata, Discrete time control system design (Prentice Hall, 1995)
12. B.A. Ogunnaike, W.H. Ray, *Process Dynamics, Modeling and Control* (Oxford University Press, Oxford/New York, 1994)

Design of Wearable Dual-Band Collar Stay Antenna for Wireless Communication

Baishali Gautam, Anindita Singha, Pooja Verma, Hashinur Islam, and Saumya Das

Abstract This article investigates the performance of a collar-stay-shaped rigid wearable dual-band antenna for the purpose of wireless communications. The collar position of formal shirt has been chosen for placing the antenna. Because of its rigid structure, it can overcome the problem of frequency resonance shifting and downfall of signal strength associated with a flexible antenna due to body bending and cloth crumpling. It finds applications in 4.34 and 5.96 GHz with bandwidths 120 MHz and 84 MHz, respectively. This antenna yields a maximum gain of 4.51 and 5.37 dBi at 4.34 GHz and 5.96 GHz, respectively.

Keywords Wearable antenna · Rigid structure · Collar of formal shirt

1 Introduction

In recent times, several researchers have started working on the development of wearable devices for the purpose of health monitoring, tracking, direction finding, and medical imaging [1, 2]. Normally wearable devices include a transmitting system, a receiving system, and an antenna. So along with transmitter and receiver, an antenna must be placed on body for successful operation of the wearable device. But most

B. Gautam (✉) · A. Singha · P. Verma · H. Islam · S. Das
Department of Electronics and Communication Engineering, Sikkim Manipal Institute
of Technology, Sikkim Manipal University, Gangtok, Sikkim, India
e-mail: baishaligautam10@gmail.com

A. Singha
e-mail: aninditasingha21@gmail.com

P. Verma
e-mail: puzavermaparhwara@gmail.com

H. Islam
e-mail: hashinur0001@gmail.com

S. Das
e-mail: saumya.das.1976@gmail.com

© Springer Nature Singapore Pte Ltd. 2020
R. Bera et al. (eds.), *Advances in Communication, Devices and Networking*,
Lecture Notes in Electrical Engineering 662,
https://doi.org/10.1007/978-981-15-4932-8_34

people do not feel comfortable to wear an antenna until they have some severe health issues. Therefore, the position of wearable antenna should be chosen such that users feel comfortability in wearing it. For that purpose, lightweight, low-profile antenna is designed over fabric materials like jeans, cotton, polyester, etc. [3–6]. A copper tape of the patch is pasted on fabric material to achieve the desired frequency of operation. However, the copper pasting leads to uneven layer formation of air and glue between the substrate and the patch which leads to differences in simulation and experimental results [7]. In addition to that, body bending and cloth crumpling cause drift in resonance frequency and downfall in signal strength [8]. To overcome these issues, a rigid but comfortable in wearing antenna has been developed which has been discussed in some research articles [9, 10]. Formal shirts require collar stays to keep the collar straight. These collar stays are made up of plastic material. In this research work, a collar-stay-shaped FR4 substrate has been considered for the development of an antenna. Thus, this particular antenna has a dual purpose. It could be used in communication besides its use as a collar stay for keeping the collar straight. Both HFSS and CST-MWS software platforms have been explored for carrying out the simulation work.

2 Antenna Construction

FR4 substrate ($\varepsilon_r = 4.4$, loss tangent tan$\delta = 0.008$) with thickness 1.6 mm has been chosen for developing the proposed collar stay antenna for 4.34 and 5.96 GHz. This collar stay antenna is a combination of three parts—a triangular part, a rectangular part, and a semi-circular part as shown in Fig. 1. And the one stub has been added over the semi-circle to get the perfect match. The conductive patch on the substrate and the ground part below the substrate is made of copper material.

3 Results and Discussion

HFSS and CST-MWS both software platforms confirm the dual-band resonance frequencies at 4.34 and 5.96 GHz as shown in Fig. 2 with reflection coefficient measurement. VSWR in Fig. 3 also establishes a value of less than 2 in the desired bands. The proposed antenna exhibits −10 dB impedance bandwidth of 140 and 84 MHz for 4.34 GHz and 5.96 GHz, respectively. The peak gain is observed as 4.51 and 5.37 dBi at 4.34 GHz and 5.96 GHz, respectively. Antenna efficiencies are reached at 43.3 and 29.4% at 4.34 GHz and 5.96 GHz, respectively. The poor efficiency maybe because of the use of FR4 material as a substrate. Almost omnidirectional radiation pattern is achieved for both resonant frequencies as shown in Figs. 4 and 5.

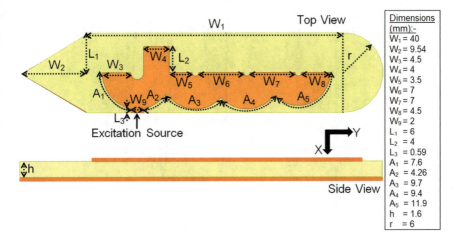

Dimensions (mm):-
$W_1 = 40$
$W_2 = 9.54$
$W_3 = 4.5$
$W_4 = 4$
$W_5 = 3.5$
$W_6 = 7$
$W_7 = 7$
$W_8 = 4.5$
$W_9 = 2$
$L_1 = 6$
$L_2 = 4$
$L_3 = 0.59$
$A_1 = 7.6$
$A_2 = 4.26$
$A_3 = 9.7$
$A_4 = 9.4$
$A_5 = 11.9$
$h = 1.6$
$r = 6$

Fig. 1 Geometry of the proposed collar stay antenna

Fig. 2 Return loss of the proposed collar stay antenna

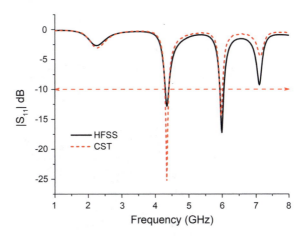

4   Conclusion

Collar stay of formal shirts is used as a substrate for developing a dual-band microstrip antenna for wearable applications. It is found that the position of this rigid antenna does not cause uneasiness in wearing and is also not vulnerable to resonance frequency drifting due to body bending and cloth crumpling. The proposed antenna yields satisfactory peak antenna gains of 4.51 and 5.37 dBi with almost omnidirectional radiation pattern at desired frequencies of 4.34 GHz and 5.96 GHz, respectively.

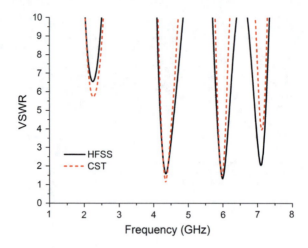

Fig. 3 VSWR of the proposed collar stay antenna

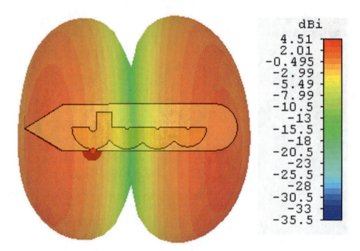

Fig. 4 Radiation pattern of collar stay antenna at 4.34 GHz

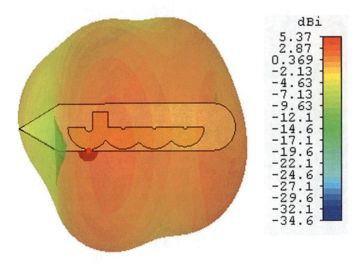

Fig. 5 Radiation pattern of collar stay antenna at 5.96 GHz

References

1. S. Das, H. Islam, T. Bose, N. Gupta, Coplanar waveguide fed stacked dielectric resonator antenna on safety helmet for rescue workers. Microw. Opt. Technol. Lett. **61**(2), 498–502 (2019)
2. S. Das, H. Islam, T. Bose, Compact low-profile body worn and wrist worn lightweight antenna for ISM and GPS band navigation and medical applications. Microw. Opt. Technol. Lett. **60**(9), 2122–2127 (2018)
3. V.K. Singh, S. Dhupkariya, N. Bangari, Wearable ultra wide dual band flexible textile antenna for wimax/wlan application. Wirel. Pers. Commun. **95**(2), 1075–1086 (2017)
4. S. Sankaralingam, B. Gupta, Determination of dielectric constant of fabric materials and their use as substrates for design and development of antennas for wearable applications. IEEE Trans. Instrum. Meas. **59**(12), 3122–3130 (2010)
5. S.J. Ha, Y.B. Jung, D.H. Kim, C.W. Jung, Textile patch antennas using double layer fabrics for wrist-wearable applications. Microw. Opt. Technol. Lett. **54**(12), 2697–2702 (2012)
6. M.A. Chung, C.F. Yang, Built-in antenna design for 2.4 GHZ ism band and GPS operations in a wrist-worn wireless communication device. IET Microw., Antennas Propag. **10**(12), 1285–1291 (2016)
7. B. Roy, A. Bhatterchya, S. Choudhury, Characterization of textile substrate to design a textile antenna in *2013 International Conference on Microwave and Photonics (ICMAP)*, (IEEE, 2013), pp. 1–5
8. Q. Bai, J. Rigelsford, R. Langley, Crumpling of microstrip antenna array. IEEE Trans. Antennas Propag. **61**(9), 4567–4576 (2013)
9. S. Das, H. Islam, T. Bose, N. Gupta, Ultra wide band CPW-Fed circularly polarized microstrip antenna for wearable applications. Wirel. Pers. Commun. **108**(1), 87–106 (2019)
10. B. Gautam, P. Verma, A. Singha, H. Islam, O. Prakash, S. Das, Design of multiple collar stay antennas for wireless wearable compact devices. Microw. Opt. Technol. Lett. 1–7 (2019)

Grid Localization Method for Spatial Touchless Interaction Applied in Wearable Devices

Lan Thi Phan and Chai-Jong Song

Abstract Touchless interaction is one of the fields that attracts a lot of interest from researchers and users for its conveniences, especially in the field where sterilization should be kept strictly or natural action in midair is preferred. In this work, we present the successful estimation for the touchless spatial interaction using the ToF sensor for wearable devices. The work is based on the control of a simple graphical user interface (GUI) by estimating the human finger position in the space of an invisible 2 dimensions' grid of 4×4 cells in front of sensors array, sensors field of view (FoV) of 270. The positioning process is based on a grid method, in which the human finger located at each cell of a grid and its position will be derived successfully by an algorithm of sensors data processing. The latest version of the commercialized ToF sensor from Microelectronics company was used in this work, named of VL53L1X. So far, this is the most flexible and convenient device due to its small size and lightweight, which could be installed easily for wearable applications. The accuracy of the touchless interaction system was 93% in real-time for object estimation and 97.1% of reliability for finger estimation.

Keywords Touchless interaction · Wearable · Grid localization · Position estimation · Time of flight sensor

1 Introduction

1.1 The Touchless Interaction as a NUI

The improvement of human–machine interface has been changed continuously from command-line interface (CLI) to Graphical user interface (GUI) and Natural user

L. T. Phan · C.-J. Song (✉)
Information & Media Research Center KETI, Seoul, South Korea
e-mail: chaijongsong@gmail.com

L. T. Phan
e-mail: Orchidna@gmail.com

© Springer Nature Singapore Pte Ltd. 2020
R. Bera et al. (eds.), *Advances in Communication, Devices and Networking*,
Lecture Notes in Electrical Engineering 662,
https://doi.org/10.1007/978-981-15-4932-8_35

interface (NUI) in order to eliminate the limits of technologies and give more convenience as natural, easy, fun, and intuitive behaviors to the user when interacting with the machine. Touchless such as human eye movement, speech, hand movement are attractively growing field in NUI, as Microsoft's Kinetics console, Leap Motion, Real Sense, for example. The work for touchless interaction using hand movement or motion are in focus with abundant quantitative and qualitative reports rather than the rest interfaces due to the more convenient to users with more benefit in some professional works [1, 2]. For example, in the operating room [3] and interventional radiology where the surgeons need to interact with scanning images of X-ray, ultrasound, the contamination may happen in case of mechanical touch with keyboard or mouse [4]. The contribution of touchless interaction is making a revolution in HMI and in the development of technologies in this field.

1.2   Related Works and Our Selection of ToF Sensors

As author's knowledge up to now, the works for touchless interaction popularly used the 3D recognition system such as depth camera [5–7] or leap motion [8]. However, they are not suitable for wearable devices due to their size and weight. And so far, there are not many academic reports about touchless interaction for wearable devices. In the process of looking for devices suitable for the wearable projector, we took a look at commercialized proximity sensors for touchless interaction, such as IR and laser sensors. On IR sensors, the estimation is based on the intensity of reflected IR back to the photon collector rather than real distance. For example, the proximity sensor by IR LED (TSL2672, VCNL4000, AN580) is available but the range is only around 50 cm. This is not suitable for a wearable projector with the distance operation over 1 m. Other samples use sound wave, the principle is based on the process of wave reflection when object locates on the way of sound transfer. Table 1 shows several types of available commercialized proximity sensors on the market and their important characteristics for touchless interaction and size for wearable applicability.

Among, ToF sensors received a special interest due to the compact, lightweight, not affected by object color, shape, and reliable performance. The depth camera is an example of using the ToF method for object positioning. However, the small size of the ToF sensor is the most important advantage for uses in wearable devices' applicability. In this work, we used a simple array of ToF sensors to facilitate the touchless using human finger in midair with the whole module size of 5 × 3 cm, containing 5 sensors with the size of 4.9 × 2.5 mm. The latest version of VL53L1X has been used due to the better stability in range, the user/object detection, programmable FoV, and the flexible range in short mode (~1.2 m), and long mode (~4 m). The version uses the array of single photon avalanche diode with size of 16 × 16 (SPAD) to detect laser 8× more signal than VL53L0X and the region of interest (ROI) could be selected by users [9].

The selection of various functions on the graphical user interface (GUI) has been implemented by human finger positioning in the field of view (FoV) of sensors based

Table 1 Proximity sensors and the characteristics

Sensor name	Size (mm)	Range/Error	Functions
VL53L0X/ST μelectronic (2016)	4.4 × 2.4	~1.2 m, error <3%	Ranging and gesture detection sensor by ToF method
VL6180X/ST μelectronic (2015)	4.8 × 2.8 × 1	0~ above 10 cm	Distance and signal level can be used by host system to implement gesture recognition by ToF method
MLX75023/Melexis (2013)	6.6 × 5.5 × 0.6	Range: adaptable; ~2 m	Optical time-of-flight (TOF) camera sensor; gesture recognition; automotive driver monitoring; surveillance
TSL2672/AMS (2016)	2×	<15 cm, <46 cm	Proximity sensor by IR LED
VCNL4000/Vishay (2015)	3.95 × 3.95 × 0.75	1–100 or 200 mm (best: 100–150)	Proximity function by IR LED + photo-pin-diode; use for mobile devices for touch screen locking
AS7264N/AMS	4.5 × 4.7 × 2.5	Around 100 mm	A tri-stimulus sensor provides measurements of color that closely match the human eye's response to the visible light spectrum
AN580/Silicon labs (2015)	2 × 2	15 cm	Touchless gesturing applications: page turning, scrolling on a tablet PC, GUI navigation

on a grid method. A virtual grid structure located at the core of FoV at the distance of 35 cm, composes of 4 rows and 4 columns with a unit cell of 4 × 5 cm. The Arduino code for an array of 5 ToF sensors module has been improved for sensors' signal stability. The algorithm based on grid method for row and column estimation has been done with high accuracy estimation and developed to filter the detection result for a robust performance in real-time. Accuracy of touchless interaction system in real-time was measured by a maximum of 92.125% for object estimation and over 97.1% of reliability for finger estimation in real work.

2 Methodology

This part shows the method to approach the object position using the array of ToF sensors by using the basic principles of distance estimation of ToF sensor and positioning method of ToF sensor array, in theoretical analysis. In detail, the methodology includes the ToF principle and the grid localization method.

2.1 ToF Sensor Principle

Time of flight is a popular principle in many modern equipments, especially in the tiny world of micro or nano-size. The name of the method has shown the principle correctly if we indicate the principle is for light or photon. As we know the velocity of a photon is a constant $c = 3.3 \times 10^8$ cm/s. So that if we know the time of flight of a photon since emitting point to collecting point, we could derive the distance between sensor and object by the following formation:
$D = v * t/2$.

D distance between sensor and object and v is the photon velocity, $v = c$.

The factor of 2 is because the photon has traveled from the sensor to the object and reflected from the object to the sensor. This means that the time for traveling of a photon is for double distance. The laser has a wavelength of 920 nm and the collector sensitivity is as good as such that it could detect an individual photon.

2.2 Grid Localization Method

As described in the ToF principle, every single sensor could estimate the object distance if the object locates in the FoV, meaning the object could be estimated in 1D. However, to estimate the object position better, we use an array of ToF sensor in this work for 2D position estimation. Basically, the position of the object could be estimated in a plane by cosine formula, as follows:
$c^2 = a^2 + b^2 - 2ab \cos\gamma$; where a, b, c: triangle length sides; γ angle made of a and b sides.

The formula shows that the object position could be estimated exactly with distances from 2 sensors. However, the sensor signal does not match the real distances of the object, causes inaccuracy in the algorithm for object position estimation. Therefore, we design an array of 5 sensors and modify the object position approach method in another way. The grid localization method described in the followings shows the modification of the original method to adapt the new conditions and this is the method applies in our work for object position estimation.

The basic approach we used here includes:

(1) A grid for object localization. The grid must be placed in the FoV area for a good signal from sensors, the size of the cell should be large enough for accurate position estimation. Depends on the number of functions, the grid will be designed with the proper number of cells.

(2) When the object located at each cell in the grid, the signal from sensors will be different, reveal different characteristics along the rows and columns in the grid. And this is the main point to derive the object position estimation in the grid.

The relative distance of the sensor's output should be considered carefully and processed in this work as sensor signals do not match the distance of the object in the real. So that we use an array of correlating factors to estimate the object position in each cell. The correlating array indicates from an initial data, which give the correspondence between object position at cells and the output signal data.

2.3 Our Criteria

- Object position is detected in a grid of 4 × 4 cells; the position of the object is indicated by column and row in the grid (Figs. 2 and 3). The ToF array contains 5 sensors alight in one row on the board.
- The time for position estimation is in real-time.
- The estimation accuracy is over 90%.

3 Experiments

To obtain the data process signal to derive the algorithm of object position touchless estimation, we setup the experimental environment with an invisible grid in FoV. The experiments were conducted in a space starting from the ToF sensor to a white background at a distance of 3 m. The space of sensor FoV is totally empty except the investigated object movements for the detection. The background color is fixed as white for a uniform performance of the sensor, otherwise, the sensor working could be different along with the different background color.

The light is room lighting intensity and it was confirmed that lighting intensity could be neglected in this work. The output does not change much in case of dark or strong lighting. There is no object in FOV of ToF sensor array (0-3m) to prevent the noise or interference to ToF sensor performance.

Field of view (FoV) is an important area for this work as the object could be detected only if it gets located in this area. FoV of a sensor is the cone with the top located at emitter/collector, cone angle is 27°. And the FoV of the system is the total space made by the individual sensor (total 5 sensors). However, the grid position is located inside the FoV area of all sensors for a good signal. The test was conducted while there is no object in FoV at a distance of 2.5–3 m. Figure 1 shows the environmental setup with an invisible grid and object position.

Object was a cylinder with diameter approximate to finger size at height of 28 cm and the final object target is human fingers. The module was an array of 5 sensors.

Sensors located in a line, parallel to ground, at a height of 23 cm. The grid size is 20.5 × 20 cm, placed at 35 cm from the sensor board. Figure 2 shows the details

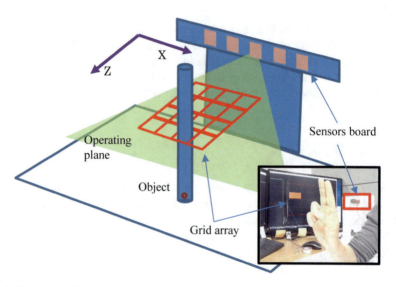

Fig. 1 Experimental setup

Fig. 2 Grid configuration in experiments and object movement. **a** Grid of 4 × 4 (red line with 1 cm width is for a separation in sensor's output when object move to next row in one column) unit cells in FoV of 5 ToF sensors array; The row (R) and column (C) of each cell was marked from 0 to 3, and presented at the center of cells. Inlet: Red point for marking and rows showing object's movement at each unit cell

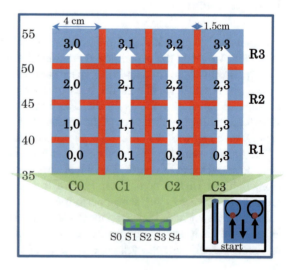

of the grid with the line boundary, distance from sensor board, row and column number, and the movement of an object in a cell in the testing process for initial data collecting.

3.1 Initial Data, Sensor Characteristics Derivation and Correlating Array (CA)

Initial data is the very first data set from trial tests to derive the characteristics of data when object locates in different positions.

Correlating array is the set of parameters that show the relative data from the sensor corresponding to the position of the object at each cell in the gird.

The correlating array was derived from the initial data. Also, the algorithm for object position was derived from these very first tests.

The test was conducted as follows:

(1) Move the object along each column, from C0, C1, C2 to C3. The test at each row was repeated 5 times to have stable and accurate data (Fig. 2, yellow arrows).
(2) At each row, the object moves in the cell up and down 3 times before moving to the next cell (Fig. 2 inlet).

The data we got from the test at C2, R2 (2, 2) cell is showed on the Fig. 3a. The up-down-up (~far-close-far) movements of object in cell 2, 2 make 3 data segments (Fig. 3a, inlet) and the distance data is not overlap with those of previous or next cell. From this range of distances, we derive the CA with minimum and maximum value of sensor at each cell, along rows and columns.

Based on the clear and stable signal of the sensor, the characteristics of sensors along rows (Table 2) and CA (Table 3) along columns (C0–C3)/rows (R0–R3) are as follows:

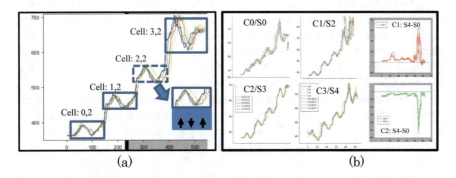

(a)   (b)

Fig. 3 Experimental setup. **a** Characteristic graph of sensor S3 signal along column C2 with the line segments at each cell (inlet). **b** The graphs of sensor signal along column C0–C3 (left) with characteristic at column 1-C1 and 2-C2 by signal S4–S0 (right)

Table 2 Sensor data characteristic

SENSORS' DATA characteristic	
C0	S0 in range (350–800), S4 > 1800
C1	S4–S0 > 0, S0, S1, S2, S3, S4 in range
C2	S4–S0 < 0, S0, S1, S2, S3, S4 in range
C3	S4 in range (350–800) S0 > 1500

Table 3 Sensor data characteristic

CA	C0		C1		C2		C3	
	CA_{min}	CA_{max}	CA_{min}	CA_{max}	CA_{min}	CA_{max}	CA_{min}	CA_{max}
R0	350	450	340	400	350	415	350	420
R1	451	520	401	499	416	500	421	514
R2	521	595	500	585	501	595	515	599
R3	596	900	586	800	596	900	600	900

3.2 Algorithm and Experimental Result for Real-Time Tests

Based on the initial data and the characteristics of the signal along columns and CA array along rows as above, the algorithm for object position estimation have been derived, as described in Fig. 4. The algorithm and initial data are for an array of 5 sensors, mean that the data will be collected from 5 sensors simultaneously and the data processing will be conducted right after that to derive the object position in real-time. The individual sensor data in 1D was collected to check if they could satisfy the 4 conditions as in Table 2, after that, based on the value of object distance from sensor which corresponds to the estimated column, the row of the sensor is derived (Table 3), and then the position of the object is estimated by column and row.

In this work, we have successfully estimated the position of an object on 3 modules of ToF array. The array contains 5 sensors as described in previous parts. The performance of each module is different due to the working situation of the individual sensor on modules. With 2 other modules, the algorithm does not change much, but for CA, the data of sensor used are M3 and M4: at (C0, C3, C1, C2)/(C0, C3, C1, C2), use (S4, S0, S3, S2)/(S4, S0, S1, S3), respectively.

For robotizing the estimation performance, we derive an algorithm based on the real-time estimation position, as in the Fig. 5. The estimation position result of the sensor is saved into an array of 10 elements including information of both column and row, called filter set (FS). The FS will be updated by time after there are 2 new signals initiated and replace the first 2 signals in current FS array. The new FS will be used to derive the final position estimation result. The result is the cell which repeats the most in FS, correspond to max GUD. This mean that instead of estimating position once by 1 sensor array signal, we use 2 sensor array signals to stabilize the result.

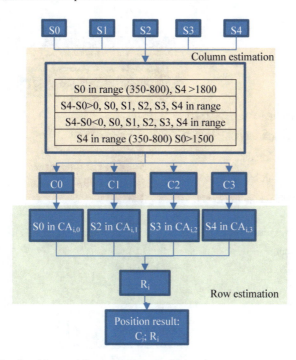

Fig. 4 Algorithm for object position estimation on grid 4 × 4

The robotizing algorithm eliminate the wrong result when the object moves between cells.

The test result using module M1 for 10 times shows that the accuracy improves from 92.13% for 2 by 1 real-time update estimation (2 sensor output data—1 position estimation) for module M1 and 97.1% for real human finger estimation.

3.3 Result Discussion and Algorithm for Robust and Stable Real-Time Tests

The test result of module M1 shows a good performance of the estimation process, with accuracy is 93% in real-time for object estimation and 97.1% in real-time for finger estimation. To obtain this result, the sensors have been calibrated using ROI and the process to eliminate the interference between sensors' signal was checked to increase the quality of the signal for a good estimation performance.

The experiment in this work has been improved step by step for form functions along with increasing the grid resolution. Some tests with a grid size of 3 × 4 cells give an accuracy of 94.1% while the corresponding number is 93% as above for 4 × 4 grid. The result shows that even the accuracy and grid resolution is in reciprocal

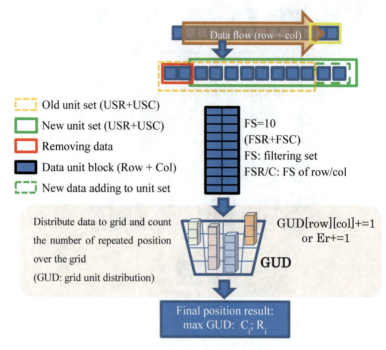

Fig. 5 Algorithm for stable estimation

ratio but it is not very significant, confirm the reliability of ToF sensors for object position in 2D estimation.

The algorithm for position estimation by grid method is quite simple with a high accuracy and high potential for other applications for NUI. The basic method is for object localization on a 2D grid in FoV of an array of 5 sensors.

The work uses the characteristic of sensor signal at every single cell in the grid to derive the algorithm of position estimation, due to a similarity in working condition, the characteristic show a principle to estimate the position of the cell by estimating the column first and the row in the second step.

4 Conclusion

The work has been conducted for estimating position using an array of ToF sensors-VL53L1X with an excellent accuracy so far: 97.1% or real-time estimation of fingers. The grid unit size is 40×35 mm in the range of 250–410 mm far from sensor's position. The algorithm of grid-based approach, tests and data analysis have described to derive the optimum parameters for position estimation. Depending on the desired resolution, the accuracy for recognition probability could be changed inversely proportional to the size of the unit square. The work has the potential for larger applications in touchless HMI. In this work, the grid scale is 4×4, meaning that there are 16 functions that could be used for any NUI with a very small and lightweight ToF sensor board, this size and weight is proper for any wearable applications. However, a stable operation should be considered more in the aspects of systematical error, in addition to the mutual impact of lasers coming from scattering, ambient light.

Acknowledgements This work was supported by IITP of MSIP (Ministry of Science and ICT) grant funded by the Korea government in 2017 [No. 2017-0-01664, Development of context adaptive wearable projection device and services based on 2D MEMS HD optical engine and space interaction].

References

1. K. O'Hara, G. Gonzalez, A. Sellen, G. Penney, A. Varnavas, H. Mentis, A. Criminisi, R. Corish, M. Rouncefield, N. Dastur, T. Carrell, Touchless interaction in surgery. Commun. ACM **57**(1), 70–77 (2014)
2. Touch Display Research, Touchless HMI sensor market 2015 report (2015)
3. A. Hurstel, D. Bechmann, Approach for intuitive and touchless interaction in the operating room. J. Multidiscip. Sci. J. **2**, 50–65 (2019)
4. F. Pereme, J.Z. Flores, M. Scavazzin, F. Valentini, J.-P. Radoux, Conception of a touchless human machine interaction system for operating rooms using deep learning, in *Proceedings of SPIE 10679, Optics, Photonics, and Digital Technologies for Imaging Applications V* (2018), p. 106790R
5. T. Kopinski; U. Handmann, Touchless interaction for future mobile applications, in *2016 International Conference on Computing, Networking and Communications (ICNC), Workshop on Computing, Networking and Communications (CNC)* (IEEE, 2016)
6. N. Zengeler, T. Kopinski, U. Handmann, Hand gesture recognition in automotive human-machine interaction using depth cameras. Sensors **19**, 59 (2019)
7. P. Bach-y-Rita, S.W. Kerce, Sensory substitution and the human-machine interface. Trends Cogn. Sci. 541–546 (2003)
8. G.M. Rosa, M.L. Elizondo, Use of a gesture user interface as a touchless image navigation system in dental surgery: case series report. Imaging Sci. Dent. **44**(2), 155–160 (2014)
9. STMicroelectronics, Long distance ranging Time-of-Flight sensor based on ST Flight Sense technology (2017)

Design of UWB Bandpass Filter Including Defective Ground Structure

Hrishikesh Thakur, Hashinur Islam, Saumya Das, Sourav Dhar, and Mruthyunjaya HS

Abstract An Ultra Wide Band (UWB) bandpass filter (BPF) including defected ground structure (DGS) has been developed in this work. Two different simulations have been utilized to design the required filter. Simulation results of software on filter performance have been compared to come into a conclusion. Return loss exceeds 10 dB and the insertion loss falls less than 1.3 dB everywhere in between 3.04 and 10.6 GHz. This filter can find applications in indoor and outdoor communications for the use of radar technology, remote sensing, cancer detection, patient monitoring system, etc.

Keywords Ultra wideband (UWB) technology · Filter designing · Defected ground structure (DGS)

1 Introduction

The Ultra Wide Band (UWB) filter finds various applications and advantages in various radar technologies to medical fields for its high-speed data, small size, low-cost, and low complexity for fast communication in todays life [1–3]. UWB is also

H. Thakur (✉) · H. Islam · S. Das · S. Dhar
Department of Electronics and Communication Engineering, Sikkim Manipal Institute of Technology, Sikkim Manipal University, Gangtok, Sikkim, India
e-mail: hrishikeshthakur715@gmail.com

H. Islam
e-mail: hashinur0001@gmail.com

S. Das
e-mail: saumya.das.1976@gmail.com

S. Dhar
e-mail: sourav.dhar80@gmail.com

M. HS
Department of Electronics and Communication Engineering, Manipal Institute of Technology, Manipal Academy of Higher Education, Manipal, India
e-mail: mruthyu.hs@manipal.edu

© Springer Nature Singapore Pte Ltd. 2020
R. Bera et al. (eds.), *Advances in Communication, Devices and Networking*,
Lecture Notes in Electrical Engineering 662,
https://doi.org/10.1007/978-981-15-4932-8_36

very useful in indoor communication as it can coexist with other communications. This is because of the very low signal strength of the UWB signal. But this requires excellent filter performance of wireless devices. As per the definition of US Federal Communication Commission (FCC), the UWB band must lie in between the frequency band of 3.1–10.6 GHz [4]. Various Researchers have developed various Ultra Wide Band Pass filters for different purposes. In [5], the authors proposed a hybrid microstrip coplanar waveguide technology for establishing a UWB filter structure. Multiple Resonant Structure including a coplanar waveguide has brought the desired UWB band. In [6], a UWB bandpass filter has been developed exploring a stepped impedance resonator technology. Odd-even mode analysis is taken into consideration. In [7], a band-notched UWB filter is designed using defected ground structure (DGS) which plays a role in restricting unwanted frequencies. The DGS is applied to suppress the upper passband spectrum of the designed filter [8].

In this work, a UWB filter is designed by creating a symmetric patch and defective ground structure. The defected ground structure is obtained by taking out slots from the conductive ground plane. The simulation work is carried out in Electromagnetic Simulation Software, HFSS, and CST-MWS. Results from different software confirm the UWB band for the use of bandpass filter.

2   Filter Configuration

In this UWB filter designing, a rectangular substrate of dimension 15 mm × 12.6 mm × 0.635 mm and relative permittivity 10.8 has been considered to achieve the required bandwidth. Symmetrical dumbbell-shaped copper patches are taken into consideration along with symmetrical defective ground structure as shown in Fig. 1. Every dimension of the UWB filter design is given in detail in Fig. 1. Design parameters have been optimized to achieve the desired characteristics of the filter.

3   Results and Discussion

The Ultra Wide Band filter is designed and implemented using ANSYS HFSS and CST-MWS software. The filter performance is measured mainly in terms of return loss and insertion loss. Return loss is the ratio of reflected power to the incident power and it should be more than 10 dB for the desired band whereas insertion loss is the ratio of the transmitted power to the incident power and it should be less than 3 dB for satisfactory filter performance. From the software results, it is confirmed that the return loss exceeds 10 dB and the insertion loss falls less than 1.3 dB everywhere in between 3.04 and 10.6 GHz. The difference between simulation results may be because of different computational electromagnetics algorithms used by different software. But both the results confirm the attainment of UWB band for the use of wireless communication (Fig. 2).

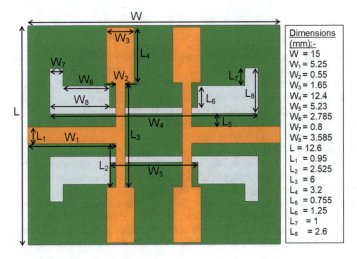

Fig. 1 Geometry of the proposed UWB bandpass filter

Fig. 2 S parameter results of UWB bandpass filter

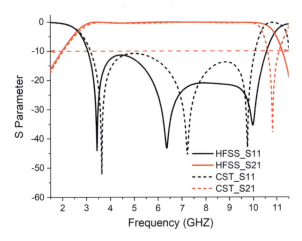

4 Conclusion

An Ultra-Wide Band (UWB) filter exploring defective ground structure has been implemented in this work. Two different simulations have been utilized to design the required filter. Simulation results of software on filter performance have been compared to come into a conclusion. It is observed that return loss exceeds 10 dB and the insertion loss falls less than 1.3 dB everywhere in between 3.04 and 10.6 GHz. This filter can be instrumental for the application of radar technology, remote sensing, cancer detection, patient monitoring system, etc.

References

1. A.N. Ghazali, M. Sazid, S. Pal, A dual notched band UWB-BPF based on microstrip-to-short circuited CPW transition. Int. J. Microw. Wirel. Technol. **10**(7), 794–800 (2018)
2. A.N. Ghazali, M. Sazid, S. Pal, Dual band notched UWB-BPF based on hybrid microstrip-to-CPW transition. AEU-Int. J. Electron. Commun. **86**, 55–62 (2018)
3. A.N. Ghazali, A. Singh, Band notched UWB-BPF based on broadside coupled microstrip/CPW transition. IETE J. Res. **62**(5), 686–693 (2016)
4. U.F.C. Commission et al., Revision of part 15 of the commissions rules regarding ultra-wideband transmission systems. First Report and Order, ET Docket, pp. 98–153 (2002)
5. B. Sahu, S. Singh, M.K. Meshram, S. Singh, Super-compact ultra-wideband microstrip band-pass filter with improved performance using defected ground structure-based low-pass filter. J. ElEctromagnEtic WavEs Appl. **32**(5), 635–650 (2018)
6. D. Sarkar, T. Moyra, L. Murmu, An ultra-wideband (UWB) bandpass filter with complementary split ring resonator for coupling improvement. AEU-Int. J. Electron. Commun. **71**, 89–95 (2017)
7. Z. Sakotic, V. Crnojevic-Bengin, N. Jankovic, Compact circular-patch-based bandpass filter for ultra-wideband wireless communication systems. AEU-Int. J. Electron. Commun. **82**, 272–278 (2017)
8. J.B. Jadhav, P.J. Deore, A compact planar ultra-wideband bandpass filter with multiple resonant and defected ground structure. AEU-Int. J. Electron. Commun. **81**, 31–36 (2017)

Design of F-Shaped Reconfigurable Bandstop Filter for Interference Resistant UWB Communication

Hashinur Islam, Sourav Dhar, Saumya Das, Tanushree Bose, and Bijay Rai

Abstract Indoor UWB (3.1–10.6 GHz) communication may disturb the existing primary users of WiMax, WLAN/ DSRC. Thus, there is a strong need for bandstop filters to provide free passage to the primary users. Reconfigurability is an essential requirement of cognitive radio implementation. A robust F-shaped reconfigurable bandstop filter design is presented in this work and the design is validated using HFSS, CST-MWS, and ADS platforms.

Keywords F-shaped · Reconfigurable filter · Bandstop filter · UWB communication

1 Introduction

The frequency band from 3.1 to 10.6 GHz has been approved by the Federal Communications Commission (FCC) as an unlicensed ultra-wideband (UWB) radio technology for indoor communication applications in the year 2002 [1, 2]. This UWB covers existing frequencies such as popular WLAN (5.2 GHz) and its short-range

H. Islam · S. Dhar (✉) · S. Das · T. Bose
Department of Electronics and Communication Engineering, Sikkim Manipal Institute of Technology, Sikkim Manipal University, Gangtok, Sikkim, India
e-mail: sourav.dhar80@gmail.com

H. Islam
e-mail: hashinur0001@gmail.com

S. Das
e-mail: saumya.das.1976@gmail.com

T. Bose
e-mail: tanushree.contact@gmail.com

B. Rai
Department of Electrical and Electronics Engineering, Sikkim Manipal Institute of Technology, Sikkim Manipal University, Gangtok, Sikkim, India
e-mail: bijay.r@smit.smu.edu.in

© Springer Nature Singapore Pte Ltd. 2020
R. Bera et al. (eds.), *Advances in Communication, Devices and Networking*,
Lecture Notes in Electrical Engineering 662,
https://doi.org/10.1007/978-981-15-4932-8_37

335

version of the DSRC (5.8 GHz), as well as WiMAX (3.5 GHz). These communication systems populated with the primary users and possibly interfere with UWB signals of interest. Thus, the coexistence of the UWB indoor communication system along with the existing communication system will cause distortion in each other's signals. A possible solution to deal with this problem is to introduce multiple notches within the passband of the UWB bandpass filter (BPF). To achieve interference resistant UWB communication systems, one of the primary concerns is the design of a reconfigurable and compact bandstop filter with multiple bands rejection capabilities over the whole operating band.

The microstrip filters possess attractive physical features like low-cost, small in size, and simpler structure. Due to these exciting characteristics reconfigurable bandpass or band-reject filters are expected to become an important device for the future generation multiband and multimode wireless radios [3].

An electronically controllable miniaturized reconfigurable bandpass filter for WiMax and WLAN applications has been proposed in [4]. By using a pair of C-shaped slots in an H-shaped strip in the proposed bandpass filter, two new resonances are achieved and it is demanded that the structure is simpler than the contemporary filter structures given in different literature. A wide bandstop switchable microstrip filter with an ultra-compact dimension of 47.616 mm^2 has been presented in [5]. Switchablility of this filter is achieved with the use of PIN diode switches. Bandpass and lowpass filters are achieved depending on the switch status. In the article [6], a UWB-BPF is developed with reconfigurable notch capability for WiMAX and WLAN band. The inclusion of rectangular resonator and stepped impedance resonator in the UWB-BPF design brought the notch characteristics in the response. The conductance of the microstrip line is varied to adjust the desired notch band. Reconfigurability has also been explored to achieve four UWB operational modes by including a transversal signal interference structure integrated with a switch controlled current paths [7]. The bandwidth of BPF has been controlled by placing transmission zeros with signal interference effect and line stubs. An interference-resistant UWB filter is also developed by exploring a microstrip CPW broadside coupling structure [8]. Reconfigurable bandstop characteristics within the UWB band is achieved for RF identification (RFID) communication (6.8 GHz) and WLAN communication (5.725–5.825 GHz). Reconfigurability is also explored for converting bandpass filters to bandstop filters [9]. Second and third-order conversion of bandpass and bandstop filter is developed in the range 2–2.7 GHz.

Referring to literature it can be said that the reconfigurable notches within the UWB range may be designed by different approaches and these filters are essential for building interference-resistant UWB filter.

In the current work, the design of a reconfigurable band-reject filter with multiple notches within the UWB band (3.1–10.6 GHz) has been proposed and simulated in different platforms like HFSS, ADS, and CST-MWS. This type of filter can be integrated with UWB-BPF in a cascaded connection to make the UWB communication interference less. WLAN (5.2 GHz), DSRC (5.8 GHz), and WiMAX (3.25 GHz) are three significant frequencies where the notch characteristics have been developed in this work.

2 Filter Configuration

The geometry of F-shaped reconfigurable filter is developed on FR4 substrate ($\varepsilon_r = 4.3$, loss tangent $\tan\delta = 0.025$) with a height of 0.8 mm is shown in Fig. 1. The overall size of the filter is 20×20 mm and the dimensions of every part of the F-shaped structure are mentioned in Fig. 1. CST-MWS, HFFS, and ADS electromagnetic simulators are used to simulate the proposed reconfigurable bandstop filter.

To design the F-shaped reconfigurable bandstop filter two diodes are incorporated for band switching purpose and different operational states of the proposed filter are as follows:

State 1:
When the diode D1 is ON and D2 is in OFF condition the F-shape filter gives frequency range 4.8–6.05 GHz to reject WiMAX, DSRC, and WLAN band application.

State 2:
When both the diodes are in ON condition, the F-shape filter gives frequency ranges of 3.1–3.63 GHz and 5.37–6.16 GHz. The multiband obtained for this frequency range corresponds to WiMAX, WLAN, DSRC, and ISM band applications.

Percentage bandwidth (%BW) of the bandstop filter can be calculated by the following formulas [3]

$$\%BW = \frac{f_H - f_L}{f_C} \times 100\% \qquad (1)$$

Fig. 1 Geometry of the F-shaped reconfigurable bandstop filter

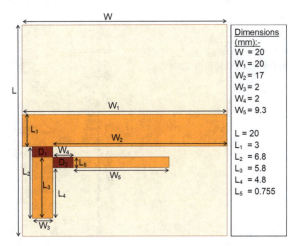

Dimensions (mm):-
W = 20
W_1 = 20
W_2 = 17
W_3 = 2
W_4 = 2
W_5 = 9.3

L = 20
L_1 = 3
L_2 = 6.8
L_3 = 5.8
L_4 = 4.8
L_5 = 0.755

where f_H=upper frequency, f_L=lower frequency and f_C= center frequency which is defined by

$$f_C = \frac{f_H + f_L}{2} \tag{2}$$

3 Results and Discussion

The designed F-shaped reconfigurable bandstop filter gives two different states using two diodes shown in Table 1.

The simulated S-parameter of the F-shaped filter is validated using three different electromagnetic simulators CST-MWS, HFSS, and ADS. The simulated insertion loss is realized greater than 20 dB for State 1 and return loss is below 2 dB displayed in Fig. 2. The 3 and 10 dB percentage bandwidth are calculated using the Eq. (1) shown in Table 1.

Table 1 Switching states of F-shaped reconfigurable bandstop filter

States	D1	D2	Frequency range (GHz)	Operating frequency (GHz)	3 dB BW (%)	10 dB BW (%)	Applications
1	ON	OFF	4.8–6.05	5.2, 5.5, 5.7, 5.8	55, 52, 50.4, 49.5	24, 22.7, 22, 21.5	WiMAX, DSRC and WLAN
2	ON	ON	3.1–3.63, 5.37–6.16	3.2, 3.5, 5.5, 5.7, 5.8	43, 39.4, 40.5, 39, 38.45	16.6, 15, 14.4, 13.9, 13.6	WiMAX, ISM, DSRC and WLAN

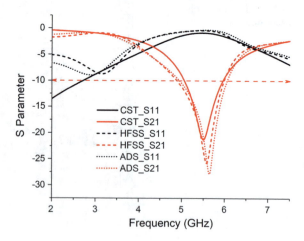

Fig. 2 Simulated S-parameters when diode D1 in ON and D2 in OFF state

Fig. 3 Simulated
S-parameters when both
diodes are in ON state

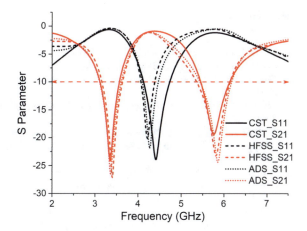

The simulated return loss for both upper and lower bands is below 2 dB and return loss is realized to be greater than 20 dB in all bands for State 2 shown in Fig. 3. The 3 and 10 dB percentage bandwidth calculated using Eq. (1) for all bands shown in Table 1.

4 Conclusion

In this research outcome, an F-shaped reconfigurable bandstop filter with two diodes has been simulated and validated using three different electromagnetic simulators CST-MWS, HFSS, and ADS. The F-shaped reconfigurable bandstop filter provides insertion loss always more than 20 dB and return loss below 2 dB in all the states. The results obtained in all the simulations strongly validate the objective of band rejection to provide interference-free passage to the primary users of WiMax and WLAN/ DSRC.

In the present work, only simulation results are shown and verified using different software (HFSS, CST-MWS, and ADS). The fabrication of this filter will be taken as the future scope of this work.

Acknowledgements The authors would like to thank Sikkim Manipal University, Sikkim, India for providing TMA Pai University Research Seed Grant-Major (Grant/ Award Number: 176/SMU/REG/TMAPURF/26/2019) for this research work.

References

1. U.F.C. Commission et al., Revision of part 15 of the commissions rules regarding ultra-wideband transmission systems. First Report and Order, ET Docket, 98–153 (2002)
2. G.R. Aiello, G.D. Rogerson, Ultra-wideband wireless systems. IEEE Microw. Mag. **4**(2), 36–47 (2003)
3. J.S.G. Hong, M.J. Lancaster, *Microstrip Filters for RF/Microwave Applications*, vol. 167 (Wiley, 2004)
4. J. Mazloum, A. Jalali, M. Ojaroudi, Miniaturized reconfigurable band-pass filter with electronically controllable for wimax/wlan applications. Microw. Opt. Technol. Lett. **56**(2), 509–512 (2014)
5. P. Kumar Singh, A. Kumar Tiwary, N. Gupta, Ultra-compact switchable microstrip band-pass filter-low-pass filter with improved characteristics. Microw. Opt. Technol. Lett. **59**(1), 197–201 (2017)
6. J. Sahay, D. Goutham, S. Kumar, A novel compact ultrawide band filter for reconfigurable notches. Microw. Opt. Technol. Lett. **57**(1), 88–91 (2015)
7. C. Shi, W. Feng, R. Gómez-García, X. Zhang, Y. Zhang, W. Che, Ultra-wideband reconfigurable filter with electronically-switchable bandpass/bandstop states. In: *2018 2nd URSI Atlantic Radio Science Meeting (AT-RASC)* (IEEE, 2018), pp. 1–4
8. Q. Zhao, C. Hua, M. Liu, Y. Lu, Compact UWB filter with a reconfigurable notched band. In: *2017 International Applied Computational Electromagnetics Society Symposium (ACES)* (IEEE, 2017), pp. 1–2
9. T. Yang, G.M. Rebeiz, Bandpass-to-bandstop reconfigurable tunable filters with frequency and bandwidth controls. IEEE Trans. Microw. Theory Tech. **65**(7), 2288–2297 (2017)

Gene Structure Analysis System

Debasmita Chakraborty, Shivam Kashyap, Udbhav Singh,
and Md Ruhul Islam

Abstract In today's world, to be able to find out the statistical significance between nucleotides or protein sequence databases is an important part of the discovery of the details of genome and correlation patterns. A very important aspect of the development of society as a whole depends on the discovery of 'what was' and 'what could be'. And it cannot be done without taking a deeper look into the genetics of mankind. The development in the field of genetics has not only helped the world of medicine in excel in various aspects but also helped mankind to find out the solution to many unsolvable health issues. Hence, 'what could be' can be derived from 'what was' and to do that, we need to take a closer look at the data that we already have, i.e. our genes. This system will receive necessary data from the given databases and will allow users to choose from a list of suitable algorithms and hence provide all the subsequent results that can be figured out from the processed data. The interface will be able to implement and compare the results of two given sets of algorithms and display the most likely output. To implement the algorithms, a certain type of Machine Learning techniques such as K-mean clustering, Neural Network Approach for Pattern Recognition, and Agglomerative clustering is also used. The resultant data can be instrumental in shaping the future. Based on the genetic algorithms and the comparison between their results, we can make various deductions which can help in the invention of new drugs, or simply understanding our own body.

D. Chakraborty · S. Kashyap · U. Singh · M. R. Islam (✉)
Department of Computer Science and Engineering, Sikkim Manipal Institute of Technology,
Sikkim Manipal University, Gangtok, India
e-mail: ruhulislam786@gmail.com

D. Chakraborty
e-mail: debikolkata2016@gmail.com

S. Kashyap
e-mail: kashyap03label@gmail.com

U. Singh
e-mail: udbhavsingh1997@gmail.com

© Springer Nature Singapore Pte Ltd. 2020   341
R. Bera et al. (eds.), *Advances in Communication, Devices and Networking*,
Lecture Notes in Electrical Engineering 662,
https://doi.org/10.1007/978-981-15-4932-8_38

Keywords Genome · Genetics · Machine learning techniques · K-Mean clustering · Neural network · Pattern recognition · Agglomerative clustering bioinformatics

1 Introduction

The trend of making software usable even without making it thoroughly complex has become a common practice these days. But, when it comes to making software that can compare gene sequence and work on a giant set of data, the accuracy is always put under the radar. The section of society, which needs these results to work on various research subjects, hence, mostly struggle to carry out the comparison and analyse the data. The process of making things available to all without the complexity is being exploited in almost all major fields of life. Because at the moment, the availability of compared and analysed data is hard to get and harder to make in a common interface, there is an utmost need to address this problem.

The proposed Gene Structure Analysis System is a must for the growing needs in the field of Bioinformatics [1] to answer biological questions. The result of the quantitative analysis [2] of data can be utilised to interpret the regulation of cells, target drug, drug formulation and disease detection. Also to keep a constant check on the constantly changing needs in the field of medical and health research [3], this system will be of immense use.

To help the users to make a faster and more efficient new approach, Gene Structure Analysis System is proposed. It will take lesser time and will produce results of higher accuracy. The analysed result can be further used or implemented in various research purposes.

2 Proposed Solution Strategy

This paper uses the comparative result of the study of two algorithms namely UPGMA and Needleman-Wunsch. It will be implemented using clustering functions available in the NET environment to form a cluster of similar nodes or trees [4]. In this, the user will be providing his dataset in the form of an ARFF file. The interface will then implement both the algorithms consecutively and compare [5] both the results. The final result of the comparison will then be displayed on the screen for the user. The user will also have the choice to store this result for further use in the form of a CSV file. The system will also record the estimated time taken for performing all the computations [6]. The timestamps will also be available for use and can be checked for time complexity. This interface doesn't store data unless the user wants to store it in the form of a CSV file, hence, it is completely safe to use.

3 Design Strategy for the Solution

The below flowchart depicts the various actions taking place in the system and the conditions responsible for the actions. It also serves the purpose of giving a clear picture of the working of the system (Figs. 1 and 2).

The algorithms that will be implemented by the system are explained as follows:

Unweighted Pair Group Method with Arithmetic Mean Algorithm:
Step 1: Construct the distance matrix by computing mismatch values between given different DNA sequences.

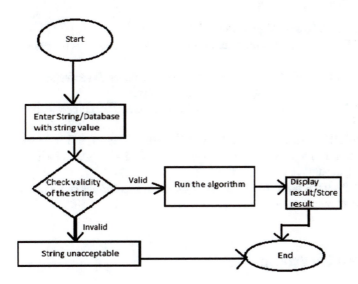

Fig. 1 Flowchart of solution strategy

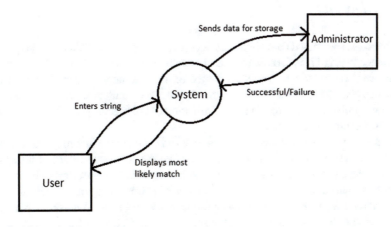

Fig. 2 0th Level of the system

Step 2: Keep repeating step 3 and 4 until you have only two clusters.

Step 3: Cluster the pair of leaves (taxa) which have the shortest distance.

Step 4: Recalculate a new average distance with the new cluster and other taxa, and make a new distance matrix, i.e.

$$d_{(ABC)} = 1/2\{d_{AC} + d_{BC}\} \tag{(1)}$$

where

d = distance

AB = Combined taxa

C = New taxa

Needleman-Wunsch Algorithm:

Step 1: Create a Matrix with columns of value A + 1 and rows of value B + 1.

For match = +1; Mismatch = −1; Gap = −2 (User driven values).

Step 2: Steps for subsequently filling the matrix:

Step 2.1: Every position Ax,y has to be the HIGHEST score available at position x, y.

Step 2.2: To calculate Ax, y, Ax, y = HIGHEST [Ax−1, y−1 + sx,y (match value or mismatch value in the diagonal) Ax, y−1 + w (gap value in 1st); Ax−1, y + w (gap value in 2nd sequence)] (2)

Step 3: Trace back: Consider the position of the current cell and look back at immediate predecessors.

4   Working Principle

4.1   Unweighted Pair Group Method with Arithmetic Mean Algorithm

In UPGMA, sequenced data alignment is considered. From the data that is acquired, a distance matrix [7] is formulated based on match and mismatch score between the sequences. The matrix is further simplified to find out the phylogenetic tree of the sequence given for analysis. From the tree that will be acquired, we can figure out the mutation probability, the percentage of mutation of the species [8]. Given following is a short example of the same.

In general terms a distance matrix is required such as one that might be created for species named A, B, C and D. Assuming that the combined values of the space between each of the species are specified in the given matrix as follows (Table 1):

In this matrix, $d_{(AB)}$ represents the number of nonmatching nucleotides divided by the total areas where a match can be found, between species A and B, while $d_{(AC)}$ is the value of the distance between the species A and C, and so on.

Table 1 Distance matrix

Species	A	B	C
B	$d_{(AB)}$	–	–
C	$d_{(AC)}$	$d_{(BC)}$	–
D	$d_{(AD)}$	$d_{(BD)}$	$d_{(CD)}$

This process is continued until only one single value remains in the matrix. A practical example using UPGMA with actual sequence data will help illustrate the general approach just described. Consider we're having alignment between six different DNA sequences as follows:

A: ATCGTGGTACTG
B: CCGGAGAACTAG
C: AACGTGCTACTG
D: ATGGTGAAAGTG
E: CCGGAAAAGTTG
F: TGGCCCTGTATC

The pairwise distance matrix that summarises the number of nonmatching nucleotides between all possible pairs of sequences shown in the Fig. 3.

Now we'll have the new matrix as follows (Fig. 4):

Fig. 3 Distance matrix

	A	B	C	D	E	F
A	------	9	2	4	9	11
B		------	9	6	3	11
C			------	5	9	11
D				------	6	10
E					------	10
F						------

Fig. 4 Distance matrix

	A/C	B	D	E	F
A/C	------	9	4.5	9	11
B		-----	6	3	11
D			----	6	10
E				-----	10
F					-----

Continuing with the calculations, we reach the final epoch where we have only one value left in the matrix, as follows:

	A/C/D/B/E	F
A/C/D/B/E	–	5.5
F		–

Hence, the Newick format for these set of strings will be ((((A, C), D), (B, E)), F). From this, we can deduce the ancestors and predecessors of a certain species and also calculate the change that it has undergone is a certain period.

4.1.1   Needleman-Wunsch Algorithm

Needleman-Wunsch algorithm is used the most when we consider the quality of global alignment as the most important factor. In this algorithm, a set of any two strings of any certain length is compared to find out the most optimal alignment which will give the maximum match score. There are 3 components to be considered: match score, mismatch score and gap. The acquired strings are aligned in the form of a matrix and for each of the scenarios of the matrix, a score is assigned. The algorithm breaks down a large chunk problem into simpler ones and the results are used to recreate a solution to the bigger problem [9].

5   Implementation & Result

The system that is proposed will take inputs in the form of strings or database containing string values. It will then implement both the algorithms one by one. In the Needleman-Wunsch algorithm, the output will be in the form of an optimal alignment of the strings using gap values efficiently that will give the highest match value, meaning, the string value that will be acquired will be of species which shares the maximum similarity both structurally and functionally [10]. In UPGMA, the output is in the form of hierarchical clusters formed using the agglomerative clustering technique. The clusters show the phylogenetic tree formation visually. It also shows the classes of the clusters depicting the levels of the phylogenetic tree. The distance between various levels is also displayed in the resultant graph. The result of UPGMA can be used to calculate the mutation percentage and the probability of mutation in the future.

It is experimentally seen that the time taken to reach the complete execution of Needleman-Wunsch [11] is higher than UPGMA due to more complex calculations happening in the system. The Needleman-Wunsch algorithm can only work on any given two sequences of string whereas, UPGMA can consider as many set as the user wants. But, Needleman-Wunsch algorithm is considered especially when quality is to be considered over quantity [12]. Besides, supporting our claim, the abstraction level of the system is quite high because the user only needs to provide with the database and the result will be displayed. Hence, needing no manual work, making it user friendly in nature (Figs. 5, 6, and 7).

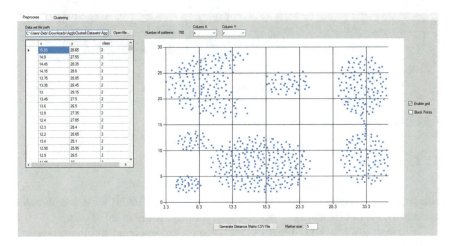

Fig. 5 UPGMA in the form of clusters

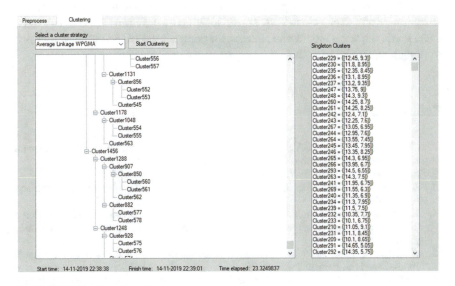

Fig. 6 UPGMA using clustering

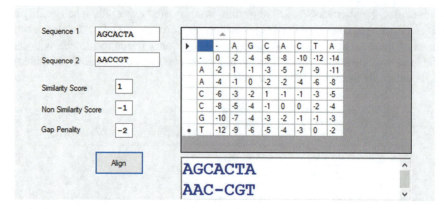

Fig. 7 Needleman-Wunsch algorithm implementation

6 Conclusion

This paper successfully performs the task of comparison and analysis of user-directed datasets to find out similarity or dissimilarity among the given set of data.

This paper can also be of great use for research and implementation purposes. It can be used as a base paper to be worked upon on a larger scale with bigger databases and even more advanced algorithms [12].

The complexity of this paper has been scaled down to suit the basic need of practitioners and even researchers. It can scale up as per need. All that has to be done, is add the necessary modules. Also, the compactness of the interface makes it working simpler and smooth.

References

1. S. Needleman, C. Wunsch, A general method applicable to search for similarities in the amino acid sequence of two proteins. J. Mol. Biol. **48**(3), 443–453 (1970)
2. M.R. Anderberg, *Cluster Analysis for Application* (Academic Press, 1973)
3. D. Sankoff, J.B. Kruskal, Macromolecules: *The Theory and Practice of Sequence Comparison* (Addison-Wesley, 1983)
4. T. Zhang, R. Ramakrishnan, M. Livny, BIRCH: an efficient data clustering method for very large databases, in *Proceedings of the 1996 ACM SIGMOD International Conference on Management of Data* (1996), pp. 103–114
5. A. Monge, C. Elkan, The field matching problem: algorithms and applications. Proc. Knowl. Discov. Data Min. **1996**, 267–270 (1996)
6. R. Durbin, S. Eddy, A. Krogh, G. Mitchison, *Biological Sequence Analysis* (Cambridge University Press, Cambridge, 1998)
7. K. Rose, Deterministic annealing for clustering, compression, classification, regression, and related optimization problems. Proc. IEEE **86**(11), 2210–2239 (1998)

8. S. Datta, J. Arnold, Some comparisons of clustering and classification techniques applied to transcriptional profiling data, in *Advances in Statistics, Combinatorics and Related Areas*, ed. by C. Gulati, Y.X. Lin, S. Mishra, J. Rayner (World Scientific, 2002) pp. 63–74

9. S. Datta, S. Datta, Comparisons and validation of statistical clustering techniques for microarray gene expression data. Bioinformatics **19**, 459–466 (2003). https://doi.org/10.1093/bioinformatics/btg025

10. S. El-Metwally, O. Ouda, M. Helmy, Next-generation sequencing technologies and challenges in sequence assembly. Springer Sci Bus **7**, 16–25 (2014)

11. M. Fakirah, M.A. Shehab, Y. Jararweh, M. Al-Ayyoub, Accelerating needleman-wunsch global alignment algorithm with GPUs, in *2015 IEEE/ACS 12th International Conference of Computer Systems and Applications (AICCSA)* (IEEE, 2015), pp. 1–5

12. Y. Jararweh, M. Al-Ayyoub, M. Fakirah, L. Alawneh, B.B. Gupta, Improving the performance of the needleman-wunsch algorithm using parallelization and vectorization technique by, published by Springer US. Online ISSN: 1573-7221. First published online: 21 Aug 2017

Prototype of Accident Detection Alert System Model Using Arduino

Nihit Kumar, Niroj Dey, Ritushree Priya, and Md. Ruhul Islam

Abstract As the usage of vehicles is increasing day by day, the hazards due to vehicles has also increased [1]. The main cause for accidents is high-speed, drink and drive, over stress work life, using a cellphone while driving, red light Jumping, overtaking in wrong manner, avoiding safety gears, etc. [2] According to the National Crime Records Bureau (NCRB) 2016 report around 148,707 road accident deaths are happening every year [3]. People are losing their life due to the fact that the medical rescue teams are not arriving on time at the place of the accident because there is no proper alert system which can notify the rescue team about the accident as soon as it happens. This paper describes the model for the accident detection notification system which will notify the emergency services about the accident as soon as it takes place. Once the vehicle meets with an accident, the sensor present inside the vehicle will detect it immediately and sends a message(SMS) to the emergency services with the accident coordinates. If the people inside the vehicle are safe a reset button is provided which can be tapped to terminate the message from being sent to the emergency services.

Keywords Road accidents · Accident detection · Microcontroller · SMS notification

N. Kumar · N. Dey · R. Priya · Md. Ruhul Islam (✉)
Department of Computer Science and Engineering, Sikkim Manipal Institute of Technology,
Sikkim Manipal University, Gangtok, India
e-mail: ruhulislam786@gmail.com

N. Kumar
e-mail: nihitkumar39@gmail.com

N. Dey
e-mail: deyneeraj666@gmail.com

R. Priya
e-mail: ritushree8335@gmail.com

© Springer Nature Singapore Pte Ltd. 2020   351
R. Bera et al. (eds.), *Advances in Communication, Devices and Networking*,
Lecture Notes in Electrical Engineering 662,
https://doi.org/10.1007/978-981-15-4932-8_39

1 Introduction

Due to the deficiency in the availability of emergency rescue facilities in our country, the lives of the people are under great risk. [1] An automatic accident detection and notification system model for vehicles is presented in this project will transmit the alert message and the coordinates of the accident to the emergency contact number within a few seconds of the accident. This model has a vibration detection sensor that can detect accidents [2] and will send an alert message to the contacts and emergency facilities that are entered in the program of the microcontroller(Arduino) via GSM module and trace the location via GPS module [3]. Then a notification message contains the Google map link with the geographical coordinates of the accident sent to the emergency facilities.

When an accident occurs a huge amount of vibration is produced which is detected via. Vibration sensor, the sensor transmits the data to the microcontroller [4]. The microcontroller sends the alert message automatically to the predefined emergency contacts of the victim. The message is sent using the GSM module and the coordinates of the accident are detected with the help of the GPS module. Hence with this project implementation, we can detect the position of the vehicle where the accident has occurred and can save the valuable life of the victim.

2 Problem Definition

According to the National Crime Records Bureau (NCRB) 2016 report around 148,707 road accident deaths are happening every year [3]. Due to the late arrival of emergency facilities at the accident spot, the people are losing their life. The main reason for the delay in emergency rescue facility is that they are not being notified about the accident on time as there is no proper system to detect the accident coordinates and send it immediately to them.

3 Proposed Solution Strategy

An accident detection notification model is proposed which will have the following components:

A vibration sensor (SW-420) which is used to measure the vibration caused during the collision of the vehicle, An ultrasonic sensor (HC-SR04) which is used to detect the presence of a human inside the vehicle. If the sensor does not detect the presence of a human the whole system will not be activated. An Arduino UNO R3, a microcontroller that is used to read the vibration frequency and it also activates the GPS module to fetch the coordinate of the accident. A NEO-6 m GPS module is used which will provide us the coordinate of the accident and sends the location

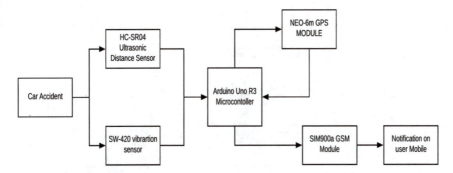

Fig. 1 Block diagram of the system

coordinates to the Arduino. The Arduino sends the GPS coordinate to the emergency facilities in the form of a Google maps SMS link via GSM module(SIM900A) [5] (Fig. 1).

The above figure shows the block design layout of the system from car accident detection to the notification about the accident to the user.

4 Design Strategy for the Solution

The above flowchart clearly depicts the various actions taking place in the system and the conditions responsible for the actions. It also serves the purpose of giving a clear picture of the working of the system (Figs 2, 3, and 4).

5 Working Principle

5.1 Vibration Sensor (SW-420)

The vibration sensor Sw-420 is a vibration detection sensor that has a comparator LM393 and a potentiometer for adjusting sensitivity [6]. A signal indication LED is also provided on the board. This sensor module gives a logic low output when there is no vibration and a logic high which ranges from 2.5 to 5v is detected depending on the intensity of vibration (Fig. 5).

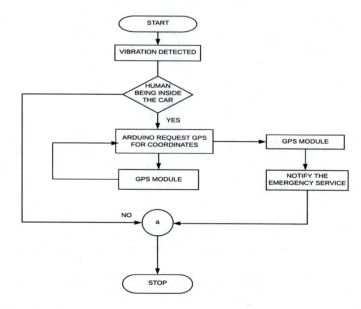

Fig. 2 Flowchart of solution strategy

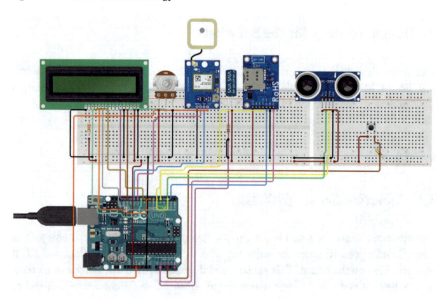

Fig. 3 Circuit diagram of the model

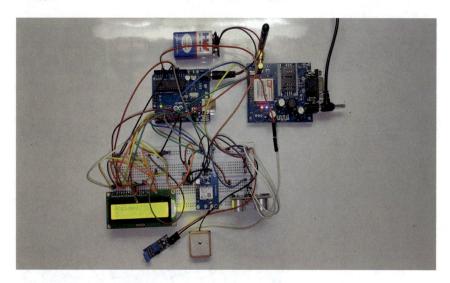

Fig. 4 Final product of the circuit

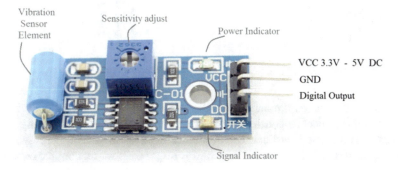

Fig. 5 Vibration sensor (SW-420)

5.2 Ultrasonic Sensor Module (HC-SR04)

The HC-SR04 ultrasonic sensor uses sound navigation ranging technology It is highly accurate and provides a stable reading. The form factor varies from 2–400 cm or 1–13 feet. It is capable of working in sunlight or with black material, although soft materials like cloth can be difficult to be sensed by the sensor. It has both ultrasonic transmitter and receiver module [7]. This module will detect the presence of a human inside the vehicle during the accident. If a human is present at the driving seat inside the vehicle, it will send a message to the microcontroller (Fig. 6).

Fig. 6 Ultrasonic sensor
(HC-SR04)

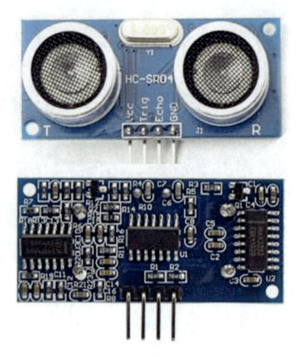

5.3 Arduino UNO R3

Arduino is working as the main brain for our project. It is a central coordination system in this project responsible for all the movements and activities going in this project. It is a single board microcontroller having a set of general purpose input–output pins (GPIO) [5, 7]. In our project it is responsible for reading the data from the vibration sensor and ultrasonic sensor, fetching the coordinates from the GPS module, and sending the fetched coordinates to the emergency services using a GSM module (Fig. 7).

Fig. 7 Arduino UNO R3

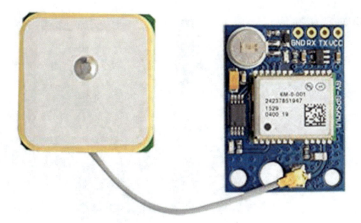

Fig. 8 GPS module (NEO-6 m)

5.4   GPS Module (NEO-6 M)

NEO-6 M is a global positioning system (GPS) module [8]. It is less in cost and high in performance GPS module. It comes with a ceramic patch antenna, a on-board memory chip, and a backup battery that can be used with many microcontrollers available in the market [8]. This module will fetch the coordinate of the accident and it will send back those coordinates to the microcontroller (Fig. 8).

5.5   GSM Module (SIM900a)

The SIM900A is a Dual-band GSM in a SMT module which is an industry-standard interface, the SIM900A delivers GSM/GPRS 900/1800 MHz performance capabilities for voice call, SMS, Data surfing. The small form factor provides low power consumption and is capable of running in between 3.5 and 5 v [6, 9]. This module is responsible for sending the coordinates to the Emergency services via SMS (Fig. 9).

6   Implementation & Result

See (Table 1).

Fig. 9 SIM900a GSM module

Table 1 Implementation test case with their results

S. No.	Test case	Result
1.	Testing Arduino by blinking in-built LED	LED is linking and all the GPIO pins are functioning well
2.	Testing the vibration sensor	Getting the reading between 2.5 and 5.0v depending on the impact of the force
3.	Testing the ultrasonic sensor	Reading the distance between the object and the sensor
4.	Testing GPS module	Receiving the coordinates of the current user location
5.	Testing GSM module	Sending the message to the registered user phone number

7   Conclusion

This project will be successful in minimizing the time taken by the Emergency service to reach the point of accident and to save the life of the victim. It provides a convenient alert notification to the emergency services and ensures that none of the precious life is lost which can be saved by the arrival of the medical team on time.

This model can work as the base model for newer technologies to be built over it. In the future [8] the accuracy of detecting the vibration can be improved by using industrial grade vibration detecting sensors and adapting the methods of Artificial Intelligence [10].

References

1. B. Madhu Mitha, G. Jayashree, S. Mutharasu, Vehicle accident detection system by using GSM and GPS gowshika. Int. Res. J. Eng. Technol. (IRJET). **06**(01) (2019) e-ISSN: 2395-0056 www. irjet.net p-ISSN: 2395-0072
2. Traffic accidents, NCRB 2016 Report, Chapter 1A: Traffic Accidents, Government of India
3. http://jhtransport.gov.in/causes-of-road-accidents.html
4. T. Kalyani, S. Monika, B. Naresh, M. Vucha, Accident detection and alert system. Int. J. Innov. Technol. Explor. Eng. (IJITEE). **8**(4S2) (2019). ISSN: 2278-3075
5. K. Shah, V. Parle, S. Bairagi, Accident detection and conveyor system using GSM and GPS module. Int. J. Comput. Appl. (0975-8887)
6. R. Ganiga, R. Maurya, A. Nanade, Accident detection system using Piezo Disk Sensor. Int. J. Sci., Eng. Technol. Res. (IJSETR). **6**(3) (2017). ISSN 2278-7798
7. http://www.circuitstoday.com/interface-gsm-module-with-arduino
8. U. Sengupta, P. Kumari, T. Paul, M.R. Islam, *Chapter 61 Smart Water Pump Controller* (Springer Science and Business Media LLC, 2019)
9. http://randomnerdtutorials.com/guide-to-neo-6m-gps-module-with-arduino/
10. E. Mathe, E. Spyrou, Connecting a consumer brain-computer interface to an internet-of-things ecosystem, in *Proceedings of the 9th ACM International Conference on PErvasive Technologies Related to Assistive Environments - PETRA '16* (2016)

Embedded Automobile Based Vehicular Security System

Saptarshi Chakravarty, Aman Srivastava, Amrita Biswas, Kathyaini Rao, and Gargee Phukan

Abstract Electronics embedded systems have become an essential and integral part of modern vehicles. This paper deals with the design and development of an embedded system, which shall be programmed to prevent/control the theft of a vehicle. The developed instrument is a basic embedded system based on Bluetooth and fingerprint technology. The first module is placed in the engine of the vehicle and the other module is placed on the dash of the vehicle, both the MCU's are interfaced with a bluetooth module. The ignition relay of the vehicle can be programmed in such a way that it responds to commands/fingerprint received from the transmitter MCU only. The Bluetooth transmitter is identified by the MCU using the hardware address which is unique for each transmitter. In addition to the above features, an extra feature of facial recognition can also be introduced to ascertain more security of the vehicle.

Keywords Bluetooth · Fingerprint · Interfacing · VSS (Vehicle Security Systems) · ISS (Intelligent Security System)

S. Chakravarty · A. Srivastava · A. Biswas (✉) · K. Rao · G. Phukan
Electronics and Communication Engineering, Sikkim Manipal Institute of Technology, 737136
Gangtok, Sikkim, India
e-mail: amritabis@gmail.com

S. Chakravarty
e-mail: chakravartysaptarshi@gmail.com

A. Srivastava
e-mail: amansriv.srivastava6@gmail.com

K. Rao
e-mail: kath210798@gmail.com

G. Phukan
e-mail: gargee161011@gmail.com

© Springer Nature Singapore Pte Ltd. 2020
R. Bera et al. (eds.), *Advances in Communication, Devices and Networking*,
Lecture Notes in Electrical Engineering 662,
https://doi.org/10.1007/978-981-15-4932-8_40

1 Introduction

Nowadays, vehicle thefts are increasing at an alarming rate all over the world. So, in order to prevent this from happening, most of the vehicle owners have started using Anti-theft control systems. But these anti-theft vehicular systems are very expensive in the commercial market and do not ensure 100% safety even after installation. Here, we made an attempt to develop an instrument using Bluetooth technology which is a simple and low-cost vehicle theft control embedded system. Before we begin the project it is very essential to know about the basics of an embedded system and how it functions [1]. To obtain a clearer knowledge of the internal embedded system of a vehicle with regard to the industrial requirements we have taken the help of the book [2] which provided us with key insights related to the basic embedded system and some significant knowledge with reference to our project.

The developed instrument is an embedded system based on bluetooth technology. Reference [3] was helpful to deliver important and useful information regarding the bluetooth module interface. Different projects and papers were studied and verified from [4], to come up with a better and innovative solution to this problem. The instrument is finally installed in the dashboard of the vehicle, and the interfacing bluetooth module is also connected to the Arduino MCU, after which the car's ignition is powered only if the fingerprint of the user matches or the bluetooth device confirms pairing with the authorised device.

2 Working

Phase 1: Establishing a connection between Arduino and fingerprint sensor, and also interfacing it with a Bluetooth device for future prospects.
Firstly, the fingerprint sensor has a flash memory that can take up to a maximum of 200 fingerprints.

The user's fingerprint is stored in the sensor and only that particular fingerprint and incase an extra one are interfaced through Arduino.

The Arduino has the sole authority in determining the eligibility of the fingerprint.

Phase 2: Connection between the Arduino of both the control modules.
The Arduino which possesses the fingerprint credentials establishes a connection with the other Arduino interfaced with relay.

Phase 3: The Arduino in the second control module is used for driving the relay.
As soon as the connection is established between the two Arduinos, the MCU turns on the ignition by activating the relay and allowing the battery to discharge.

The DC fan rotates depicting that the car has acquired ignition.

3 Flowchart

See (Fig. 1).

Fig. 1 Flowchart depicting the entire working

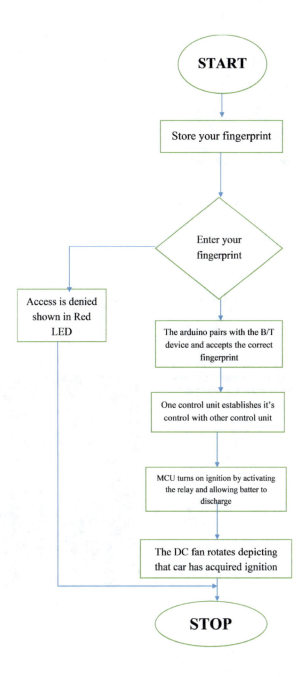

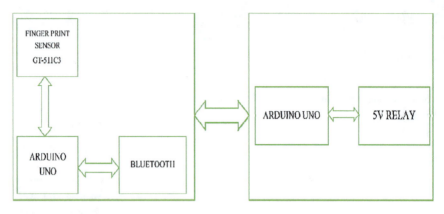

Fig. 2 Centralised embedded system

4 Block Diagram

See (Fig. 2).

5 Images of the Prototype

See (Figs. 3, 4, 5, and 6).

Fig. 3 Fingerprint sensor
ready to accept fingerprint

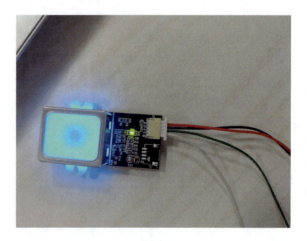

Fig. 4 Fingerprint sensor interfaced with Arduino microcontroller

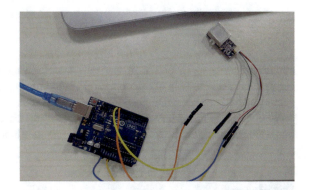

Fig. 5 Stepper motor signifying the engine of the vehicle and the 5 V ignition relay

Fig. 6 The bluetooth module

6 Conclusion

The basic objective throughout the project has been to take precautionary measures regarding vehicles. To obtain it, we must avoid the theft of vehicles. With bringing in fingerprint and bluetooth communication modes, vehicular theft management becomes smarter and much more feasible. The user can protect their cars in a much more modern and smarter way. Also, by identifying the driver digitally, chances of this technology being misused or stolen become minimal.

References

1. R. Kamal, *Embedded System: Architecture, Programming and Design*, 2nd edn. (Tata McGraw Hill Publisher, 2008)
2. N. Navet, F. Simonot-Lion, *Automotive Embedded Systems Handbook (Industrial Information Technology)* (2009), pp. 1–1, 1–3, 3–1, 4–1, 7–1, 8–1, 11–1
3. H.H.B. Aziz, N.H.A. Aziz, K.A. Othman, Mobile phone car ignition system using Embeded-Blue 506 Bluetooth technology, in *2011 IEEE International Conference on Control and System Graduate Research Colloquium* (IEEE, 2011), pp. 1–4
4. CCMTA Best Practice Models for Combating Auto Theft, Version 6.1, Anti AutoTheft Project Group (2006)
5. http://www.howstuffworks.com
6. http://www.wikipedia.org
7. http://www.atmel.com
8. http://www.simcom.com
9. http://www.datasheetcatalog.com

High Gain X-band Rectangular Microstrip Antenna Using Fractal Slot Geometry

Tanmoy Sarkar, Abhijyoti Ghosh, Subhradeep Chakraborty, L. Lolit Kumar Singh, and Sudipta Chattopadhyay

Abstract This paper presents an investigation on fractal slots on a rectangular patch antenna (RMA) and its suitable position on the patch for improved input and radiation characteristics of the RMA. The study has been carried out on RT duroid 5870 substrate ($\varepsilon_r = 2.33$, height $= 1.575$ mm) using Ansoft high-frequency structure simulator (HFSS) software. Different fractal slot geometries are incorporated at the corners of the patch to achieve improved input and radiation performance. Stable radiation pattern with co-polarization gain of 9.58 dBi with 3-db beam width of 53° along with 24% active patch area reduction has been achieved through the final structure.

Keywords Rectangular microstrip antenna · Fractal geometry · Gain · Beamwidth

1 Introduction

Demand of small handheld devices is very much popular in current era. To accumulate the complex circuitry in a closed cluster is a challenging job for manufacturing industry. Antenna is one of the vital components of handheld communication devices. The performance of the communication device largely depends on the performance of the antenna associated with the device. In this context, microstrip antennas are very much popular in these handheld communication devices due to its tininess, lightweight, simple design, ease of implementation, and easy compatibility with microwave-monolithic-integrated circuits. Thus, microstrip antennas find increasing number of new applications day by day. However, microstrip antennas in

T. Sarkar
Radionics Lab, Department of Physics, The University of Burdwan, Burdwan, India

A. Ghosh (✉) · L. Lolit Kumar Singh · S. Chattopadhyay
Department of ECE, Mizoram University, Aizawl, Mizoram, India
e-mail: abhijyoti_engineer@yahoo.co.in

S. Chakraborty
Microwave Devices Area, CSIR-CEERI, Pilani, India

© Springer Nature Singapore Pte Ltd. 2020
R. Bera et al. (eds.), *Advances in Communication, Devices and Networking*,
Lecture Notes in Electrical Engineering 662,
https://doi.org/10.1007/978-981-15-4932-8_41

their conventional structure suffer from some limitations such as low gain (3–4 dBi), narrow impedance bandwidth (2–4%), and poor co-polarization to cross-polarization isolation (polarization purity) (10–13 dB) predominantly in the H-plane [1]. Different work has been carried out by antenna research community to overcome the limitations of such antennas in such a way that it can be associated with the modern handheld communication devices even more effectively. Use of fractal shape geometry for miniaturization of microstrip antenna is popular in antenna research. Different types of fractal shapes like Minkowski-island-shaped fractal [2] and Sierpinski-carpet-shaped fractal [3] have reported to achieve miniaturization. Microstrip antenna miniaturization with fractal EBG and SRR loads for linear and circular polarizations is reported in [4] with unstable reflection coefficient profile. In [5], the miniaturization of a microstrip patch antenna has been carried out with a Koch fractal structure. In this report, the shifting in reflection coefficient is high over different iteration processes and the gain is also very low. A miniaturized concentric hexagonal fractal rings based monopole antenna has been documented in [6] without any evidence of improved gain. In [7], an E-shape fractal geometry based multiband patch antenna has been studied, where the radiation pattern changes with multiple nulls with number of iterations. Square shape fractal geometry [8] has been inserted for miniaturization in a loop antenna in [8]. Multiple order-2 Koch fractal structure has been implemented in [9] for the purpose of miniaturization but failed to improve other characteristics of the antenna. A Jerusalem cube fractal base patch antenna has been reported in [10]. A high VSWR in higher frequency and high directive pattern has been observed in [10] which is not suitable for omnidirectional communication. In all the above cases, the antenna input and output parameters have altered due to structural changes.

In the present investigation, a simple single layer single element rectangular microstrip antenna with square type fractal cuts imported at the corners of the patch is proposed for improved radiation and input characteristics. The proposed structure provides high gain along with reduction of active patch area (miniaturization) simultaneously which is very rear with fractal-geometry-based microstrip patch antenna. The fractal cut has been introduced at all the corners of a rectangular patch antenna for stable radiation pattern. The simulation has been carried out in different iteration steps and improved input and radiation performance has been observed.

2 Antenna Structure

At the beginning, a 8×12 mm^2 co-axial feed microstrip patch antenna has been designed over a RT duroid substrate (66×66 mm^2)with permittivity $(\varepsilon_r) = 2.33$ and height $(h) = 1.575$ mm using commercially available High-Frequency Structure Simulator (HFSS v.14). The antenna length is oriented along xz-plane and width along yz-plane which is shown in Fig. 1. The complete parameters of the conventional structure are given in Table 1. Afterward square-shaped fractal geometry has been introduced at the four corner of the patch in three different iterations. The gradual incorporation of the fractal geometry is shown in Fig. 2.

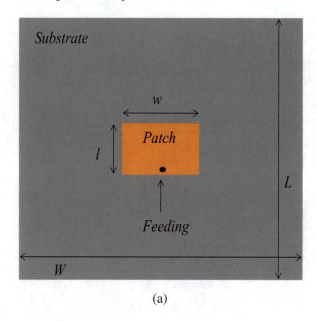

(a)

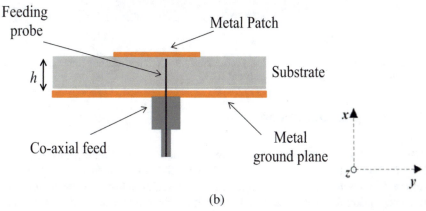

(b)

Fig. 1 Schematic diagram of conventional rectangular patch antenna **a** top view, **b** side view

Table 1 Detail parameters of the conventional RMA

L (mm)	W (mm)	l (mm)	w (mm)	f_r (GHz)	ε_r	h (mm)
66	66	8	12	10.86	2.33	1.575

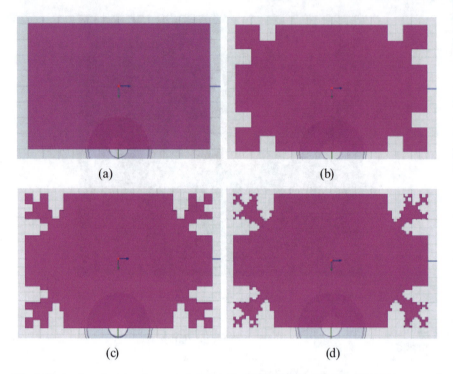

Fig. 2 Evaluation of proposed structure **a** conventional RMA, **b** first iteration (first ITE), **c** second iteration (second ITE), **d** third iteration (third ITE)

3 Result and Discussion

The simulation has been carried out using commercially available High-Frequency Structure Simulator (HFSSv14). Figure 3 shows the reflection coefficient profile of different iterations along with conventional antenna. From Fig. 3, it can be observed that convention RMA resonates at 10.86 GHz while the present antenna (3rd ITE) operates around 10.5 GHz. This confirms that the active patch area of the present antenna reduces by 24% compared to conventional RMA.

Radiation pattern for different iteration steps with conventional antenna is also compared and shown in Fig. 4. It can be seen that the radiation pattern for all the structures is very stable and symmetric in both E-plane and H-plane. The gain of the antenna is around 9.58 dBi which is 3 dB higher than conventional RMA. The polarization purity is also moderate.

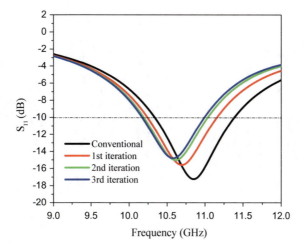

Fig. 3 Reflection coefficient profile at different iterations

The electric field magnitude over the substrate in conventional RMA as well as the present antenna is shown in Fig. 5. From the distribution, it is confirmed that each of the antenna has been excited by its dominant resonant TM_{10} mode and no other higher mode has not come into picture after perturbation of the structure. The entire radiation pattern shows a good broad side radiation with better co-cross isolation.

4 Conclusion

In this paper, a fractal slot geometry based RMA has been studied. The response has been obtained using structural changes on a conventional patch antenna. The input and output parameters like reflection coefficient, radiation pattern, and electric field distribution have been observed using HFSS simulation software. The antenna has excited by its dominant TM_{10} mode which is unaltered after perturbation using fractal slots. Fractal slots from four corners have less disturbance in modal variations. So any rectangular patch antenna miniaturization process using fractal slots should be incorporated at the corner only. A stable radiation with co-polarization gain of 9.58 dBi with 53° beam width (3-dB) in both principle planes along with 24% miniaturization has been achieved.

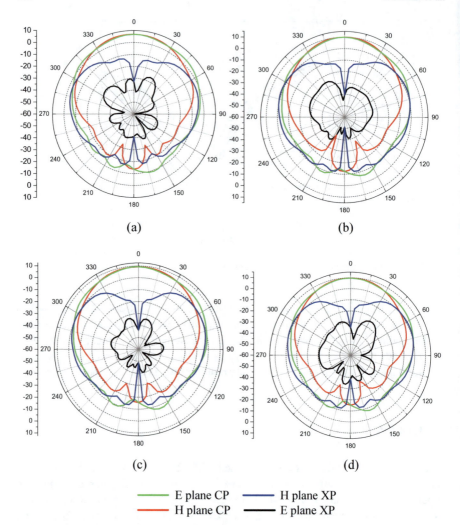

(a)

(b)

(c)

(d)

E plane CP
H plane CP
H plane XP
E plane XP

Fig. 4 Radiation pattern for different types of antennas **a** conventional RMA, **b** first iteration (first ITE), **c** second iteration (second ITE), **d** third iteration (third ITE)

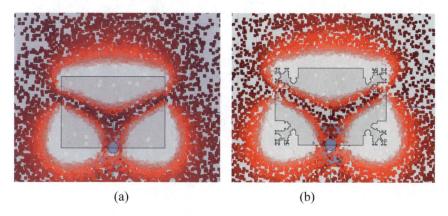

(a)   (b)

Fig. 5 E-field distribution **a** conventional RMA, **b** third Iteration (third ITE)

References

1. R. Garg, P. Bhartia, I. Bahl, A. Ittipiboon, *Microstrip Antenna Design Handbook* (Artech House, Norwood, 2001)
2. J.P. Gianvittorio, Y. Rahmat-Samii, Fractal antennas: a novel antenna miniaturization technique, and applications. IEEE Antennas Propag. Mag. **44**, 20–36 (2002)
3. A. Rakholiya, N. Langhnoja, A. Dungrani, Miniaturization of microstrip patch antenna for mobile application. Int. Res. J. Eng. Technol. **4**, 917–921 (2017)
4. M.S. Sedghi, M. Naser-Moghadasi, F.B. Zarrabi, Microstrip antenna miniaturization with fractal EBG and SRR loads for linear and circular polarizations. Int. J. Microw. Wirel. Technol. **9**, 891–901 (2016)
5. E.A.M. Souza, P.S. Oliveira, A.G. D'Assunção, L.M. Mendonça, C. Peixeiro, Miniaturization of a microstrip patch antenna with a koch fractal contour using a social spider algorithm to optimize shorting post position and inset feeding. Int. J. Antennas Propag. 1–10 (2019)
6. K. Djafri, M. Challal, R. Aksas, F. Mouhouche, M. Dehmas, Miniaturized concentric hexagonal fractal rings based monopole antenna for LAN/WiMAX application. Radio Eng. **28**, 39–44 (2019)
7. A. Janani, A. Priya, Design of E-shape fractal simple multiband patch antenna for S-band LTE and various mobile standards. Int. J. Eng. Sci. **3**, 12–19 (2013)
8. H.M. Elkamchouchi, M.N. Abd El-Salam, Square loop antenna miniaturization using fractal geometry, in *Proceedings of the Twentieth National Radio Science Conference (NRSC'2003)* (IEEE Cat. No.03EX665), Cairo, Egypt (2003), pp. 1–8
9. K. Garg, Miniaturization of microstrip patch antenna using fractal geometry. Int. J. Adv. Res. Comput. Commun. Eng. **5**, 180–184 (2016)
10. C. Padmavathi, Miniaturisation of patch antenna using novel fractal geometry. Int. J. Electron. Commun. Eng. Technol. **7**, 63–74 (2016)

Implementation of IoT for Studying Different Types of Soils for Efficient Irrigation

Sankha Subhra Debnath, Abhinaba Dutta Choudhury, and Amit Agarwal

Abstract In the twenty-first century, one of the most important issues that remains is the water resource, especially freshwater and water used for irrigation. India as a third world country, agriculture remains the largest sector that contributes to the GDP of the country. Most of the freshwater goes for the irrigation sector. Even with all the development in technologies we have seen that comparatively the agricultural industry still is unable to make the most out of it and the outreach is still poor. Still today a large sector is being operated manually and the traditional techniques like drip irrigation, sprinkler irrigation are still widely used. With the help of large-scale automation along with IoT implementation it is possible to make a paradigm shift in the agriculture Industry of the country to yield the best result. This paper focuses primarily on the making of a smart irrigation system that focuses on the optimal usage of water and the study of soil samples along side with the implementation of IoT. Thus, this paper discusses not only to contribute for minimizing human efforts but also in studying and understanding the different types of soils required and their moisture holding properties for agricultural fields in the system.

Keywords Agriculture · Irrigation · Automation · Arduino · IoT · Analytics · Water management · Soil · Database

1 Introduction

The traditional method used by the gardeners usually wastes plenty of water and the labour required is also quite huge. The importance of a proper irrigation system is that, if you water the plants timely, then the yields may double if a steady supply of water is provided to the plant. The statistics that has been found from the survey shows us that 85% of freshwater resources are being used for agriculture. With growing demand and more production, more percentage of the freshwater will get utilized. The advancement of IoT and its implementation in agriculture sector with

S. S. Debnath · A. D. Choudhury · A. Agarwal (✉)
ECE Department, Sikkim Manipal Institute of Technology, Majitar, East Sikkim 737136, India
e-mail: amiteng2007@gmail.com

© Springer Nature Singapore Pte Ltd. 2020
R. Bera et al. (eds.), *Advances in Communication, Devices and Networking*,
Lecture Notes in Electrical Engineering 662,
https://doi.org/10.1007/978-981-15-4932-8_42

network and sensor technology has given us the insight of how well we can control and minimize water wastage [1, 2].

The IoT is a technology undergoing huge research and development at a very large number of sectors. The IoT technology enables any mobile devices or computers to communicate to each other [3, 4]. It is like the Internet of all electronic smart objects connected to each other. They can not only just communicate with each other via different paths and remotely but also the exchange and transfer of information occur in a very efficient and large scale over remote distances. In this case, for this project the sensors included are soil sensors, moisture and humid sensors that will be connected throughout the system.

The paper aims to develop a smart irrigation system with real-time IoT analytics for the studying of the soil that farmers will be using for irrigation. The sensors used along with microcontroller hold the central part of the hardware system. The system will be connected to an IoT platform by using the ESP8266 [5]. The real-time data will be collected from the sensors over the period that will enable us to study about the soil used in the irrigation system which will result in more optimal water management for the whole system.

2 Problem Statement

In today's busy world, it is hard to maintain a watering schedule for gardens, farms and fields because it takes a lot of time and human efforts. In traditional watering method, plenty of water gets wasted and overwatering severely damages plants. Nowadays, some systems use technology to reduce the number of workers and time required to water the plants. Even with such systems, the control is very limited, and many resources are still wasted. An extensive study of soil along with the irrigation can help us to implement smart irrigation in much more optimal way.

These days labour is becoming more and more expensive. As a result, if no effort is invested in optimizing these resources, there will be more money involved in the same process. Technology is probably a solution to reduce costs and prevent loss of resource, labour expenditure and understanding more optimal irrigation. The traditional irrigation systems shall be reviewed in order to move forward to a much more optimal smart irrigation industry in the coming decades.

3 Literature Survey

The starting of the project has been carried out by studying and analysing the current methods and approaches, analysing the requirements and finally by making a design for the system.

The basic structure of the system in this paper comprises the soil sensor, temperature and humidity sensors, and ESP8266 to connect to the ThingSpeak IoT platform for data analytics of the sensors. The data from the sensors are being fed into the IoT platform directly where we can visually represent the values of soil moisture and humidity for a large period [6, 7].

This paper on [8] 'Smart Irrigation system with IoT analytics for soil study' has been studied thoroughly for understanding the irrigation mechanism which is capable of controlling the pumping motor according to the dryness and moist value of the soil and at the same time all the values of the soil data moisture will be fed into the IoT platform for further analytics.

In the old paper, 'Arduino based Automatic irrigation System Using IoT', we have looked into the understanding of current approaches and the requirements needed for developing the system [9].

The paper 'IoT based Smart irrigation System using Node Mcu' describes how the sensors are being interfaced with the environment and how the sensing values are being taken from analog to digital thus contributing in IoT analytics.

4   Proposed System

The whole system has been divided into two major parts: the Hardware design, IoT implementation and analytics. The control unit was achieved with the help Arduino UNO R3 microcontroller. The sensors connected to the microcontrollers are soil sensor, temperature and humidity sensors that are connected with the motor pump along with the L293D motor driver. The ESP8266 Wi-Fi module was connected with the Arduino microcontroller which was configured via network to the IoT platform, namely, Thingspeak is an open-source IOT application and API to store, analyse and extract data from the sensors and objects connected. It uses the HTTP and MQTT protocols both via LAN and the net.

With the help of ESP8266 connected to the Thingspeak we have managed to send the sensing values of soil sensors and the temperature sensors to the server where we had created multiple channels where the sensing values of all the sensors can be visualized separately according to our requirements. The data is being uploaded at a very frequent time scale over a period of time that can help us to collect enough data so that we can further proceed for analysing. This procedure is being used for collecting data of a number of soil samples, hence giving us a dataset of different soil samples over a same period of time for further analysing.

4.1   Block Diagram of the System

See Fig. 1.

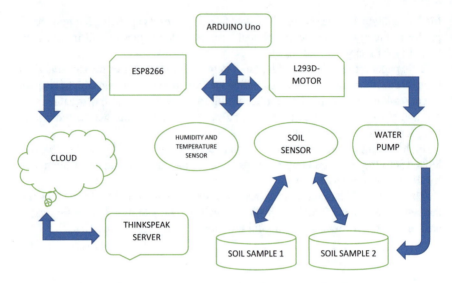

Fig. 1 Block diagram of smart irrigation system with IoT

4.2 Hardware Design

The major components of the hardware design comprised of the Arduino UNO R3, soil sensors, humidity and temperature sensors, water pump and ESP8266. The Arduino compiler software has been used to upload the programming for the soil sensor, motor driver and ESP8266 (Fig. 2).

Software used for uploading programmes to Arduino UNO R3 is Arduino IDE. All the Arduino programmes for the soil sensor, humidity sensor and the ESP module had been compiled using Arduino IDE to obtain a working functional physical model of the irrigation system (Fig. 3).

4.3 IoT Implementation and Analytics

The ThingSpeak IoT platform has been used for the purpose of IoT analytics. The ESP8266 module has been configured with the Thingspeak server and different channels were created for different sensors that are present. With the presence of an active Internet connection we are able to send the sensing values from the soil sensors from different soil samples present in the system over a specific period of time at a particular repetitive interval so as to create a dataset for particular soil samples over time. The same dataset can be fed to mobile device by creating a simple app that enables us to monitor the data of the sensors over time (Figs. 4, 5, 6 and 7).

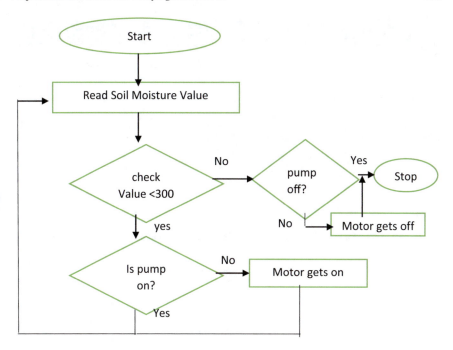

Fig. 2 Flow chart for soil moisture sensor

Fig. 3 System setup

5 Smart Irrigation System in Action

When both the hardware system and the Thingspeak server are connected, we are able to get a working model of the irrigation system where water is given to the soil sample at intervals of time regularly when the moisture level is dry. The sensing values are being sent to the IoT server continuously with intervals from where we can visualize the data. The power supply associated with the motor driver and the ESP8266 ranges between 9 and 12 V in the prototype version that has been developed with the help of Arduino UNO R3.

Fig. 4 Graph relation between moisture sensor level and pump status for soil sample 1

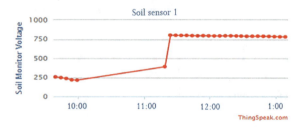

Fig. 5 Sensing values for soil sample 1

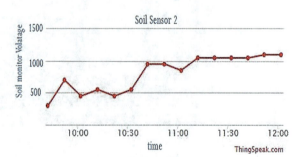

Fig. 6 Sensing values for soil sample 2

Two soil sensors are used to make the system more complex and we obtain the sensing values for two soil samples simultaneously by creating two different channels in the Thingspeak server. This helps in visualizing the data together and makes it easier for analysis (Figs. 8 and 9).

Entries: 749

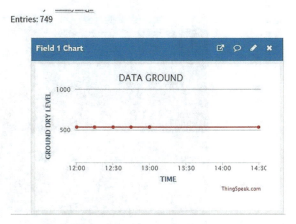

Fig. 7 Data entries of sensing values for a large number of times over a specific period

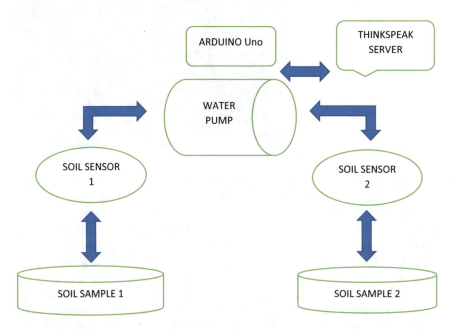

Fig. 8 Block diagram of system with two soil samples

6 Conclusion

The smart irrigation system with IoT analytics has been designed and tested successfully. The system has been tested to function automatically. The water pump is working for the soil sensors whenever the sensing values are changing as per the dryness and moisture level is defined. The same data of the sensing values are sent

Fig. 9 Working model

to the Thingspeak IoT platform with the help of ESP8266 that has been configured with the Arduino UNO R3. The data of the soil sensor 1 and soil sensor 2 along with the humidity and temperature sensor is obtained in Thingspeak by creating multiple channels and obtaining graphs and timelines as per our needs to compare and study the different soil samples that have been used in the system. We see from the graphs that for two different soil samples, the moisture level and the dryness are totally different for a taken period of time. For sample 1, we are seeing a steep rise of dryness after a certain time. For Soil sample 2, we can see that the moisture is gradually decreasing for the period of time. Both data are being compared for understanding their nature of moisture and dryness level. The data are also extracted in sheet forms that are used for further studies and comparison.

The smart irrigation industry will be the future of the agricultural sector in the coming decades as automation is becoming more relevant in every aspect of the industry and production. With the help of a working smart irrigation system that is capable of studying and collecting data of the soils in which particular agriculture is being done can be a huge help in the agriculture sector to understand the water management required for various types of soils and their properties. With a enhanced large-scale model of the whole prototype it is possible to layout a smart irrigation system that can cover large fields of different types of soil in order to automate the water system in the fields and at the same time for collecting a large number of data for different types of soil that are present in the field for different yield products.

References

1. S. Darshna, T. Sangavi, S. Mohan, A. Soundharya, S. Desikan, *Smart Irrigation System* (2015), pp. 31–36
2. S. Rawal, *IOT based Smart Irrigation System* (2017), pp. 7–11
3. V.N.R. Gunturi, *Micro Controller Based Automatic Plant Irrigation System* (2013), pp. 194–198
4. S.R. Kumbhar, A.P. Ghatule, Microcontroller based controlled irrigation system for plantation, March 2013, in *IMECS* (2013)
5. C. Arun, K. Lakshmi Sudha, *Agricultural Management Using Wireless Sensor Networks—A Survey* (2012), pp. 76–80
6. A. Abdullah, S.A. Enazi, I. Damaj, AgriSys: a smart and ubiquitous controlled-environment agriculture system, in *2016 3rd MEC International Conference on Big Data and Smart City (ICBDSC)*, Muscat (2016), pp. 1–6
7. J. John, V.S. Palaparthy, S. Sarik, M.S. Baghini, G.S. Kasbekar, Design and implementation of a soil moisture wireless sensor network, in *2015 Twenty First National Conference on Communications (NCC)*, Mumbai (2015), pp. 1–6
8. P. Naik, A. Kumbi, V. Hiregoudar, N.K. Chaitra, H.K. Pavitra, B.S. Sushma, J.H. Sushmita, P. Kuntanahal, *Arduino Based Automatic Irrigation System Using IoT* (2017), pp. 881–886
9. P.B. Chikankar, D. Mehetre, S. Das, An automatic irrigation system using ZigBee in wireless sensor network, in *2015 International Conference on Pervasive Computing (ICPC)*, Pune (2015), pp. 1–5

Design and Implementation of a Warfield Unmanned Ground Vehicle to Carry Payload

Sankha Subhra Debnath, Sameer Hussain, Aditya Gaitonde, and Amit Agarwal

Abstract The main objective of this project is to design and implement a Warfield-based unmanned ground vehicle that is capable of carrying medical supplies, scouting, spying, and information transfer from the enemy territory to the fellow soldiers. The UGV can be used where it is difficult for a human foot soldier to access. The transport payload is another important aspect of this project. For usage in real-time scouting, the UGV is also capable of retrieving video feeds in night mode and can detect the presence of metals in contact with the vehicle. Both Arduino and Raspberry Pi 3 have been used in the control unit of the UGV. Various sensors like ultrasonic sensor, infrared sensor, motion sensor, and GPS module are used to get more information of the environment and behavior of the UGV. WI-FI module is used for wireless communication between the UGV and foot soldiers which is a less cost and reliable form of communication. The whole project is oriented toward the development of a UGV in wartime or emergency situation, it is also capable of surveillance and exploration.

Keywords Unmanned ground vehicle · ROBOT · Payload · Sensor · Wireless · Microcontroller

1 Introduction

What is an UGV? UGV stands for unmanned ground vehicle. An unmanned ground vehicle is a vehicle that operates both manually from remote distance and autonomously with the help of highly complex smart system embedded into it [1, 2]. There is no onboard human presence on such vehicles. With more advancement in robotics, humans have found a way to work in any environment

S. S. Debnath · S. Hussain · A. Gaitonde · A. Agarwal (✉)
Electronics & Communication Department, Sikkim Manipal Institute of Technology, Sikkim Manipal University, Majitar, East Sikkim 737136, India
e-mail: amiteng2007@gmail.com

© Springer Nature Singapore Pte Ltd. 2020
R. Bera et al. (eds.), *Advances in Communication, Devices and Networking*,
Lecture Notes in Electrical Engineering 662,
https://doi.org/10.1007/978-981-15-4932-8_43

without its presence [3, 4]. The UGV can operate in a real-world and real-time environment in terrains and landscapes that are difficult for human scouting and penetration. Such UGVs are capable of getting information from its environment and navigating by its complex smart system [5, 6]. Such UGVs can also be controlled manually from remote distance and can be taken control by both the sender and the destination host [7]. There has been an enormous development of the UGV industry and its operation in the current timeline. The UGVs are based on land-based operation. It has its counter parts like UAV and UUV. The major applications of UGVs found its way for military, exploration, surveillance, hospital, and commercial usage (few) [8, 9].

Our current project deals with the design of the UGV and implements it in rough terrain and warfield-like situation where it can take a significant amount of payload from the source to destination, that is, from foot soldiers to soldiers taking the real military account. We are able to understand how the payload works and how much is required for different types of operations. It can also save lives by giving the location of the soldiers or those who have the receiver transponder of the UGV as data will be sent continuously. We can also track the location down for such system. The payload can also be used to carry medical supplies and foods from a source to a remote area with less access to resources and Internet connections. The project also aims at making a prototype of the whole UGV cost-efficient and development of the payload capacity of the system.

2 Design and System Overview

2.1 Physical Body and Design

The base of any UGV is the main physical body that serves as the founding structure for carrying and fitting all the modules, power supply, control unit, communication system, and most importantly the payload structure. We have a body of aluminum for the chassis and four tubeless wheels that are capable of a minimum of 8 kgs payload each with four DC motors connected to four wheels (Fig. 1).

2.2 Control Unit Design

The control system is the major part of the UGV. The whole system is divided into three parts—the base station, the onboard mechanisms of the UGV, and the destination user end.

At the base station, the main user is able to manually control all the operations by WI-FI and LAN communication. It is important to have a working server for the control and communication. At the receiver destination end, we have a display and

Fig. 1 UGV chassis and
base structure

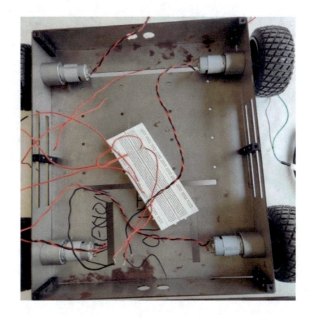

WI-FI connection to get the signal of the UGV and the live video feed from the camera onboard. The base station also comprises WI-FI and LAN connections and the display setup for tracking of the UGV and receiving its various data along with its location.

2.3 Communication System Onboard

The communication to the base station and the vehicle and the vehicle to the receiver destination can be established at a particular range or area. We are using an IEEE 802.11 wireless Wi-Fi module for establishing a LAN connection. Wireless camera modules are used for live video feeding to the base station. A GPS module is used for establishing route to the receiving end. In time of no LAN network or communication via Internet, we are using the magnetometer that analyzes the latitude and longitude of the route established by making various points along the path and storing the values. This database is used by the UGV for moving from destination A to destination B.

2.4 Operations Onboard

The UGV is capable of carrying out various range of tasks onboard autonomously without the presence of human. The major operations focused in this project are

Fig. 2 Flowchart for the
GPS operation

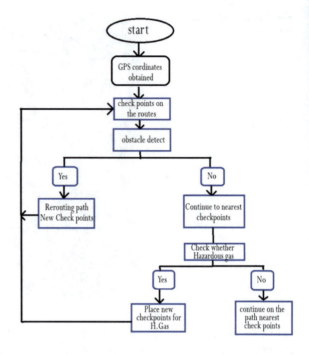

detecting hazardous gas in the environment, obstacle detection, metal detector underneath the UGV, live video feeding within the range of Wi-Fi camera, and GPS tracking. The main sensors used for the obstacle detection are the infrared as well as the ultrasonic sensors at a very close distance to avoid collisions. The Arduino mega microcontroller is being used for making the hazardous gas circuit and the metal detection circuit. MQ-4/MQ-7 sensors are used in the hazardous gas system. The ESP module is used for the video feeding from the UGV to the base station.

2.5 Algorithm for GPS Tracking and Checkpoints

The GPS module present onboard the UGV is responsible for the tracking the location of the UGV. The data is sent to the base station (Fig. 2).

2.6 Power Requirements

A lithium polymer battery of 11–12 V with 5200 mAh has been used initially. Such batteries are used because these batteries are rechargeable and much more efficient than other normal batteries. We are using battery or source current for the base station PC for display and the server. The PC power supply is connected to a 110 V AC power.

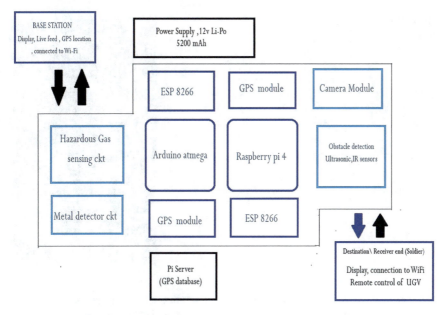

Fig. 3 Block diagram of the system

We have to power both the Arduino and Raspberry Pi with 12 V battery onboard the chassis. Additional backup power supply is kept onboard in case of emergency.

3 System Block Diagram

See Fig. 3.

4 Preliminary Tests

An initial prototype has been built where the payload for the structure has been tested for several types of terrains. We have also checked for two sets of payload, respectively, 6 kg and 10 kg. The base chassis has the capability to climb up to a steep path of 40°–45° inclination with the current power of 11.2 V with 5200 mAh. The Wi-Fi module has been configured with the Arduino for the initial data exchange. Onboard circuit for hazardous gas sensing has been tested out separately. The ultrasonic sensors and the IR sensors are placed on all four sides of the chassis and are connected with Arduino for the obstacle detections. The obstacles detected are above the level of the wheels of the body. The Raspberry Pi 4 has been configured with the Raspberry Pi camera module. The GSM module has been placed onboard

Fig. 4 Payload of a 6 kg
stool on test

and is connected with the Arduino. On the base station end, we have set up a display
where we can get the live feed of the camera onboard the UGV and with the help of the
GPS module onboard and implementation of our algorithm for checkpoints we can
start getting the route datasets. We are using an accelerometer and a magnetometer.
The magnetometer will keep a track on the latitude and longitude of the UGV path.
The whole structure as mentioned above has been assembled and several round up
tests have been done separately for each part to ensure its working initial prototype
(Fig. 4).

5   Conclusion

A prototype of an unmanned ground vehicle onboard with GPS module, metal
detector, hazardous gas sensor circuit, ESP module, and microcontroller (Arduino
and Raspberry Pi 4) has been assembled. The mechanism of the body has been tested
in rough and smooth terrain with payload of around 10 kg on the structure. All the
components are assembled using a specific blueprint so as to make the most efficient space usage inside the chassis. The power supply has been assembled along
with the components. The LiPo battery is to be charged fully before one round trip
from source to destination. The self-sustainable power backup can be achieved with
further implementation of solar charge and power storage. The GPS tracking algorithm allows both the base station and the receiver end at the destination to keep a
track of the UGV. The display used at both ends allows the users to have a visual
representation of the GPS display at real time of the UGV. The LAN server with the
help of Raspberry Pi created onboard serves as the communication of the UGV. We
can control the UGV remotely in the range of the LAN as the ESP module has been

configured. The current project focuses extensively on the payload structure and how much payload can be transported from point A to point B with the help of the UGV. The security is also an issue for such system for which more modified structures are required for payload security. The whole project will undergo much more sophisticated processes and assembling with the implementation of all the modules and the communication systems more smoothly. There is a huge research and development scope for the payload structures for miniature autonomous UGVs.

References

1. J. Khursid, H. Bing-rong, Military robot-a glimpse from today and tomorrow, in *8th International Conference on Control, Automation, Robotics and Vision*, Kunming, China (2004)
2. M.E. Purdy, *Presentation Slide "Ground Robotics Technology"* (Joint Ground Robotics Enterprise, Department of Defense, June 2007)
3. S. Ioannou, K. Dalamagkidis, K.P. Valavanis, E.K. Stefanakos, P.H. Wiley, On improving endurance of unmanned ground vehicles: the ATRV-Jr case study, in *14th Mediterranean Conference on Control and Automation, MED'06*
4. J.A. Kramer, R.R. Murphy, UGV acceptance testing, in *Proceedings of SPIE—The International Society for Optical Engineering, v 6230 I, Unmanned Systems Technology VIII* (2006), p. 62300
5. S.S. Mehta, et al., Adaptive vision-based collaborative tracking control of an UGV via a moving airborne camera: a Daisy chaining approach, in *Proceedings of the 45th IEEE Conference on Decision & Control*, San Diego, CA, USA, December 13–15 (2006), pp. 3867–3872
6. S.J. Lee, D.M. Lee, J.C. Lee, Development of communication framework for unmanned ground vehicle, in *Proceedings of the International Conference on Control Automation and Systems*, Oct. 14–17, Seoul, South Korea (2008), pp. 604–607
7. X. Feng, et al., Enhanced supervisory control system design of an unmanned ground vehicle, in *Proceedings of the 2004 IEEE International Conference on Systems, Man and Cybernetics*, October 10–13, The Hague, The Netherlands (2004), pp 1864–1869
8. J. Ortiz, et al., Description and tests of a multisensorial driving interface for vehicle teleoperation, in *Proceedings of the 11th International IEEE Conference on Intelligent Transportation System*, October 12–15, Beijing, China (2008), pp. 616–621
9. L. Lo Belloy, et al., Towards a robust real-time wireless link in a land monitoring application, in *Proceedings of the 11th IEEE Conference on Emerging Technologies and Factory Automation, EFTA'06*, September 20–22, Prague, Czech Republic, pp. 449–452

Maximum Power Point Tracking of Photovoltaic Systems Using Fuzzy-PID-Based Controller

Raval Parth Pradip, Saurabh Bhowmik, Noimisha Hazarika, Chitrangada Roy, and Anirban Sengupta

Abstract In this paper, a fuzzy-PID controller for DC–DC boost converter at variable irradiation level in the solar photovoltaic (PV) system has been designed. The problem that PV system has is that the output voltage depends on weather conditions, mainly temperature and irradiance level, which eventually changes the load operating point. Classical controllers such as PI and PID have been used previously that have undesirable transient response. Here, a hybrid fuzzy logic based PID controller is analyzed and designed to control the DC–DC converter acting as a regulator. This proposed model eliminates overshoots and oscillations of the output voltage and ensures maximum amount of power extraction from PV system. The transient response parameters and steady-state errors are analyzed to conclude the benefits of such a control system. The achieved results at different levels of irradiation at constant temperature are shown. Analysis of results supports the advantage and validity of the hybrid fuzzy-PID controller.

Keywords Hybrid fuzzy-PID controller · Maximum power point tracking (MPPT) · PID controller · Fuzzy logic controller

R. P. Pradip · S. Bhowmik · N. Hazarika · C. Roy · A. Sengupta (✉)
Department of EEE, Sikkim Manipal Institute of Technology, Sikkim Manipal University, Majitar, Rangpo, Sikkim, India
e-mail: anirban.s@smit.smu.edu.in

R. P. Pradip
e-mail: parthraval31@yahoo.com

S. Bhowmik
e-mail: nick2604888@gmail.com

N. Hazarika
e-mail: noimisha_201600097@smit.smu.edu.in

C. Roy
e-mail: chitrangada.r@smit.smu.edu.in

© Springer Nature Singapore Pte Ltd. 2020   393
R. Bera et al. (eds.), *Advances in Communication, Devices and Networking*,
Lecture Notes in Electrical Engineering 662,
https://doi.org/10.1007/978-981-15-4932-8_44

1 Introduction

The increasing rate of exhaustion of fossil fuel and its adverse effect on the environment are compelling the world to seek for alternative energy sources that are renewable and sustainable at the same time. The advancement in technology in the recent times has not only made it easier to search for new and better alternatives to the conventional sources of energy, but also improved the efficiency of the renewable energy sources by a thousand folds. Of all, solar energy is considered as one of the best sources of renewable energy due to its availability and environment friendliness [1]. Solar cells are becoming an important part of electric power systems, specifically in smart grids. There is no doubt that photovoltaic (PV) systems will have great significance in the future of energy systems if not the center spot among all. In fact, solar energy systems provide the advantage of lower fuel costs and lower maintenance over other energy systems. PV systems have been used for many different applications including smart grids. A PV system can produce wide ranges of voltages and currents at terminal output; however, a PV output is inconsistent due to unregulated sun power. A PV cell, therefore, must generate a constant DC voltage at the desired level for the application, regardless of variations of light illumination and temperature. Different MPPT techniques have been designed and presented by researchers in order to operate the PV system in its maximum power point.

In [2], Mohammad Junaid Khan and Lini Mathew designed and compared the simulation results of Fuzzy Logic Controller (FLC)-based and Perturb and Observe (PO)-based MPPT for obtaining the maximum power from solar PV array even under variable irradiation conditions. The comparative analysis showed that FLC-based controller gave better and more accurate output response under changing irradiation level than PO-based controller, hence increasing the efficiency and providing stable output power and reducing the fluctuations [2]. In [3], the authors presented a new digital control strategy using FLC-based PO scheme for maximum power extraction in PV system along with a dual MPPT controller. The proposed scheme consists of a dual-axis solar tracker to keep track of the maximum power from sun and a FLC-based PO controller to keep system power at maximum. Results showed elimination of oscillations and enhancement of operating point convergence speed. Boumaaraf, Talha, and Bouhali proposed a new control technique for tracing the MPP of a grid-connected PV system using Artificial Neural Networks (ANN) and analyzed its performance in MATLAB Simulink software [4]. Simulation results showed that the system was able to adjust according to the neural network parameters for faster response and also provide energy to the grid with unity power factor and lower harmonics distortion of the output [4]. Fatemi, Shadlu, and Talebkhah presented a new three-point P&O method for MPPT of solar PV output by generating controlled gate pulses for the DC–DC boost converter used between the PV panel and the load and compared its simulation results with that of conventional Hill Climbing (HC) method [5]. Results clearly showed better response in three-point

P&O technique than HC-based controller under both constant and varying temperature and irradiation levels [5]. An Adaptive Neuro-Fuzzy Inference System (ANFIS)-based MPPT controller was developed in MATLAB/Simulink for a PV solar system consisting of PV array, DC–DC boost converter, and the ANFIS-based controller by Mohammed, Devaraj, and Ahamed in [6] and its results were compared with those without MPPT technique and with incremental-conductance-based MPPT technique. Simulation results showed that ANFIS-based controller could change the duty cycle faster and more accurately according to changes in temperature and irradiation levels [6]. In [7], the authors gave a detailed study of operation and comparison between two of the most commonly reliable MPPT techniques, namely, Perturb and Observe (P&O) method and Incremental-Conductance (IC) method. Comparisons were done on various parameters like complexity, flexibility, consistency, availability, etc. and it was concluded that selection of MPPT technique purely depends on the capital on hand, application, location, environmental conditions, etc., as both methods have their own pros and cons [7].

Many controllers such as P, PI, and PID [8] are used to control the performance of the power electronics converter in the PV system. But, the linear control technique is not sufficient to deal with the variation in different electrical quantities like current and voltage. Therefore, nonlinear control techniques such as FLC controlling technique can be implemented to increase the performance of the converter.

In this paper, a hybrid fuzzy logic based PID controller is analyzed and designed to control the DC–DC converter acting as a regulator between the PV array and the load. The transient response parameters and steady-state errors are analyzed to conclude the benefits of such a control system.

2 Problem Formulation

The proposed system consists of PV array and boost converter, which is controlled by hybrid fuzzy-PID controller and a DC load. A PV system can produce wide ranges of voltages and currents at terminal output; however, a PV output is inconsistent due to unregulated sun power. A PV cell, therefore, must generate a constant DC voltage at the desired level for the application, regardless of variations of light illumination and temperature. In this paper, a control system has been designed to obtain linear output voltage of the boost converter with changing irradiation level and thus the solar cell's output voltage.

2.1 Modeling of PID Controller for Boost Converter

The classical controller is used instead of fuzzy logic control for the photovoltaic model with boost converter to compare its performance with FLC. The Proportional–Integral Derivative (PID) controller is used as classical controller [8]. The

PID controller can be expressed as

$$U(t) = K_p e(x) + K_i \int_0^x e(x) \mathrm{d}x + K_d \frac{\mathrm{d}e(x)}{\mathrm{d}x} \tag{1}$$

where $e(x)$ is the input error that is calculated by comparing the reference and output voltage.

2.2 Modeling of Fuzzy Logic Controller for Boost Converter

The output voltage of a DC–DC buck converter is controlled by the FLC. The FLC uses linguistic variables as input instead of numerical variables. The error $e\_1(y)$ and change in error $\Delta e\_1(y)$ are calculated from the two equations below. The error is the difference between the buck–boost output voltage and reference voltage, while the change in error is difference between the present error and pervious error.

$$e\_1(y) = V_{\mathrm{ref}} - V_{\mathrm{out}} \tag{2}$$

$$\Delta e\_1(y) = e\_1(y) - e\_1(y-1) \tag{3}$$

2.3 Hybrid Fuzzy-PID Controller

The classical controller requires accurate information of the variables of the system in order perform sensitively, whereas the fuzzy logic controller does not require exact information for system within variations in the system variables. Combining these two controllers, to control and minimize the steady-state error of the system, while the fuzzy logic control cannot eliminate steady-state error. The block diagram of hybrid fuzzy-PID controller for any process is shown in Fig. 1.

In Fig. 1, error signal is the difference between the reference and the output voltages. Fuzzy logic control in this figure is the proposed fuzzy-based control method and the PID is used to control the output voltage. The design of the whole module system developed for this study is shown, which includes the boost converter design with pulse generator, PID controller, and hybrid fuzzy-PID controller. MATLAB/SIMULINK has been used to simulate the design of the proposed model.

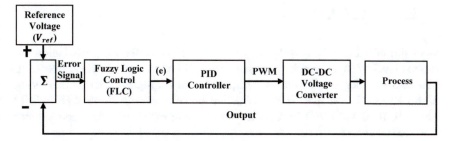

Fig. 1 Block diagram of hybrid fuzzy-PID controller for any process

3 Flowchart of the Proposed Controller

In Fig. 2, flowchart of the proposed hybrid fuzzy-PID controller is shown.

Fig. 2 Flowchart of design
of the hybrid fuzzy-PID
controller

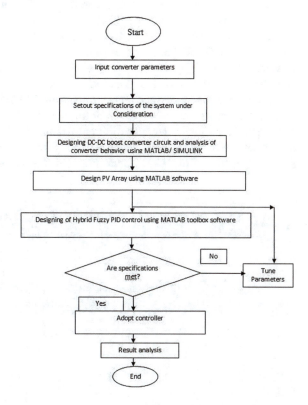

4 Results and Analysis

Modeling of a PV system with DC–DC boost converter for fixed irradiation and varying level of irradiation is proposed with PID controller and hybrid fuzzy-PID controller in order to drive the switching of MOSFET [8]. For each of these there are two results, one at 1000 W/m^2 irradiation and another with different [700, 1100, 900, 750, 1000, 400] W/m^2 irradiation levels along with final output power graph have also been observed and analyzed.

Figure 3 shows output waveform at irradiation of 1000 W/m^2 using PID controller [8]. Figure 3c shows output voltage of the PV system which is around 31 V. Here, 3(B) and 3(c) show output current of DC–DC converter which is increased from 4 to 11 A and thereafter it becomes almost linear and output of boost converter, which is increased from 12 to 32 V and thereafter is almost linear.

Figure 4 gives three-output waveform at different levels of solar irradiation [700, 1100, 900, 750, 1000, 400 W/m^2] using PID controller [8].

Figure 5 gives output waveform at constant irradiation level 1000 W/m^2 in presence of hybrid fuzzy-PID-based MPPT controller. Graph (A) gives the output voltage of the solar cell which is at constant irradiation level of 31 V. Graph (B) shows the output of DC–DC boost converter using hybrid fuzzy-PID controller in boost converter. Graph (C) gives the current drawn from DC–DC converter, it is almost constant (6.5 A) with time at irradiation of 1000 W/m^2. Here, output voltage of the boost converter is increased and is near to constant (34 V) at irradiation of 1000 W/m^2.

Figure 6 shows output waveform at different levels of irradiation using hybrid fuzzy-PID-based controller. For different irradiation levels, graph (A) shows the output voltage of the solar cell which is varying from 22 to 34 V. Graphs (B) and (C) show output current of DC–DC converter, which is almost constant with respect to

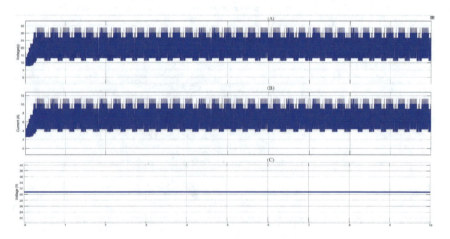

Fig. 3 **A** Output voltage of boost converter, **B** output current of boost converter [8], **C** output of solar panel at 1000 W/m^2 irradiation that is fed to input of boost converter

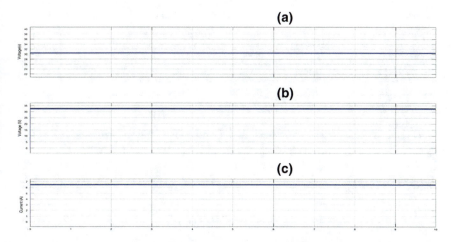

Fig. 4 **A** Output voltage of boost converter, **B** output current of boost converter [8], **C** output of solar panel at [700, 1100, 900, 750, 1000, 400] W/m² irradiation that is fed to input of boost converter

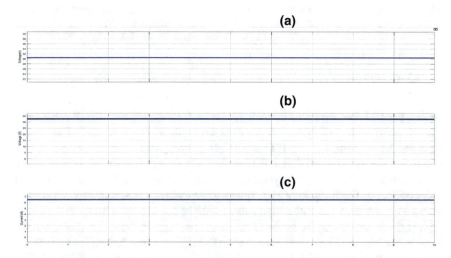

Fig. 5 **A** Output voltage of solar panel at 1000 W/m² irradiation, **B** output voltage of the boost converter, and **C** output current of boost converter

time at constant irradiation level in the range of 6.5–7.1A and the output of DC–DC boost converter with respect to time using hybrid fuzzy-PID controller, respectively. Here, output voltage of the boost converter is increased and is almost constant from 33 to 36 V at different levels of irradiation that ensure maximum power extraction from the solar system.

(a)

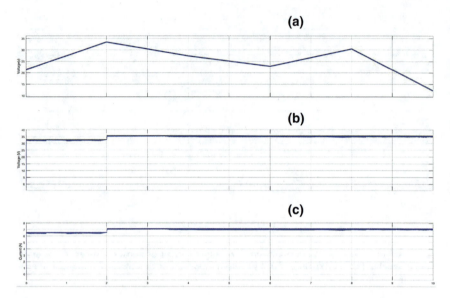

(b)

(c)

Fig. 6 **A** Output of solar panel from (22 to 34) V at [700, 1100, 900, 750, 1000, 400] W/m² irradiation that is fed to input of boost converter, **B** output voltage from (33 to 36) V of boost converter, **C** Output current from (6.5 to 7.1) A of boost converter using hybrid fuzzy-PID controller

5 Conclusion

In this paper, boost converter modeling and performance analysis in the PV solar system using PID and hybrid fuzzy-PID controllers are presented. The results obtained using hybrid fuzzy-PID controller is compared with conventional PID controller. Results showed that in PID control, the system had a very short rise time with overshoot. However, the system took long time to settle into the steady-state value of 36.5 V. The fuzzy logic controller had the system arrive at the desired voltage after a long rise time and with a high voltage oscillation. In hybrid fuzzy-PID control, with variable irradiation the system had a short rise time without overshoot, and the system had less voltage oscillations than PID controller. The hybrid fuzzy-PID has showed better performance in variable irradiation as compared to PID controller under same system conditions. Therefore, hybrid fuzzy-PID controller can be seen as a more effective and accurate controller for the MPPT application in solar PV systems.

References

1. F.M. Guangul, G.T. Chala, Solar energy as renewable energy source: SWOT analysis, IN *2019 4th MEC International Conference on Big Data and Smart City (ICBDSC)*, Muscat, Oman (2019), pp. 1–5
2. M.J. Khan, L. Mathew, Artificial intelligence based maximum power point tracking algorithm for photo-voltaic system under variable environmental conditions, in *2017 Recent Developments in Control, Automation & Power Engineering (RDCAPE)*, Noida (2017), pp. 114–119
3. R. Sankar, S. Velladurai, R. Rajarajan, J.A. Thulasi, II. PV system description: maximum power extraction in PV system using fuzzy logic and dual MPPT control, in *2017 International Conference on Energy, Communication, Data Analytics and Soft Computing (ICECDS)*, Chennai (2017), pp. 3764–3769
4. H. Boumaaraf, A. Talha, O. Bouhali, Maximum power point tracking using neural networks control for grid-connected photovoltaic system, in *4th International Conference on Power Engineering, Energy and Electrical Drives*, Istanbul (2013), pp. 593–597
5. S.M. Fatemi, M.S. Shadlu, A. Talebkhah, Comparison of three-point P&O and hill climbing methods for maximum power point tracking in PV systems, in *2019 10th International Power Electronics, Drive Systems and Technologies Conference (PEDSTC)*, Shiraz, Iran (2019), pp. 764–768
6. S. Sheik Mohammed, D. Devaraj, T.P. Imthias Ahamed, Maximum power point tracking system for stand alone solar PV power system using adaptive neuro-fuzzy inference system, in *2016 Biennial International Conference on Power and Energy Systems: Towards Sustainable Energy (PESTSE)*, Bangalore (2016), pp. 1–4
7. T. Jayakumaran, et al, A comprehensive review on maximum power point tracking algorithms for photovoltaic cells, in *2018 International Conference on Computation of Power, Energy, Information and Communication (ICCPEIC)*, Chennai (2018), pp. 343–349
8. S. Bhowmik, R.P. Pradip, C. Roy, Design of PID controller for maximum power point tracking for PV energy systems, in *Advances in Communication, Devices and Networking. Lecture Notes in Electrical Engineering*, ed. by R. Bera, S. Sarkar, O. Singh, H. Saikia, vol. 537 (Springer, Singapore, 2019)

Exploring the Electrical Behavior of High-K Triple-Material Double-Gate Junctionless Silicon-on-Nothing MOSFETs

Ningombam Ajit Kumar, Aheibam Dinamani Singh, and Nameirakpam Basanta Singh

Abstract As semiconductor devices advance to sub-45 nm range, there has been a great deal of development in the device area along with some critical issues. One such issue is the decrease in device threshold voltage which is due to a decrease in control by the gate upon the channel region and an increase in drain/source charge sharing. This case is generally termed as short channel effects (SCEs). In this paper, a study is demonstrated to reduce the said effect on a junctionless silicon-on-nothing (SON) double-gate (DG) Metal Oxide Semiconductor Field Effect Transistor (MOSFET) by fusing the idea of multi-material gate technology and high dielectric oxide technique. This paper presents the analysis of significant device parameters such as potential and the threshold voltage of the proposed device built on ATLAS simulations data. A comparison with other device structures is also carried out. The effects on changes in various device parameters are also studied.

Keywords High-K dielectric · Hafnium dioxide · Junctionless silicon-on-nothing (JLSON) · Triple material

1 Introduction

Semiconductor device industries have been following Moore's law, doubling the processing power every 2 years. This increase in processing power has been achieved by the downscaling of devices [1]. As the downscaling reaches the nano levels, it

N. A. Kumar (✉) · A. D. Singh
Department of Electronics and Communication Engineering, North Eastern Regional Institute of Science and Technology, Nirjuli, Arunachal Pradesh, India
e-mail: ningombama@gmail.com

A. D. Singh
e-mail: ads@nerist.in

N. B. Singh
Department of Electronics and Communication Engineering, Manipur Institute of Technology, Imphal, Manipur, India
e-mail: basanta_n@rediffmail.com

© Springer Nature Singapore Pte Ltd. 2020
R. Bera et al. (eds.), *Advances in Communication, Devices and Networking*,
Lecture Notes in Electrical Engineering 662,
https://doi.org/10.1007/978-981-15-4932-8_45

increases the short channel effects (SCEs) resulting in performance degradation. As a result of this scaling challenges in bulk conventional MOSFETs, several non-conventional geometry MOSFET structures had been proposed [1].

To overcome the SCEs, various technologies have been put up by researchers like multi-material gate, round gate, junctionless MOSFET, etc. The junctionless transistor has been in the spotlight for its robust discretization over SCEs [2]. As the name signifies, a junctionless transistor has no junctions between channel and source/drain region. Due to its unique property of having no junctions, it provides a device with a lower thermal budget and solution to large doping concentration requirements [3–5]. SOI is among the structures that have been well received by researchers and device engineers due to its inherent advantages [2]. Among the SOI technology, SON is the most improved architecture. The most significant advantages of SON are: (i) Reduction in minimal channel length due to reduced drain/source-channel coupling through the oxide layer [6, 7], (ii) higher circuit speed as a result of reduced parasitic capacitance between the substrate and drain/source from the presence of nothing/air layer below the Silicon film, and (iii) zero charge accumulation in the air-gap [8, 9]. Architecturally, SON is the same with SOI. The only difference is that in SON air replaces the buried oxide. SON also exhibits a fabrication property, allowing fabrication of exceptionally thin-film silicon and buried oxide of 5–20 nm and 10–30 nm, respectively, which can give quasi-total suppression of Short Channel Effects resulting in good electrical performance [10, 11]. Numerous structural models have also been proposed to reduce SCEs such as the retro-gate channel doping [9], super-halo gate doping [12], etc. Among this, the notable ones are the multi-material gate engineering (in which multiple gate electrodes with decreasing order of their work functions from source to drain are placed) [13–17], and high-K oxide layer (in which the conventional silicon dioxide layer is replaced by a high dielectric material such as hafnium dioxide) [18]. In multi-material gate engineering, gate electrode at the source end controls the subthreshold characteristic. While, gate electrode at the drain end lowers the electric field at the drain end, thereby lowering SCEs. Hafnium dioxide (dielectric constant = 22–25; bandgap = 5.6 eV) replacing the oxide layer provides higher thermal stability with lower bulk trap density, and relatively thicker physical gate oxide [18].

Thus, we can imagine that with little improvement to the present fabrication techniques, a high-K junctionless triple material double gate (JLTMDG) SON MOSFETs will be practical for fabrication in the coming future. In this paper, the oxide layer has been replaced by high-K hafnium dioxide for JLTMDG SON MOSFETs. The surface potential and the threshold voltages are compared with high-K JLTMDG SOI, high-K JLTMDG SOI MOSFETs. Also, the behavior with the changes in device parameters is also studied using 2-D ATLAS simulation [19].

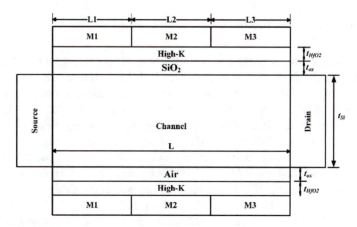

Fig. 1 Schematic diagram of high-K JLTMDG SON MOSFET

2 Structure of High-K TMDG SON MOSFET

Figure 1 presents the schematic of the proposed high-K JLTMDG SON MOSFET, where t_{ox}, t_{ox1}, t_{si} and L are high-K thickness, oxide thickness, silicon thickness, and channel length, respectively. The x-axis conveys the SiO_2/Si interface and the y-axis conveys the Source/Si interface. The channel is divided equally into 3 regions each having a length L_1, L_2, L_3 for 3 different gate materials.

3 Simulation Parameters

Table 1 shows the different parameters taken into consideration while simulating the proposed device. Several ATLAS models like the Auger recombination model, Standard Shockley-Read-Hall, Drift-diffusion, and Lombardi (CVT) models are incorporated in the simulation process for all the device structures.

Table 1 Structural parameters of high-K JLTMDG SON MOSFET

Device parameter	Symbol	Value
Channel length	L	30 nm
Channel thickness	t_{Si}	10 nm
High-K thickness	t_{ox1}	2 nm
SiO_2 thickness	t_{ox}	2 nm
Doping concentration	N_a	1×10^{19} cm^{-3}
Air thickness	t_{air}	2 nm

4   Results and Discussion

Figure 2 presents the comparison of central potentials across channel length of high-
K JLTMDG SOI with high-*K* JLTMDG SON MOSFETs. It is seen in the figure that
high-*K* JLTMDG SON MOSFET expresses the closest minimum surface potential
with a characteristic stepping profile toward the source side. This helps in channel potential protecting it from the fluctuating drain bias voltage which lowers the
Drain Induced Barrier Lowering (DIBL) effects. Moreover, in high-*K* JLTMDG
SON MOSFET, the barrier height between the source-to-channel and the drain-to-
channel is better in contrast to high-*K* JLTMDG SON MOSFET. Hence, will exhibit
improvement in circuit performance. Figure 3 presents the surface potentials across
the channel length of high-*K* JLTMDG SOI for different values of V_{gs}. It is seen
from the figure that as the gate to source voltage increases the potential barrier

Fig. 2 Change in central
potential across the channel
length in high-*K* JLTMDG
SOI and high-*K* JLTMDG
SON MOSFETs

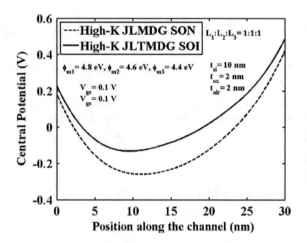

Fig. 3 Change in central
potential across the channel
length in high-*K* JLTMDG
SON MOSFET with a
different value of V_{gs}

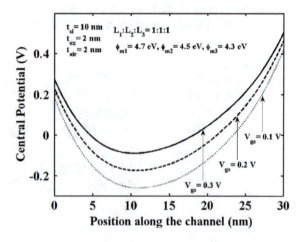

height decreases. This is expected as a vertical electric field along the channel that increases with an increase in V_{gs}. As V_{gs} increases the MOSFET becomes more enhanced allowing more current to flow from source to drain.

Figure 4 presents the change of threshold voltage concerning the gate oxide work function of the devices. It is seen from the figure that JLTMDG SON has a threshold voltage lower than JLTMDG SOI. As work function increases the voltage required to trigger the gate electrode increases resulting in a lower vertical electric field thereby increasing the threshold voltage. Figure 5 presents the change in threshold voltage concerning the thickness of the oxide layer. It clearly shows that the threshold voltage increases with increasing oxide thickness. This increase in threshold voltage is the result of the decrease in the capacitance across the gate oxide.

Figure 6 presents the current-voltage characteristic of high-K JLTMDG SON MOSFET for the different drain to source voltages. The figure clearly shows that

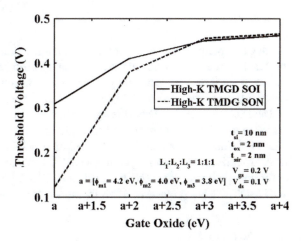

Fig. 4 Change in threshold voltage with gate oxide work function for high-K JLTMDG SON and SOI MOSFET

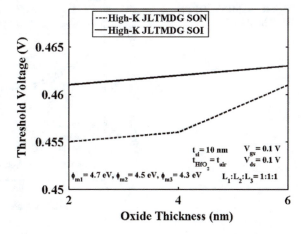

Fig. 5 Change in threshold voltage with a thickness of the oxide layer for high-K JLTMDG SON and SOI MOSFET

Fig. 6 I-V graph for high-K JLTMDG SON MOSFET with a change in the drain to source voltage, V_{ds}

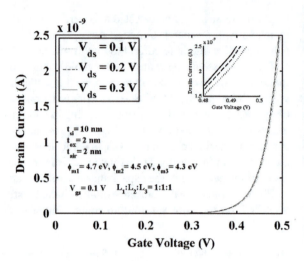

as the gate voltage increases, there is an increasing drain current. As drain voltage increases there is an increase in the horizontal electric field and the depletion region widens towards the source resulting in increased drain current as shown in the figure. Figure 7 presents the electric field deviation across the channel in High-K JLTMDG SOI and SON MOSFETs. From the figure, it clearly shows that the source end electric field is higher than the drain end electric field for both JLSON MOSFET and JLSOI MOSFET. And, the drain end electric field of JLSON MOSFET is less than that of JLSOI MOSFET. The lowering of the electric field towards the drain terminal ensures lower hot-carrier electrons (HCEs) in JLSON MOSFET. Hence, High-K JLTMDG SON MOSFET has preferable HCEs immunity. Moreover, a better constancy of the

Fig. 7 Electric field deviation long the channel in high-K JLTMDG SOI and SON MOSFETs

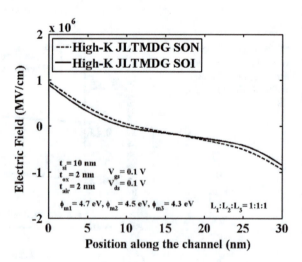

electric field across the channel is achieved due to the higher electric field around the source end.

5 Conclusion

In this paper, studies on the impact of multi-material gate and replacing high dielectric in place of the conventional oxide layer in a junctionless double gate SON MOSFET is presented. We observed a reduction in threshold voltage for high-K JLTMDG SON as compared to high-K JLTMDG SOI MOSFET as a result of lower surface potential minima. It is also observed that the SON structure has better immunity to DIBL and HCEs providing better protection of channel potential from the fluctuating drain bias voltage. So, we can conclude that the JLSON structure with multi-gate engineering and high-K oxide is a better replacement of conventional JLSOI structure.

Acknowledgements Ningombam Ajit Kumar thankfully acknowledges the financial support obtained from Visvesvaraya Ph.D. Scheme, Ministry of Electronics Information Technology (MeitY), Government of India in the form of Fellowship.

References

1. The International Technology Roadmap for Semiconductor, Emerging Research Devices (2009)
2. R. Trevisoli, R.T. Doria, M. Souza, M.A. Pavanello, Substrate bias influence on the operation of junctionless nanowire transistors. IEEE Trans. Electron Dev. **61**(5), 1575–1582 (2014). https://doi.org/10.1109/TED.2014.2309334
3. C.-J. Su, T.-I. Tsai, Y.-L. Liou, Z.-M. Lin, H.-C. Lin, T.-S. Chao, Gate-all-around junctionless transistors with heavily doped polysilicon nanowire channel. IEEE Electron Dev. Lett. **32**(4), 521–523 (2011). https://doi.org/10.1109/LED.2011.2107498
4. D. Ghosh, M.S. Parihar, G.A. Armstrong, A. Kranti, High-performance junctionless MOS-FETs for ultralow-power analog/RF applications. IEEE Electron Dev. Lett. **33**(10), 1477–1479 (2012). https://doi.org/10.1109/LED.2012.2210535
5. V. Kilchytska, T.M. Chung, B. Olbrechts, Ya. Vovk, J.-P. Raskin, D. Flandre, Electrical characterization of true silicon-on-nothing MOSFETs fabricated by Si layer transfer over a pre-etched cavity. Solid State Electron. **51**(9), 1238–1244 (2007). https://doi.org/10.1016/j.sse.2007.07.021
6. M.-H. Han, C.-Y. Chang, H.-B. Chen, J.-J. Wu, Y.-C. Cheng, Y.-C. Wu, Performance comparison between bulk and SOI junctionless transistors. IEEE Electron Dev. Lett. **34**(2), 169–171 (2013)
7. Y. Tian, W. Bu, D. Wu, X. An, R. Huang, Y. Wang, Scaling capability improvement of silicon-on-void (SOV) MOSFET. Semicond. Sci. Technol. **20**(2), 115–119 (2005). https://doi.org/10.1088/0268-1242/20/2/002
8. S. Monfray, F. Boeuf, P. Coronel, G. Bidal, S. Denorme, T. Skotnicki, Silicon-on-nothing (SON) applications for low power technologies, in *2008 IEEE International Conference on Integrated Circuit Design and Technology and Tutorial* (2008), pp. 1–4. https://doi.org/10.1109/ICICDT.2008.4567232

9. J. Pretet, S. Monfray, S. Cristoloveanu, T. Skotnicki, Silicon-on-nothing MOSFETs: performance, short-channel effects, and backgate coupling. IEEE Trans. Electron Dev. **51**(2), 240–245 (2004). https://doi.org/10.1109/TED.2003.822226

10. M. Jurczak, T. Skotnicki, M. Paoli, B. Tormen, J. Martins, J.L. Regolini, S. Monfray, et al., Silicon-on-nothing (SON)-an innovative process for advanced CMOS. IEEE Trans. Electron Dev. **47**(11), 2179–2187 (2000). https://doi.org/10.1109/16.877181

11. N.A. Kumar, A.D. Singh, N.B. Singh, Partially depleted and fully depleted silicon on insulator: a comparative study using TCAD. Int. J. Electron. **6**(3), 3 (2017)

12. S. Mudanai, W.-K. Shih, R. Rios, X. Xi, J.-H. Rhew, K. Kuhn, P. Packan, Analytical modeling of output conductance in long-channel halo-doped MOSFETs. IEEE Trans. Electron Dev. **53**(9), 2091–2097 (2006). https://doi.org/10.1109/TED.2006.880371

13. M.J. Kumar, A. Chaudhry, Two-dimensional analytical modeling of fully depleted DMG SOI MOSFET and evidence for diminished SCEs. IEEE Trans. Electron Dev. **51**(4), 569–574 (2004). https://doi.org/10.1109/TED.2004.823803

14. N.A. Kumar, A.D. Singh, N.B. Singh, Threshold voltage and subthreshold slope comparison of silicon on insulator (SOI) and silicon on nothing (SON) MOSFET using TCAD. Int. J. Electron. **6**(3), 3 (2017)

15. H. Kaur, S. Kabra, S. Haldar, R.S. Gupta, An analytical threshold voltage model for graded channel asymmetric gate stack (GCASYMGAS) surrounding gate MOSFET. Solid State Electron. **52**(2), 305–311 (2008). https://doi.org/10.1016/j.sse.2007.09.006

16. P.K. Tiwari, S. Dubey, M. Singh, S. Jit, A two-dimensional analytical model for threshold voltage of short-channel triple-material double-gate metal-oxide-semiconductor field-effect transistors. J. Appl. Phys. **108**(7), 074508 (2010). https://doi.org/10.1063/1.3488605

17. N.A. Kumar, A.D. Singh, N.B. Singh, Examining the short channel characteristic and performance of triple material double gate SON MOSFETs with grading channel concentration. J. Nanoelectron. Optoelectron. (2019). https://doi.org/10.1166/jno.2019.2667. American Scientific Publisher

18. S.K. Sarkar, P.K. Dutta, N. Bagga, K. Naskar, Analysis and simulation of dual metal double gate son MOSFET using hafnium dioxide for better performance, in *Michael Faraday IET International Summit 2015*, vol. 69, no. 4 (2015). https://doi.org/10.1049/cp.2015.1665

19. ATLAS User's Manual, (2015) SILVACO Int., Santa Clara, CA, USA

A Literature Review of Current Vision Based Fall Detection Methods

Amrita Biswas and Barnali Dey

Abstract Falls, especially among the elderly, is a serious issue. With the increase in the number of nuclear families, very often the elderly are left alone at home. Personal caregiver is not always affordable. In such cases, a home installed automatic fall detection system could prove to be very useful. A lot of research has been going on in this field and a large number of work has been published especially over the last five years. In this paper, we have comprehensively summarized the latest research going on in this field and the shortcomings that are there which need to be overcome.

Keywords Fall detection · Depth · 2D camera · Vision based · Database · Survey

1 Introduction

Falls, especially among the elderly, is a serious cause of concern and very often lead to major injuries and maybe even death. Deaths from unintentional injuries are the seventh main cause of death among the elderly and the largest percentage of those deaths are caused by falls. On an average one in four U.S. residents aged ≥ 65 years (older adults) report falling each year and emergency visits due to falls are estimated at approximately 3 million per year [1, 2]. Falls can also indicate serious health concerns like blood pressure, heart attack, etc. The time for which a person has remained fallen on the ground is also vital. The sooner a fallen person is attended the better it is. Due to the susceptibility of the elderly to falls very often they have to be accompanied by nurses and other support staff increasing the pressure on manual labour. As a result, the demand for some intelligent monitoring system for detecting falls has gained primary importance. Over the years a lot of researchers have proposed various fall detection methods.

A. Biswas (✉) · B. Dey
Electronics and Communication Engineering Department, Sikkim Manipal Institute of Technology, Rangpo, Sikkim, India
e-mail: amritabis@gmail.com

B. Dey
e-mail: barnali.d@smit.smu.edu.in

© Springer Nature Singapore Pte Ltd. 2020
R. Bera et al. (eds.), *Advances in Communication, Devices and Networking*,
Lecture Notes in Electrical Engineering 662,
https://doi.org/10.1007/978-981-15-4932-8_46

The methods have been divided into mainly vision-based and sensor-based approaches. The four main approaches towards data acquisition for fall detection is as follows:

1. RGB Camera-based.
2. Depth Camera-based.
3. Wearable Sensor-based.
4. Ambience/Fusion-based.

Details have been shown in the block diagram in Fig. 1.

Vision-based approaches make use of standard RGB cameras for data acquisition. The number of cameras installed varies. Generally, image processing algorithms for estimation of the pose is used to detect falls. Since pose estimation is essentially better in 3D, multiple cameras are sometimes used which help in better pose analysis of the person. The sequence of images from a single camera and multiple cameras are also used as set up for pose estimation [3].

Wearable sensor-based methods involve the use of sensor-based devices that have to be constantly worn by the subject which is a constraint. A subject may be reluctant to use a wearable device all the time. However, these devices are cheap and easy to install [4, 5].

Ambience based approaches also use sensors. These sensors need not be worn by the users but are installed elsewhere like the floor, walls, bed, etc. Different types of sensors have been used by researchers. Some are based on sensing the pressure or vibrational data on the floor, some are based on audio signals received by

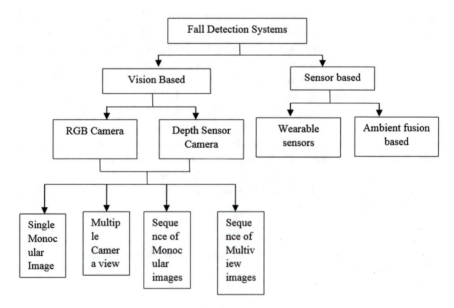

Fig. 1 Classification of fall detection systems

microphones, etc. However, this approach has higher false alarm rates and separate hardware installation for only fall detection is required [6, 7].

As far as data acquisition for fall detection is concerned vision-based approaches like the use of RGB camera and depth camera are the most convenient because it works unobtrusively in the background, is easy to install and set up, and can have a wide area of coverage. Also, the video surveillance cameras are widely used these days for monitoring and security purposes. The same set up could be programmed for fall detection as well. One of the issues with vision-based device is the privacy of the user especially in private areas like the bathroom where due to the slippery floor the probability of fall is even more. Few researchers have tried to address this issue as well to maintain the privacy of the subjects being monitored [8–10].

In this survey paper, we have discussed the recent work done on fall detection using the camera/vision-based approaches, their shortcomings and scope for future work. We will also discuss the publicly available databases used for carrying out research in the field of fall detection.

The latest survey paper on fall detection that we have come across has been published by Zhang et al. [9] in 2015. They have covered some approaches to fall detection over the period 2005 to 2014. However, a significant amount of research in the field of vision-based fall detection has been carried out over the past five years. In my work, I have concentrated on work done after 2014 and the related work not covered by Zhong et al.

Operation of Vision Based fall detection systems can be divided into four basic parts: Data Acquisition, Preprocessing, Feature Extraction and Classification.

Data acquisition is carried out using 2D cameras (vision-based) or 3D sensor cameras.

The preprocessing step generally involves steps like frame extraction, background subtraction, illumination normalization, noise removal, and human detection.

Feature Extraction will involve the removal of redundant data and extraction of features from the image that will be necessary for fall detection also leading to dimensionality reduction of the data set. Various methods like Convolutional Neural Networks, HOG-based approaches, ad hoc segmentation, threshold and shape features and Discrete Wavelet Transforms have been used.

Classification step is carried out to decide whether the detected pose is fallen or not fallen. Since fall detection is essentially a 2 class problem various researchers have used SVM for classification. Other methods like Neural Networks, and k Nearest Neighbours have also been used.

In Sect. 2 we have discussed the publicly available fall databases and the latest work done on fall detection using 2D cameras and depth sensors have been discussed in Sects. 3 and 4, respectively. Section 5 has a brief discussion on the popular methods used and their shortcomings and conclusion in Sect. 6.

2 Public Fall Datasets

Several fall databases are available publicly for testing and validation of fall detection algorithms. They are as follows:

TST Fall Detection Dataset V1

The TST database has been constructed using depth data collected from the Microsoft Kinect Sensor. Top view configuration has been used to collect data. Four subjects participated in the experiments to simulate some fall actions. The database contains 20 tests details of which are: Tests from 1 to 10 contain data of two or more people walking in the monitored area. Tests from 11 to 20 contain depth images of a person simulating falls. [11]

TST Fall Detection Dataset V2

This database contains depth frames and skeleton joints recorded using the Microsoft Kinect v2. It also contains acceleration samples collected from an IMU during the simulation of ADLs and falls. The dataset contains depth frames and skeleton joints collected using Microsoft Kinect v2 and acceleration samples provided by an IMU during the simulation of Daily activities and falls. The simulation is carried out by 11 subjects. A total of 264 sequences are available and for every recording depth frames of resolution 512×424 is available along with two acceleration data sequences obtained from IMUs. The IMUs are connected to the waist and right wrist of the subjects [12].

SDU FALL

SDU Fall Database has also used the Kinect Sensor to record data. RGB, depth videos and 20 skeleton joint positions were recorded. The resolution of all the videos is 320×240 pixels per frame and frame rate of 30 fps. Twenty subjects performed various postures which include six actions 10 times each, namely: bending, squatting, sitting, falling down, lying, walking, etc. Some recordings were made under certain specific conditions like carrying or not carrying large objects, changing illumination conditions and different positions with respect to the camera [13].

FALL Database

The database consists of falls and activities of daily living performed by two subjects. Each subject performed all activities twice. Hence, the database consists of 72 video sequences, containing 40 falls and 32 activities of daily living. The different scenarios are adopted by Noury et al. [14] and fall detection using these are implemented in [15].

High Quality Fall Simulation Dataset

This dataset consists of fall recordings captured using 2D cameras. Different recordings were made in the realistic background and the falls were made as natural as possible. Cameras were installed in the homes of the elderly in the nursing home room [16].

Multi Cam Fall Dataset
This database has recordings of 24 situations that have been recorded using 8 IP video cameras. The first 22 recordings are of fall and abrupt situations and the last 2 recordings contain only abrupt situations [17].

UP-Fall Detection Dataset (Multi Modal)
UP-Fall Detection has recordings of 11 activities and 3 trials per activity. Subjects performed six simple activities and five types of falls. The volunteers constituted 17 healthy young adults without any impairment. They carried out daily activities and simulated falls and the data were collected using five wearable sensors in addition to cameras. The sensors provided accelerometer, gyroscope and ambient light data [18].

FDD Dataset
The FDD dataset has been recorded using a single camera in a real world video surveillance scenario. The recordings have a frame rate of 25 frames/s with an image resolution of 320 × 240 pixels. The recordings aptly portray the elderly home environment and simple office room settings. They also include illumination variations, occlusion problems and background texture and clutter situations. The dataset contains 191 annotated videos that permit independent feature classification. Also, the recordings were carried out at different locations like home, office, lecture room and coffee room [19].

Besides vision-based databases some databases containing falls recorded by wearable devices are also available publicly. DLR, MobiFall, SisFall, tFall, UMAFall, Cogen Labs, Gravity Project, UniMiB SHAR are some of them. For details about these databases, readers can refer to [18, 20] which contains a comprehensive summary of these databases.

3 Latest Research on Fall Detection Systems Using 2D Cameras

Nunez-Marcos et al. [21] used Convolutional Neural Networks to determine if the video frames contain a falling person. The motion in the video was represented using optical flow images, that capture the motion of adjacent video frames and ignore other details such as colour, brightness, contrast or lighting variations. Their optical flow images were fed as input to the networks and the network was trained in three phases. Fall datasets usually have a lower number of samples so transfer learning techniques were used for training the network. They applied their proposed algorithm to the URFD, Multicam and FDD database for testing their proposed algorithm.

de Miguel et al. [22] proposed a fall detector also based on optical flow images. The initially captured data is preprocessed using segmentation and background subtraction and noise removal is carried out using Kalman filters. Optical flow images are used as input to the classifier. Classification is carried out using the KNN classifier. They have developed their own database for validating their work.

Harrou et al. [23] have used the multivariate exponentially weighted moving average (MEWMA) statistic for the detection of falls. The MEWMA features offer a better response in detecting falls and has low computational cost so is simpler to implement in real time. The URFD and FDD databases were used for testing their work.

Lu and Chu [24] proposed a fall detection system for identifying falls that occur when a person is sitting down or standing up from a chair. They use the YOLOv3 algorithm to detect objects in the image frame including the subject. Followed by the Camshift algorithm to track the detected person continuously. Their application areas mainly cover nursing homes and other such public places that have surveillance systems installed. They have tested their work on videos were falls were simulated and also on videos obtained from Youtube and Giphy.

Haraldsson [25] in his master's thesis designed a fall detection where the images are sampled and their motion history is generated. Convolutional neural network is used for extracting relevant features and a neural network of densely connected neurons are used for classification. He used the FDD database for testing his proposed algorithm.

Debard et al. [26] used weighted structural intensity histograms and upper body detectors along with particle filter tracking to improve the foreground segmentation. For feature extraction they computed five parameters namely: Aspect Ratio, Change in Aspect Ratio, Centre Speed and Head speed and fall angle and used this spatial data as a feature vector. The classification was carried out using SVM.

Devishay et al. [27] proposed an image-based fall detection where feature extraction was carried out using a Histogram of Oriented Gradients and the HAAR extractors post frame extraction and background subtraction. The classification was carried out using SVM.

Shukla and Tiwari [28] proposed a fall detection system based on two features: human centroid height relative to the ground and motion history image-based fall detection. Human centroid computation is suitable when the fall is unoccluded. However, for occluded falls the motion history images are used. The first method is well suited for unclouded falls because most falls will occur near the ground. For falls whose ends are occluded by furniture or other obstacles MHI-based human fall detection is used where the angle of falling is computed and if the falling object is found below the threshold angle then the situation is classified as a fall.

Chen [29] in his work analyses the video frames using the Gaussian modelling of the background to extract positions of the moving objects. Noise removal of the processed frames is carried out using morphological filtering. Human head positions are identified in the foreground image and their median point, height and aspect ratio is computed. Relative changes in these features are used to track persons in the frame. The height and aspect ratio of the human body is used to detect falls in their work.

Belshaw et al. [30] In their work used images obtained from a camera installed in the ceiling. Their system uses visual background modelling for separation of the shadowed silhouette of a person from shadowless silhouette regions. Velocity, area and moment features of the preprocessed frames is used for feature extraction. Multilayer Perceptron Neural network was used for classification.

Liu and Zuo [31] proposed a fall detection algorithm based on three features, namely: human aspect ratio, effective area ratio and centre variation rate.

Triantafyllou et al. [32] in their work compute the trajectories of a fallen person like vertical velocity and variance of the area which constitutes the feature vector set. Hidden Markov Models are used to model the fall incident.

4 Latest Research on Fall Detection Systems Using 3D Cameras

Volkhardt et al. [33] work used a Kinect sensor connected to a mobile robot to help people fallen on the ground. They segmented objects in the frame into point clouds and layered them so that the effects of occlusion could be taken care of. They extracted four sets of features from their data namely: Geometrical and statistical features, Histogram of Local Surface Normals, Fast Point feature histogram and surface entropy and tested with four different sets of classifiers: Ada Boost, Random Forests, Nearest Neighbours and SVMs. Based on their experimental results they obtained the best results using Histogram of Local Surface Normals along with SVM as classifier.

Sree Madhubala and Umamakeswari [34] Used a Microsoft Kinect sensor to capture the normal activities of subjects and the image frames obtained were processed in Raspberry pi. The boundary of the subject in the frame was computed using the Canny edge detection method followed by morphological processing operations to correctly obtain the outline of the person. Shape detection is carried out by contour analysis and classification is done using the mean matching procedure.

Gasparrini et al. [11] used Ad hoc segmentation algorithm on the depth images obtained from the Kinect sensor. Anthropometric relationships and features are used to detect humans in the depth images. Post detection, a tracking algorithm is applied between the frames of the detected images. The frame containing the background of the scene is used as a reference frame. Fall detection is carried out using the nearness of the person detected to the floor.

Nizam et al. [35] extracted the subject and the scene details and frame wise tracking was carried out. The velocity at the tracked joints with respect to the previous location was computed and fall was detected if the position of the subject was detected near the floor after some abrupt change in the computed velocity.

Yang et al. [36] in their work first applied Median filters for foreground, as well as background processing. The outline of the moving person is computed by the background subtraction of frames. Disparity maps were computed using the histogram statistics of horizontal projections and vertical projections of the data. The disparity maps and least square method were used to estimate the floor plan and floor plan equations. Human body centroids were computed from the depth images and also the angle between the human body and the floor. When both these parameters are below certain preset threshold it is classified as an occurrence of fall.

Sase et al. [37] in their work processed depth videos to separate the frames followed by background subtraction to detect the region of interest. By comparing the region of interest of consecutive frames a threshold is computed which is used to detect falls. They tested their proposed approach on UR and SDU fall databases.

Aslan et al. [38] Modelled the falls using Curvature Scale Space features and Fisher Vector encoding on the depth data obtained using the Kinect sensors. The classification was carried out using Support Vector Machines.

Abobakr et al. [39] designed a fall detection system using random decision forests to distinguish between fall and non-fall activities. The data were acquired using a 3D sensor camera. Local variations in the depth pixel were analyzed to determine the posture. The classification was done using SVM.

Wang et al. [40] proposed a feature extraction method that combined directional gradient of depth feature (DGoD) and local difference of depth feature (LDoD) around a pixel for mutual comparison to calculate the difference between the pixels. Random forest classifier has been applied to the depth image to classify pixels on various parts of the human body. The human skeleton is detected using the gravity centre of each part.

5 Discussion on Current Methods Used in Fall Detection

Based on the literature review it can be seen that researchers have used various databases to test their proposed work and some have recorded their own videos as well. Performance metrics used for validating and evaluating their work are also different so it is difficult to make a general claim as to performance comparisons. But it can be safely said that the current research papers are on 2D camera-based fall detection systems.

Various tracking methods are being used to track the moving person in the video frames and optical flow has been used widely. However, when the distance between the camera and subject increases the optical flow tracking fails. The problem also arises when multiple moving objects are present in a single frame.

Several researchers have used computation of human body centroid and parameters like moments, velocity, median, centre variation, aspect ratio, etc. to detect falls. However, partial occlusion in such cases is a drawback.

Classification of falls have been carried out by Neural Networks and Support Vector Machines mostly. Sometimes due to lack of sufficient training data the network is unable to learn leading to performance degradation.

The depth sensors have been used in place of 2D cameras by some researchers. Kinect sensor by Microsoft has gained a lot of popularity in the field of computer vision because of its affordable cost and easy availability. The results obtained using depth sensors are relatively better than 2D sensor cameras as the depth data is robust to illumination variations.

The main challenges in fall detection have been summarized as follows:

- poor identification of pose
- the area being scanned-whether at home or in public places
- number of people in the frame
- poor lighting conditions
- quality of the image captured
- occlusion
- distance of the subjects from the camera
- usage of accessories like walking aids.

6 Conclusion

In this paper, is mostly focused on the fall detection algorithms proposed over the last five years. Though research on fall detection has been going on for almost a decade now some issues still remain due to the nature of the problem. Most of the fall databases available are recorded using actors simulating fall events as real time data is hard to get due to privacy issues. Also, the detection system should be able to discriminate between falls and other activities carried out in the day to day life. So it is important to prepare a database that includes fall, as well as daily life activities. Implementing fall detection systems using vision-based approaches is easy when a single camera needs to be installed. However, sometimes a single camera does not provide satisfactory results especially if the area of coverage is more. In that case, multiple cameras need to be installed and calibrated accordingly. Also, it has been inferred that depth camera-based approaches have the best performance in fall detection.

References

1. National Center for Health Statistics (US) (ed.), Health, United States, 2016: with chartbook on long-term trends in health, National Center for Health Statistics (US), Hyattsville. Report No. 2017-1232, Health, United States (2017)
2. G. Bergen, M.R. Stevens, E.R. Burns, Falls and fall injuries among adults aged ≥65 years—United States. MMWR Morb. Mortal Wkly Rep. **65**(37), 993–998 (2016). https://doi.org/10.15585/mmwr.mm6537a2
3. H.-B. Zhang, Y.-X. Zhang, B. Zhong, Q. Lei, L. Yang, J.-X. Du, D.-S. Chen, A comprehensive survey of vision-based human action recognition methods. Sensors (Basel). **19**(5), pii: E1005 (2019). https://doi.org/10.3390/s19051005
4. M. Saleh, R.L.B. Jeannès, Elderly fall detection using wearable sensors: a low cost highly accurate algorithm. IEEE Sens. J. **19**(8), 3156–3164 (2019). https://doi.org/10.1109/jsen.2019.28911285
5. F. Wu, H. Zhao, Y. Zhao, H. Zhong, Development of a wearable-sensor-based fall detection system. Int. J. Telemed. Appl. (2015). http://dx.doi.org/10.1155/2015/576364

6. I. Chandra, N. Sivakumar, C.B. Gokulnath et al., IoT based fall detection and ambient assisted system for the elderly. Cluster Comput. **22**(Supplement 1), 2517–2525 (2019). https://doi.org/ 10.1007/s10586-018-2329-2
7. Q. Zhang, M. Karunanithi, Feasibility of unobtrusive ambient sensors for fall detections in home environment, in *38th Annual International Conference of the IEEE Engineering in Medicine and Biology Society (EMBC)*, Orlando, FL, pp. 566–569 (2016)
8. Y. Nizam, M.N.H. Mohd, M.M.A. Jamil, A study on human fall detection systems: daily activity classification and sensing techniques. Int. J. Integr. Eng. **8**(1), 35–43 (2016)
9. Z. Zhang, C. Conly, V. Athitsos, A survey on vision-based fall detection, in *PETRA '15*, 01–03 July 2015, Island of Corfu, Greece. ACM. ISBN 978-1-4503-3452-5/15/07. http://dx.doi.org/ 10.1145/2769493.2769540,2015
10. Z. Zhang, S. Ishida, S. Tagashira, A. Fukuda, Danger-pose detection system using commodity Wi-Fi for bathroom monitoring. Sensors **19**, 884 (2019). https://doi.org/10.3390/s19040884, www.mdpi.com/journal/sensors
11. S. Gasparrini, E. Cippitelli, S. Spinsante, E. Gambi, A depth-based fall detection system using a kinect® sensor. Sensors **14**(2), 2756–2775 (2014). https://doi.org/10.3390/s140202756
12. S. Gasparrini, E. Cippitelli, E. Gambi, S. Spinsante, J. Wahslen, I. Orhan, T. Lindh, Proposal and experimental evaluation of fall detection solution based on wearable and depth data fusion, in *ICT Innovations 2015* (Springer International Publishing, 2016), pp. 99–108. https://doi.org/ 10.1007/978-3-319-25733-4_11
13. M. Yilmaz, SDU dataset (2018). https://doi.org/10.6084/m9.figshare.7166453.v1
14. N. Noury, A. Fleury, P. Rumeau, A.K. Bourke, G.O. Laighin, V. Rialle, J.E., Fall detection— principles and methods, in *Engineering in Medicine and Biology Society*, vol. 2007, pp. 1663– 1666 (2007)
15. R. Planinc, M. Kampel, Robust fall detection by combining 3D data and fuzzy logic, in *ACCV Workshop on Color Depth Fusion in Computer Vision*, Daejeon, Korea, pp. 121–132, Nov 2012
16. G. Baldewijns, G. Debard, G. Mertes, B. Vanrumste, T. Croonenborghs, Bridging the gap between real-life data and simulated data by providing realistic fall dataset for evaluating camera-based fall detection algorithms. Healthc. Technol. Lett. **3**(1), 6–11 (2016). https://doi. org/10.1049/htl.2015.0047. eCollection
17. E. Auvinet, C. Rougier, J. Meunier, A. St-Arnaud, J. Rousseau, Multiple cameras fall dataset. Technical report 1350, DIRO—Université de Montréal, July 2010
18. L. Martínez-Villaseñor, H. Ponce, J. Brieva, E. Moya-Albor, J. Núñez-Martínez, C. Peñafort-Asturiano, UP-fall detection dataset: a multimodal approach. Sensors (Basel). **19**(9), 1988 (2019). https://doi.org/10.3390/s19091988
19. A. Trapet, Fall detection dataset (2013), http://le2i.cnrs.fr/Fall-detection-Dataset?lang=en. Le2i Laboretoire Electronique, Informatique et Image UMR CNRS 6306
20. E. Casilari, J.A. Santoyo-Ramón, J.M. Cano-García, Analysis of public datasets for wearable fall detection systems. Sensors (Basel). **17**(7), 1513 (2017). https://doi.org/10.3390/s17071513
21. A. Núñez-Marcos, G. Azkune, I. Arganda-Carreras, Vision-based fall detection with convolutional neural networks. Wirel. Commun. Mob. Comput. **2017**, 16 pages (2017). https://doi.org/ 10.1155/2017/9474806. Article ID 9474806
22. K. de Miguel, A. Brunete, M. Hernando, E. Gambao, Home camera-based fall detection system for the elderly. Sensors **17**, 2864 (2017). https://doi.org/10.3390/s17122864, www.mdpi.com/ journal/sensors
23. F. Harroua, N. Zerroukib, Y. Suna, A. Houacineb, Vision-based fall detection system for improving safety of elderly people. IEEE Instrum. Meas. Mag. **20**, 49–55 (2017). http://dx.doi.org/ 10.1109/mim.2017.8121952
24. K.-L. Lu, E.T.-H. Chu, An image-based fall detection system for the elderly. Appl. Sci. **8**(10), 1995 (2018)
25. T. Haraldsson, Master's thesis on real-time vision-based fall detection with motion history images and convolutional neural networks, Computer Science and Engineering, master's level 2018, Luleå University of Technology Department of Computer Science, Electrical and Space Engineering Luleᵒa Tekniska Universitet Institutionen fᵒor System- och rymdteknik, Supervisor Dr. D. Walther, Dr. J. Hallberg (2018)

26. G. Debard, M. Mertens, T. Goedemé, T. Tuytelaars, B. Vanrumste, Three ways to improve the performance of real-life camera-based fall detection systems. Hindawi J. Sens. **2017**, 15 pages (2017). https://doi.org/10.1155/2017/8241910. Article ID 8241910
27. D. Mishra, N. Mukta, S. Rana, P. Karnawat, Computer vision based fall detection for elderly person using HOG descriptor and HAAR feature extractor. Int. J. Eng. Res. Comput. Sci. Eng. (IJERCSE) **5**(3) (2018). ISSN (Online) 2394-2320
28. P. Shukla, A. Tiwari, Vision based approach to human fall detection. Int. J. Eng. Res. Gen. Sci. **3**(6) (2015). ISSN 2091-2730
29. M.-C. Chen, A video surveillance system designed to detect multiple falls. Adv. Mech. Eng. **8**(4), 1–11 (2016). https://doi.org/10.1177/1687814016642914, aime.sagepub.com
30. M. Belshaw, B. Taati, D. Giesbrecht, A. Mihailidis, Intelligent vision-based fall detection system: preliminary results from a real-world deployment, in *RESNA Conference Proceedings*, RESNA_ICTA (2011)
31. H. Liu, C. Zuo, in *AASRI Conference on Computational Intelligence and Bioinformatics*. An improved algorithm of automatic fall detection. AASRI Proc. 353–358 (2012)
32. D. Triantafyllou, S. Krinidis, D. Ioannidis, I.N. Metaxa, C. Ziazios, D. Tzovaras, A real-time fall detection system for maintenance activities in indoor environments. IFAC Papers On Line **49–28**, 286–290 (2016)
33. M. Volkhardt, F. Schneemann, H.-M. Gross, Fallen person detection for mobile robots using 3D depth data, in *Proceedings IEEE International Conference on Systems, Man, and Cybernetics (IEEE-SMC 2013)*, Manchester, GB (IEEE Computer Society, CPS 2013), pp. 3573–3578
34. J. Shree Madhubala, A. Umamakeswari, A vision based fall detection system for elderly people. Indian J. Sci. Technol. **8**(S9), 167–175 (2015)
35. Y. Nizam, M.N.H. Mohd, M.M.A. Jamil, Human fall detection from depth images using position and velocity of subject, in *2016 IEEE International Symposium on Robotics and Intelligent Sensors*, IRIS, Tokyo, Japan. Proc. Comput. Sci. **105**(2017), 131–137, Dec 2016
36. L. Yang, Y. Ren, W. Zhang, 3D depth image analysis for indoor fall detection of elderly people. Digit. Commun. Netw. **2**, 24-34 (2016)
37. P.S. Sase, S.H. Bhandari, Human fall detection using depth videos, in *2018 5th International Conference on Signal Processing and Integrated Networks (SPIN)*, Noida, 2018, pp. 546–549
38. M. Aslan, A. Sengur, Y. Xiao, H. Wang, M. Ince, X. Ma, Shape feature encoding via fisher vector for efficient fall detection in depth-videos. Appl. Soft Comput. https://doi.org/10.1016/j.asoc.2014.12.035,2015
39. A. Abobakr, M. Hossny, S. Nahavandi, A skeleton-free fall detection system from depth images using random decision forest. IEEE Syst. J. **12**(3), 2994–3005 (2018)
40. H. Wang, F. Zhou, W. Zhou, L. Chen, Human pose recognition based on depth image multi-feature fusion. Complexity **2018**, 12 (2018), https://doi.org/10.1155/2018/6271348. Article ID 6271348

Surveillance Robocar Using IoT and Blynk App

Nitesh Kumar, Barnali Dey, Chandan Chetri, and Amrita Biswas

Abstract The motivation of the paper is the real time application of an unmanned vehicle, "robocar" using internet of things (IoT) for surveillance in harsh terrains for military purpose. In this paper, a robocar has 6 been used for the aforesaid purpose. The robocar has a 'pick and place' robotic arm designed with Arduino and controlled by the "Blynk" app which is a free mobile application software that connects the robocar with the internet and helps the user to control the device from a remote location. A significant attribute of the robocar 10 is its capability to pick and place an object of around 250 gm. Also, it has a live monitoring feature that enables the user to control the direction of its movement from anywhere by using the "Blynk" app on his mobile phone. The movement of the arms and grippers of the robocar is controlled by the rotation of the servomotor. To provide internet of things (IoT) access, NodeMCU which is a microcontroller integrated with the Wi-Fi module on a single board is used. Another important feature of the robocar is that it can go off-road as it uses a caterpillar chain for its movement that enables it to move even on rough terrain.

Keywords IoT (Internet of things) · NodeMCU (ESP8266 Wi-Fi module) · Android app (Blynk) · Servo motor · Caterpillar chain

1 Introduction

Today we all are familiar with the advancement of technology in real time application. It not only provides new possibility for extending new ideas, but also future ways to be implemented. This has resulted in an unprecedented enhancement in the development of robotics that can range from automation to surveillance. As we know, the idea of surveillance has become crucial to keep close systematic supervision that is applicable to be maintained over a person, group, property, offices, important public places, and especially in international borders. Surveillance may be done inside,

N. Kumar · B. Dey (✉) · C. Chetri · A. Biswas
SMIT, SMU, Rangpo, Sikkim, India
e-mail: barnali.d@smit.smu.edu.in

© Springer Nature Singapore Pte Ltd. 2020
R. Bera et al. (eds.), *Advances in Communication, Devices and Networking*,
Lecture Notes in Electrical Engineering 662,
https://doi.org/10.1007/978-981-15-4932-8_47

as well as outside these areas by the human or with the help of technology manifestation, in the form of robots. A robot may be defined as an extension of the human, which is capable of performing specific tasks as preprogrammed, by the user into them. This provides highly accurate results and quite easily overcomes the limitation of human beings. Robots as surveillance devices in the international borders of countries like India is gaining strong footholds as we see from recent studies that military warfare is slowly moving towards unmanned machines. The use of robots will help save human lives in such a situation [1]. Ideally, robots for the military services need to be designed as small as possible so that they are least likely to be located by the human eye and also they need to be designed in such a way that they can move in any terrain especially uncertain rugged terrains. The army in most cases uses a robot for war field that helps in reducing soldier causality [2]. However, there are certain limitations in such robots such as they have limited range of coverage as they are based on RF Technology, Zigbee and Wi-Fi, they are only capable of sensing one or two physical quantities, do not have a smooth movement in rough terrains, etc. [3]. All these limitations have motivated the authors to design a robot that not only senses but also have the ability to pick and place objects, move in rugged terrain and also use the internet for communication via the cellular phone so as to have a wide range of area coverage in real time.

2 Proposed System Architecture

The overall system reported in this paper is based on real time applications. Taking out one of the major drawbacks, we enhanced the limitation of the robocar to go on off-road with the help of the caterpillar chain [4]. The chain provides a strong grip to the car, for smooth travel in odds road. We have used Microcontroller integrated board which has a Wi-Fi module. The main advantage of this is that space consumption is reduced. The implementation and use of IoT to get control or access over any devices makes time efficiency to do the work [5]. Today real time applications such as home automation, on-off light with the help of an app, etc., are possible with the use of IoT. The open-source platform provides the user and developer, to access and to make a strong communication between hardware and software. The NodeMCU is connected to the internet with the help of hotspot or Wi-Fi access. It can go to the distance where till it can access the internet [6]. The servo robot arm movement is controlled by an android app called BLYNK. The android provides a platform for the transmission of data from smartphone to hardware. A camera that provides a good quality of video is used for viewing [7]. The main applications of these robocars are in military rescue and search operation, bomb defuse operation, carrying medic, detection of enemies present ahead, and many more. The servo arm was so designed so as to provide a movement of 180° for 'pick and place' purpose with objects of minimum weight of 250 gm that it can handle [8–10].

3 Proposed Circuit Design of the Robocar System

Figure 1 gives a picture of how the whole circuit works. The Arduino (microcontroller) is connected directly to the motor driver circuit for the controlling of the dc servo motor. Along with the side, dc source of constant power source is given to the motor driver circuit. The motor moves ahead-backward, left-right [11].

In Fig. 2, initially when the user gives command via mobile, with the help of an android app then the signal is fetched to NodeMCU. NodeMCU is a Wi-Fi-based module that provides an internet service to the microcontroller to execute the given command from the user. The microcontroller is connected through certain pins to the motor driver. According to the input, i.e., user from the android app, the microcontroller gives the driver instruction's to rotate either left-right, forward or backward [9]. The driver and all these accessories are connected through a portable power supply.

In the Fig. 3, the circuit shows the working of the robotic arm. Three servo motors are connected with the controller, when the controller receives the command from the mobile app it gives the three different movements which are twisting, rotating, and gripping for better control in picking and placing of objects [12].

Figure 4 shows the whole block diagram of the robot arm. The coder codes the program and loads it to the microcontroller. Along with it android app (BLYNK) is used, which is connected directly to the cloud source. The cloud server also provides back the information from the microcontroller to the android app. From where the user can see the output [13]. After all these, the microcontroller does the specific task given by the user to the robotic arm to move left-right, ups-downs, and covering an angle pf 180°. The power supply which is a portable battery provides a constant supply to the microcontroller [14].

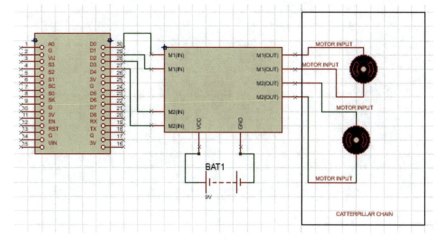

Fig. 1 Circuit diagram of the robocar

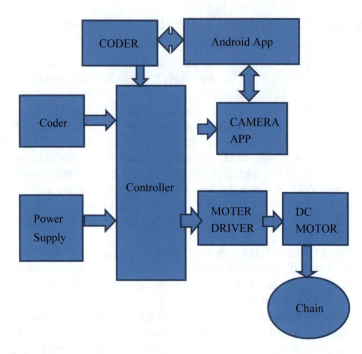

Fig. 2 Block diagram of the robocar

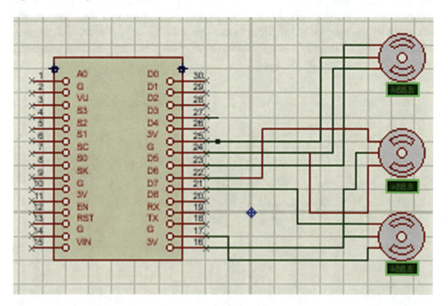

Fig. 3 Circuit diagram of the robotic arm

Fig. 4 Block diagram of the arm

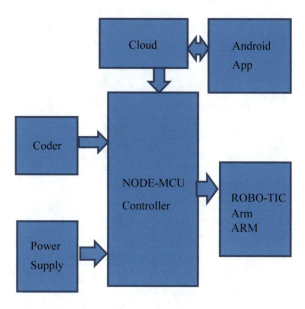

4   Result and Analysis

The result and analysis of the research paper would like to highlight its application and implementation in terms of medical, defense, surveillance, and many more. Our research paper's key implementation is of the caterpillar chain to go in off-road. We simulated the circuit on proteus and the output came as expected. The algorithm of the entire simulation of the system is depicted in Table 3 of the following section. The commands that has been given as an input to the robocar system are tabulated in Table 1. The commands that has been given as an input to the Grips and Arms controller system are tabulated in Table 2 (Figs. 5, 6, and 7).

Table 1 Input output status of the robocar system

Command	Robocar status
if (X == 128 && Y == 128)	Stop
else if (X > 103 && X < 132 && Y $\geq$ 129)	Forward
else if (X > 123 && X < 132 && Y $\leq$ 127)	Reverse
else if (Y > 123 && Y < 132 && X $\leq$ 127)	Left
else if (Y > 123 && Y < 132 && X $\geq$ 127)	Right

Table 2 Input output status of the grips and arms controller

Command	Robocar status
Slider 1	Twisting in 180°
Slider 2	Arms up/down in 180°
Slider 3	Gripping of objects

Table 3 Algorithm of the proposed robocar

Step 1	Programming should be done as per the requirement
Step 2	Configuration of circuit and module with blynk app
Step 3	Burn the program in NodeMCU
Step 4	Use the digital pins of NodeMCU as per the programming
Step 5	Connect the output pins of the NodeMCU with Motor driver module
Step 6	Output of the motor driver will be connected with Motors to drive the caterpillar chain
Step 7	Connect the IP camera with the module
Step 8	Give command from the blynk app to move car in forward, reverse, left, right and stop
Step 9	Burn the program for arms control of robocar in another NodeMCU
Step 10	Connect the digital pins of the NodeMCU with servomotors
Step 11	Give command from the blynk app to control the arms in the desired direction for gripping the objects

Fig. 5 The Blynk app platform

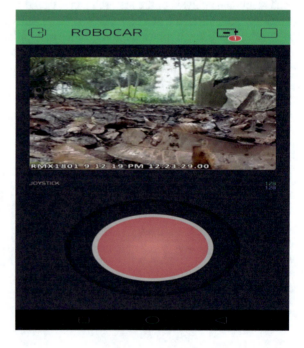

Fig. 6 The working model
of the robocar

5   Conclusion

In this article, we would like to highlight that the existing robocar could be used for the off-road purpose. With the introduction of caterpillar chains, thus it is possible which was one of the major limitations of many previous robocars developed so far. This robocar will be very useful for carrying lightweight amenities and first aid for soldiers. The servo arm is one of the major parts which is used to lift objects with 180° coverage. For an otherwise hazardous situation and in times of need and emergency, this robocar can serve as a life saver.

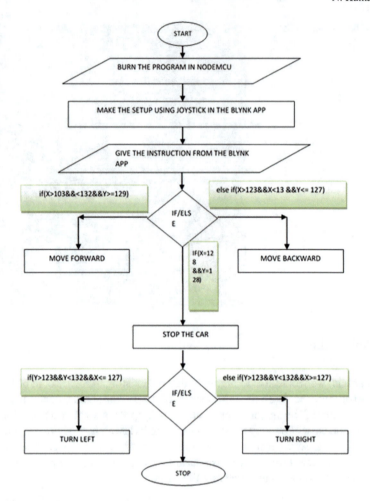

Fig. 7 Flowchart of the working of the robocar model

References

1. T. Kaur, D. Kumar, Design of cell phone operated multipurpose security robot for military applications using solar panel. Int. J. Sci. Eng. Technol. Res. **3**(16), 3472–3475 (2014). ISSN 2319-8885
2. S. Pavithra, S.A. Siva Sankari, 7TH sense-a multipurpose robot for military, in *2013 International Conference on Information Communication and Embedded systems (ICICES)* (IEEE, 2013), pp. 1224–1228
3. T. Kaur, D. Kumar, Wireless multifunctional robot for military applications, in *Proceedings of 2015 RAECS*, UIET, Panjab University, Chandigarh, 21–22 Dec 2015, pp. 1–5
4. R. Selvarasu, H.I. Kadiri, A.S. Joshi, A.B. Wani, IOT based remote access human control robot using MEMS sensor. Int. J. Comput. Sci. Mob. Comput. **5**(3), 816–826 (2016)
5. K. Gawli, P. Karande, P. Belose, T. Bhadirke, A. Bhargava, Internet of things (IoT) based robotic arm. Int. Res. J. Eng. Technol. (IRJET) **4**(3), 757–759 (2017)

6. D. Butkar Vinayak, R. Devikar Sandip, B. Jaybhaye Vikas, Android based pick and place. Int. J. Inf. Futur. Res. **2**(4) (2014). ISSN (Online) 2347-1697
7. B. Stroustrup, *The C++ Programming Language* (Addison-Wesley, 2014), p. 18
8. M. Vanitha, M. Selvalakshmi, R. Selvarasu, Monitoring and controlling of mobile robot via internet through raspberry PI board. 978-1-5090-1706-5/16/$31.00 ©2016 IEEE
9. A. Luigi Atzori, B. Antonio Iera, G. Mora, The internet of things: a survey
10. M. Summerfield, in *Rapid Gui, Programming With Python and Qt. Python is a Very Expressive Language* (2009)
11. H. Chu, M. Chien, T. Lin, Z. Zhang, Design and implementation of an auto-following robot-car system for the elderly, in *2016 International Conference on System Science and Engineering (ICSE)*, UK, Jan 2016, pp. 1–4
12. D. Kuhlman, in *A Python Book: Beginning Python, Advanced Python, and Python Exercises* (2012)
13. Sun Developer Network, C. Object-Oriented Programming, The History of Java Technology (2010)
14. B. Erich, New JavaScript Engine Module Owner, Brendaneich.Com, 16 July 2016

System Identification and Speed Control of a DC Motor Using Plug-In Repetitive Controller in the Presence of Periodic Disturbance

Rijhi Dey, Anirban Sengupta, Naiwrita Dey, and Ujjwal Mondal

Abstract DC motors are extensively used in industries, as they provide high torque. Therefore, precise speed control of DC motor has been a major interest among many researchers. Many works have been carried out with conventional PID controller for speed control of DC motor but it does not hold desired performance in the presence of periodic disturbance in the system. A plug-in Repetitive Controller (RC) is proposed in this paper. A comparative study has been carried out between both the controllers with an overview of performance of the controllers. A set of various input and output data are accumulated and further processed to determine the DC motor model, and it has been implemented in Matlab with the help of identification toolbox.

Keywords DC motor · PID · IMP · Plug-in RC

The original version of this chapter was revised: The correction in author's affiliation has been incorporated. The correction to this chapter is available at https://doi.org/10.1007/978-981-15-4932-8_60

R. Dey (✉)
Department of Electronics and Communication Engineering,
Sikkim Manipal Institute of Technology, Sikkim Manipal University,
East Sikkim, India
e-mail: rijhi.d@smit.smu.edu.in

A. Sengupta
Department of Electrical and Electronics Engineering,
Sikkim Manipal Institute of Technology, Sikkim Manipal University,
East Sikkim, India

N. Dey
Department of Electronics and Communication Engineering,
RCCIIT, Kolkata, India

U. Mondal
Instrumentation Engineering, Applied Physics, University of Calcutta, Kolkata, India

© Springer Nature Singapore Pte Ltd. 2020, corrected publication 2020   433
R. Bera et al. (eds.), *Advances in Communication, Devices and Networking*,
Lecture Notes in Electrical Engineering 662,
https://doi.org/10.1007/978-981-15-4932-8_48

1 Introduction

DC motors are used in most of the industry for its simple speed control property. The speed of the DC motor can be controlled both below and above the base speed. But the speed of the DC motor changes drastically with some sudden change of load. This sudden change of speed can be reduced with the application of PID controller [1]. But controlling the speed of the DC motor with the help of controller requires proper tuning of the coefficients of controller.

Several control techniques are available in the literature for tuning the controller parameters. Li et al. proposed a PSO-based technique for speed control of DC motor [2]. Naung et al. proposed neural network based control of DC motor [3, 4]. But all these techniques are offline in nature. Moreover, PID controllers can reduce the error signal only. They do not reduce the disturbances. This problem can be eliminated with the help of repetitive controller.

Repetitive controller deals with the periodic signal and is established on the application of Internal Model Principle (IMP). IMP is a process that simulates the response of the system in order to evaluate the outcome of a system disturbance [5]. The tasks to control the system are often repetitive in nature and can be affected by periodic disturbances. In different applications like hard disk and CD drives, robotics and inverters, we often confront a condition where the periodical reference signals/commands have to be tracked or rejected.

RC is also known as a controller which uses iterative learning technique, where the error introduced in tracking a limited time period reference signal is minimized by means of correcting the input in every iteration based on the error which is observed in the previous iterations [6–10].

In this work as a controller, a PID with a plug-in repetitive controller is used to control the voltage which is proportional to the speed of the motor. The combination design method of PID and RC system possess high performances in different areas. PID control executes better in the existence of noise and high error but tuning is not easy. RC can be easily tuned but very sensitive to noise or high error. So the advantages of both PID and repetitive controller are implemented in this work.

2 Experimental Setup of a DC Motor System

The DC motor system fundamentally performs as a transducer which converts electrical energy into mechanical energy exhibited in the configuration of rotating shaft connected to the rotor [11]. The basic portrayal of the DC motor used is depicted in Fig. 1.

The DC motor setup is modelled in an open-loop configuration. The input/output data sets are taken from the given system in open-loop configuration and in no-load condition. The input/output data records are fed through system identification toolbox of Matlab, a GUI-based outlook, to determine the model of the DC motor.

Fig. 1 DC Motor

The outcome is validated through fitness of system identification toolbox with the actual responses. The fitness percentage is calculated by the provided equation:

$$Best\ fits = [\frac{1 - (|a - a^*|)}{|a - \hat{a}|}] \times 100 \tag{1}$$

In equation number (1), 'a' is the true response, 'a*' is the anticipated model output and '$\hat{a}$' is the mean of 'a'. The best fitness is marked by 100%. The best fitness is achieved to be 85.14% as the recognized model is stable up to this fitness level with respect to frequency response as shown in Fig. 2.

The transfer function of the system given by

$$G(s) = \frac{0.1086}{s^2 + 0.2066s + 0.0005406} \tag{2}$$

3 Design of Modified Repetitive Control System

Conventional PID controller is not adequate to furnish consistent adaptation in performance when periodic disturbances are subjected to the process. The plug-in RC incorporates the factor of $\frac{1}{1-e^{-Ls}}$, which correlates to the sub-harmonic and harmonics of the period L; the controller can trace any periodic signal and minimize/remove any

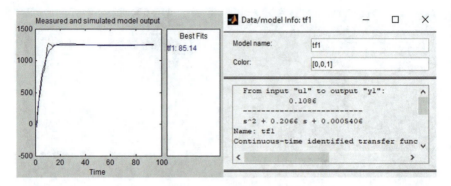

Fig. 2 Identified model is validated with the help of system identification toolbox based on DC motor system

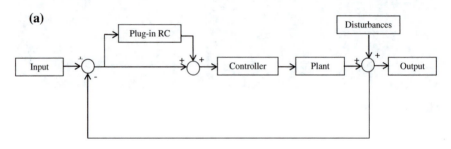

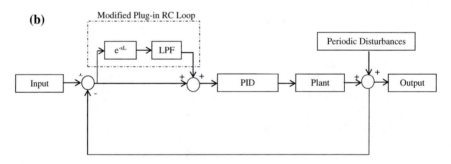

Fig. 3 **a** Block diagram of a plant with plug-in RC-based PID. **b** Proposed modified RC loop for periodic disturbance rejection

disturbance having period L [4]. This problem can be minimized by using a plug-in RC as shown in Fig. 3a. For removal of periodic interference, a modified architecture of repetitive controller with PID controller is proposed in this paper as depicted in Fig. 3b.

The periodic disturbance which is added is a sinusoidal signal having V_{P-P} of 2 V and frequency 0.2 Hz. In the modified RC loop, there is a time delay along with

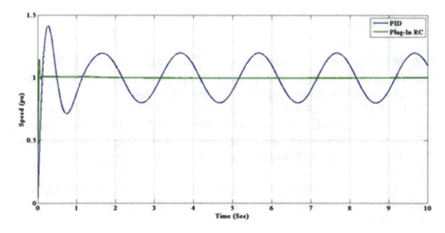

Fig. 4 Disturbance rejection performance of PID versus plug-in RC system

a low-pass filter. A 5 rad/s cut-off frequency is selected for the low-pass filter f(s) is given as

$$f(s) = \frac{5}{s+5} \tag{3}$$

Therefore, transfer function of the system is given by

$$G_{RC}(s) = \frac{1}{1 - e^{-LS}f(s)} \tag{4}$$

4 Result

A comparative analysis is done by using PID controller and plug-in RC control system using DC motor as a plant which is depicted in Fig. 4. A step input of 4 V was provided to the DC motor, with an added sinusoidal disturbance. The PID controller was unable to reject the periodic disturbances, whereas the plug-in RC was able to eliminate the disturbances and also improved the stability of the system.

The above figure clearly specifies gradual improvement in tracking ability for the motor shaft while being driven by plug-in RC system in comparison to a conventional PID scheme.

5   Conclusions

This paper includes a short explanation of modelling of a DC motor used in laboratory. Easy and apprehensive method is applied to reduce the difficulty for identifying the system, owing to that the suggested idea remains unaltered.

The addition of the appropriately implemented Low-Pass Filter (LPF) is used to decrease the consequence due to positive feedback in the loop of RC. It also improves the stability of the proposed design of the control system of plug-in RC. Comparison of eliminating added periodic disturbances by PID Controller and plug-in RC system is shown properly.

Though future improvising can be done to explore more, the conclusion that plug-in RC eliminates periodic disturbances which PID controller cannot is justified with proper validation.

References

1. V. Kumawat, K. Bhuvnesh, S. Arvind, PID controller of speed and torque of servo motor using MATLAB. Int. J. Recent Innov. Trends Comput. Commun. **1**(20) (2013)
2. H. Li, M.Y. Chow, Z. Sun, EDA-based speed control of a networked dc motor system with time delays and packet losses. IEEE Trans. Ind. Electron. **56**(5), 1727–1735 (2009)
3. Y. Naung, S. Anatolii, Y.H. Lin, Speed control of dc motor by using neural network parameter tuner for pi-controller, in *2019 IEEE Conference of Russian Young Researchers in Electrical and Electronic Engineering (EIConRus)* (IEEE, 2019), pp. 2152–2156
4. U. Bhatnagar, Application of grey wolf optimization in optimal control of DC motor and robustness analysis (2018)
5. B.A. Francis, W.M. Wonham, The internal model principle of control theory. Automatica **12**(5), 457–465 (1976)
6. U. Mondal, A. Sengupta, N. Dey, Tracking of periodic reference signal: a parameterized finite dimensional repetitive control approach. Trans. Inst. Meas. Control **40**(13), 3640–3650 (2018)
7. J. Santana, J.L. Naredo, F. Sandoval, I. Grout, O.J. Argueta, Simulation and construction of a speed control for a DC series motor. Mechatronics **12**(9–10), 1145–1156 (2002)
8. U. Mondal, A. Basu, S. Bose, A. Sengupta, U. Saha, Finite dimensional robust repetitive controller for tracking periodic reference input, in *IEEE International Conference on Electrical, Computer and Communication Technologies (ICECCT)* (IEEE, 2015), pp. 1–4
9. U. Mondal, A. Sengupta, Real time system development & speed control of a stepper motor using commercial soft tools & open source codes (2011)
10. Y. Onuki, H. Ishioka, Compensation for repeatable tracking errors in hard drives using discrete-time repetitive controllers. IEEE/ASME Trans. **6**, 132–136 (2001)
11. U. Mondal, A. Sengupta, R.R. Pathak, Servomechanism for periodic reference input: discrete wavelet transform-based repetitive controller. Trans. Inst. Meas. Control **38**(1), 14–22 (2016)

Optimum Frequency Utilization Model for Industrial Wireless Sensor Networks

K. Krishna Chaitanya, K. S. Sravan, and B. Seetha Ramanjaneyulu

Abstract In many industries, thousands of wireless sensor networks (WSNs) are needed to perform different assignments to execute a task effortlessly, in which lot of errors bound to occur due to lack of proper coordination between these sensors. In order to establish the synchronization between them and to implement the Internet of Things, sensors should be able to communicate among themselves without interference. So, in order to reduce the internetwork interference, we propose a novel method to optimize the utilization of available frequencies based on the networks' occupancy rate varying from 10 to 100%. Simulation has been carried out in a dense network scenario consisting of 50 typical networks, using GLPK 4.65 (GNU Linear programming kit) solver. By using the proposed model, we can mitigate the spectrum scarcity and reduce the starvation of networks for frequencies. The simulation results show that frequencies have been utilized optimally compared to the conventional greedy method.

Keywords Factory of Things (FoT) · GLPK · Industrial wireless sensor networks (IWSNs) · IIoT · Occupancy rate

1 Introduction

Wireless networks have gained more popularity in the recent times, especially in many industries which have lots of devices that require interconnectivity among themselves. Collection of information from those devices had become a hectic task. When compared to conventional wireless communication, sensor networks have the

K. Krishna Chaitanya (✉) · K. S. Sravan · B. Seetha Ramanjaneyulu
VFSTR, Vadlamudi, A.P, India
e-mail: krishna.kandregula444@gmail.com

K. S. Sravan
e-mail: sraonekumar92@gmail.com

B. Seetha Ramanjaneyulu
e-mail: ramanbs@gmail.com

© Springer Nature Singapore Pte Ltd. 2020   439
R. Bera et al. (eds.), *Advances in Communication, Devices and Networking*,
Lecture Notes in Electrical Engineering 662,
https://doi.org/10.1007/978-981-15-4932-8_49

advantage of supporting machine-to-machine interaction rather than human–machine interaction [1, 2]. With the introduction of Industrial Internet of Things (IIoT), finding a solution to this problem has become vital. There are different ways to collect the data from each and every sensor node among which deploying wireless sensor nodes is considered to be the most preferred one, because they can be deployed in every nook and corner of the industry by providing a cost-effective and energy-efficient solution.

Wireless sensor network technologies integrated with the Internet of Things (IoT), which results in the global interconnection of heterogeneous smart physical objects [2] with possible improvements in convenience, productivity, and energy efficiency.

Following the era of IoT, emerging technologies are being introduced inside modern factory environments which are termed as intelligent factory environments or "Factory of Things (FoT)" [2, 3]. Here, introduction of wireless sensor networks in gathering information from the moveable objects plays a key role in the efficient design of industrial processes. They can provide seamless connectivity, during the node failure, i.e., even if one node fails the other can take over. Wireless sensor nodes can be deployed in many applications like remote sensing [3], event detection, and local control of actuators.

Several studies have been conducted, where wireless sensor nodes are deployed for solving various problems in different areas. Researchers realized that sensor network designs have to meet some specific characteristics such as scalability, adaptability, and cost-effectiveness.

As we know, industrial applications don't require a huge amount of data transfer and they don't travel much distance either, yet they are time-sensitive. Typically, IEEE 802.15.4 6LoWPAN standard supports these kinds of networks which are widely used in industrial environments nowadays. Section 2 discusses related work which is happening in this area, Sect. 3 describes the proposed system model, Sect. 4 describes the simulation results that are obtained, and Sect. 5 concludes the paper.

2 Background Work

In [1], authors have proposed an algorithm which dynamically assigns sampling times and manages network performance by using various network efficiency parameters, which also decreases the power consumption by using a programmable logic circuit (PLC) that allows users to have an overall view of the network.

Many researchers have been focusing on the channel assignment strategies [3, 4]; authors have proposed an adaptive channel assignment technique, which treats the whole network as two sets of systems, namely, primary systems and the secondary systems. Here the author proposed a method for heterogeneous wireless sensor networks that can be applied to real world by using wireless sensor networks to implement IOT.

In order to make the wireless sensor nodes work for longer time, we need various energy-efficient techniques to mitigate battery constraint of the wireless sensor node and make them work efficiently.

In papers [5, 6], different techniques have been proposed where two sets of networks are considered that operate in different modes: one is ad hoc and other one is infrastructure mode. They have used the cognitive radio mechanism which senses the spectral holes and allocates the channels optimally leaving minimum spectral holes. Thus, improving spectral efficiency of the network is an important factor in our work. Here, these networks are partitioned as two sets, namely, the active set and the passive set just like incumbents and CR-CPEs. The proposed algorithm increases the efficiency by opportunistically using the primary and secondary sets. It identifies the malicious nodes that interfere the network and also energize the nodes that are going to sleep, thereby increasing the lifetime of the node.

Thereby, increasing the energy of the wireless sensor node efficiently in the industrial environment and the data transfer happens among the networks. For instance, if any node fails then troubleshooting the defected node among the other nodes is a severe problem. This problem can be mitigated by using different cloud computing techniques which have been proposed in [7–9]. In [8], various connectivity layouts for implementing Factory of Things (FoT) have been proposed.

Automation in industry is growing rapidly; however, Industry 4.0 and Industrial Internet of Things (IIoT) have bought revolutionary changes in the manufacturing sector, which incorporates new manufacturing strategies [10, 11], for the betterment of interoperability among these networks. This in turn poses new challenges to the industrial networks that are supposed to provide better performance in terms of timeliness, reliability, security, and connectivity. On the other hand, new communication technologies that are being developed for information technology (IT) came into existence, which offers promising chances for the further evolution of automation systems.

3 System Model

The network architecture considered in this paper consists of massively dense, interacting groups of wireless nodes that operate to provide seamless communication services to the user terminals. A scenario consisting of 1–50 networks in an increasing order has been considered. Figure 1 shows the networks which are trying to access the available channels depending upon their occupancy rates, i.e., (10–100%). A threshold of 15 frequencies was considered as maximum number of available frequencies that must be optimally allocated between the networks based on their occupancy rates. When the requirement exceeds the available frequencies, there occurs the scarcity of channels. Incorporating the wireless sensor nodes in industry needs some industrial IoT techniques [12].

Usually, networks at 10% occupancy require less number of frequencies and at the same time an occupancy rate of 90% requires almost all the available frequencies to

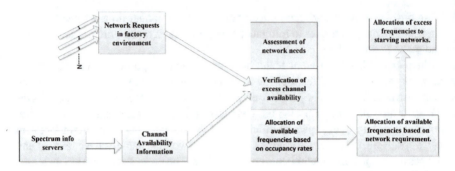

Fig. 1 System model

meet the channel requirements of the networks. As the number of networks increases gradually (i.e., 1–50), channel scarcity also increases. When the channel requirement is increased more than 15, there occurs starvation of network; this information is obtained from the spectrum info servers.

Based on the network requests and the channel availability information in the factory environment, an assessment has been done to allocate the frequencies in an optimal way. Finally, based on that assessment we can utilize the excess amount of frequencies which are available at low occupancy rate to the networks which are starving for available frequencies at high occupancy rate. The conventional greedy method of channel allocation leads to starvation of dense networks with more occupancy rate, and at the same time wastage of frequencies is occurred by allocating the whole number of frequencies to the networks at lower occupancy rates. The simulation is carried out for 200 time instances, by varying occupancy rates from 10 to 90%, for 1–50 networks gradually. A GLPK 4.65 (GNU linear programming kit) solver is used to obtain the optimal number of frequencies required by the networks. GLPK (GNU linear programming kit) is an open-source software to solve mathematical programs specifically linear programming (LP) and mixed-integer programming (MIP). GLPK consists of a set of routines written in ANSI C language.

From the obtained analysis, the average of the optimal frequencies required at varying occupancy rates by increasing the number of networks from 1 to 50 for 200 instances has been obtained and analyzed.

4 Results and Discussions

In this section, we describe the simulation results obtained by using the proposed method. Figure 2 describes the average optimum frequencies required by single user at varying occupancy rates which are obtained by using GLPK 4.65 solver.

Figure 3 describes the simulation results for 200 time instances for multiple networks at varying occupancy rates from 10 to 100% using GLPK4.65 solver. The vertical axis represents the average optimum number of frequencies required for

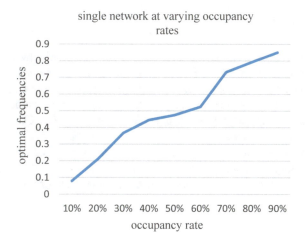

Fig. 2 Single network at varying occupancy rates

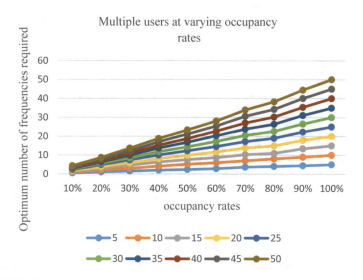

Fig. 3 Multiple users at varying occupancy rates

multiple networks at varying occupancy rates. Here we had considered the number of networks as 5, 10, 15, 20, 25, 30, 35, 40, 45, and 50 which are plotted in different colored lines as shown in Fig. 3.

5 Conclusion

In this paper, a novel method was proposed to mitigate internetwork interference by optimally allocating available frequencies among multiple networks with varying occupancy rates from 10 to 100%. The simulation results show that with varying occupancy rates for dense networks (i.e., from 5 to 50) the number of optimum frequencies required is varying, which are obtained by using GLPK 4.65 solver which shows that when we consider 15 as the maximum number of available frequencies then the network starvation is also increased. Our proposed method decreases the network starvation by smartly allocating the available channels which are available at lower occupancy rates, thereby accommodating more networks all the time.

References

1. M. Collotta, L. Gentile, G. Pau, G. Scatà, A dynamic algorithm to improve industrial wireless sensor networks management, in *IECON 2012—38th Annual Conference on IEEE Industrial Electronics Society* (Montreal, QC, 2012), pp. 2802–2807
2. O. Takyu, K. Shirai, T. Fujii, M. Ohta, Adaptive channel assignment with predictions of sensor results and channel occupancy ratio in PhyC-SN. IEEE Access **7**, 44645–44658 (2019)
3. M. Usman, D. Har, I. Koo, Energy-efficient infrastructure sensor network for ad hoc cognitive radio network. IEEE Sens. J. **16**(8), 2775–2787 (2016)
4. M. Al-Jemeli, F.A. Hussin, An energy efficient cross-layer network operation model for IEEE 802.15.4-based mobile wireless sensor networks. IEEE Sens. J. **15**(2), 684–692 (2015)
5. C. Karakus, A.C. Gurbuz, B. Tavli, Analysis of energy efficiency of compressive sensing in wireless sensor networks. IEEE Sens. J. **13**(5), 1999–2008 (2013)
6. H. Lin, L. Wang, R. Kong, Energy efficient clustering protocol for large-scale sensor networks. IEEE Sens. J. **15**(12), 7150–7160 (2015)
7. U. Wetzker, I. Splitt, M. Zimmerling, C.A. Boano, K. Römer, Troubleshooting wireless coexistence problems in the industrial internet of things, in *2016 IEEE International Conference on Computational Science and Engineering (CSE) and IEEE International Conference on Embedded and Ubiquitous Computing (EUC) and 15th International Symposium on Distributed Computing and Applications for Business Engineering (DCABES)* (Paris, 2016), pp. 98–98
8. AB Ericsson, Ericsson mobility report—on the pulse of the networked society (2015)
9. N. Baccour, A. Koubˆaa, L. Mottola, M.A. Z´uˇniga, H. Youssef, C.A. Boano, M. Alves, Radio link quality estimation in wireless sensor networks: a survey. ACM Trans. Sens. Netw. **8**(4) (2012)
10. G. EkbataniFard Hossein, Monsefi Reza, T. Akbarzadeh, R. Mohammad, H. Yaghmaee Mohammad, A multi-objective genetic algorithm based approach for energy efficient QoS-routing in two-tiered wireless sensor networks, in *2010 5th IEEE International Symposium on Wireless Pervasive Computing (TSWPC'10)* (2010), pp. 80–85
11. P. Krasiński, B. Pękosławski, A. Napieralski, IEEE 802.15.4 wireless network application in real-time automation systems, in *Proceedings of the 20th International Conference Mixed Design of Integrated Circuits and Systems—MIXDES 2013* (Gdynia, 2013), pp. 508–511
12. R. Mangharam, M. Pajic, S. Sastry, Demo abstract: embedded virtual machines for wireless industrial automation, in *2009 International Conference on Information Processing in Sensor Networks* (San Francisco, CA, 2009), pp. 413–414

A Review on Convolutional Neural Networks

Shubhajit Datta

Abstract A significant development has been seen in artificial intelligence and deep learning during the last few decades. Nowadays, it is used in different industries such as retail, hospital, and education. Convolutional Neural Network (CNN) is one of them, which is mostly used in computer vision, image classification, object detection, text recognition, etc. The study of CNN got a breakthrough after AlexNet was proposed in ImageNet completion, in 2010. This paper presents a review work on CNN. At first, it presents the details of CNN, structure of CNN, different layers associated with it, activation function, and cost function. Then it presents the development of CNN over the years. Different CNN models and their advantages over others have been discussed in this paper. Later, it discusses some additional techniques to make the model more accurate.

Keywords Convolutional neural network · Deep learning · Activation function · Cost function · Image classification · Object detection

1 Introduction

In recent days, we can find the use of Artificial Neural Networks (ANN) and deep learning [1] in different areas. Social networking site uses ANN to make user's friend suggestion, whereas Amazon uses it to recommend us books, Netflix recommends movies, etc. [2]. Convolutional Neural Network (CNN) is the widely used deep neural network in the area of image classification, object detection, text recognition, action recognition, and many more. CNN was initially inspired by visual cortex of animals [3]. For large volume images like RGB images, ANNs lead to an explosion in the number of weights which requires more memory and computation data. This problem can be solved using CNNs [4] by using the sparse connections and parameter sharing. Like ANNs, CNNs also have neurons, weights, and objectives. The major properties of CNNs are the presence of sparse connection between the layers, and the weights are

S. Datta (✉)
Surendra Institute of Engineering and Management, Siliguri, India
e-mail: shubhajitdatta1988@gmail.com

© Springer Nature Singapore Pte Ltd. 2020
R. Bera et al. (eds.), *Advances in Communication, Devices and Networking*,
Lecture Notes in Electrical Engineering 662,
https://doi.org/10.1007/978-981-15-4932-8_50

shared between output neurons in hidden layer. The main difference between ANN and CNN is that ANN learns global patterns from its inputs, whereas CNN learns local patterns which help to recognize images more accurately. To learn translation-invariant patterns and to learn spatial hierarchies of patterns are the characteristics of CNN.

2 Layers of CNN

Like ANNs, CNNs stack a sequence of layers followed by an output layer. The layers of CNN are as follows.

2.1 Convolution Layer

This is the first layer of CNN [5]. The main reason to use convolution layer in CNN is that it learns local patterns of the image, whereas dense layers learn only global patterns. In convolution layer, every output neuron is connected with the small neighborhood in the input through a weight matrix, which is called Kernel or Filter [6]. Multiple filters can be assigned for each convolution layer. Each filter is moved around the whole input and gives a 2D output. The output of each filter is stacked to get a volume output. The size of the output is calculated by $((n_x - f_x + 1) \times (n_y - f_y + 1)$ where $(n_x \times n_y)$ is the size of image matrix and $(f_x \times f_y)$ is the size of filter. In the following figure, a 5×5 image matrix is convolved with a 3×3 kernel filter. So, the size of the output is $(5 - 3 + 1) \times (5 - 3 + 1) = 3 \times 3$.

Here, first the convolutional kernel is superimposed on the top left corner of the image. Then, the corresponding elements are multiplied, and the sum of product is calculated. Then the result becomes the first element of the output feature map. By doing so, the first element of the output feature map is found $(1 \times 1 + 0 \times 1 + 0 \times 0 + 0 \times 2 + 3 \times 1 + 2 \times 1 + 4 \times 1 + 2 \times 0 + 0 \times 2) = 10$. Then the filter is slided by one pixel at a time to the right or the bottom, and the corresponding elements of the output features maps are calculated (Fig. 1).

In the above example, it has been shown that the size of the output feature map is less than the input size. So, if several convolution layers operate, the size of the feature map will be decreased very quickly. To prevent this, zero padding can be done before doing convolution. This type of convolution is called padded convolution. Padded convolution is done to preserve the size of the feature map.

But, if the size of the input feature map is very large and it should be decreased gradually, strided convolution can be used. In strided convolution, the stride value may be more than one, which means multiple pixels can be skipped.

So, the size of the output feature map is calculated by

a)

1	0	0	2	1
0	3	2	0	1
4	2	0	0	2
0	0	3	1	2
1	0	2	0	1

b)

1	1	0
2	1	1
1	0	2

*

c)

10	10	11
19	11	11
14	6	13

=

Fig. 1 Convolution operation: **a** Image matrix, **b** kernel matrix, **c** output feature map

$$\frac{(n - f) + 2P}{S} + 1$$

where n is the size of input feature map, f is the size of filter, P is the zero padding, and S is the stride value.

2.2 Pooling Layer

The next layer of CNN is pooling layer. This layer provides translational invariance by sub-sampling [7]. It reduces the size of the feature map. The mostly used pooling techniques are max pooling and average pooling. In the following example, the input of the pooling layer is a 4 × 4 feature map. During the pooling operation of stride 2, a 2 × 2 matrix is superimposed on the input and the max value should be taken as the first element of the output. Here, the max value among 1, 1, 4, and 0 is 4, so 4 is taken as the first element of the output. Then the same process is continued for all the input points and output is generated. So, basically sub-sampling is done in this layer. In average pooling, average value is taken instead of max value (Fig. 2).

1	1	3	2
4	0	8	6
3	1	1	2
1	0	2	4

⇒

4	8
3	4

Fig. 2 Max pooling operation strided by 2

2.3 Fully Connected Layer

The convolution layers and pooling layers are alternatively stacked leading to a series of fully connected layers followed by an output layer. The fully connected layers of CNN work like the layers of ANN exactly [8]. The neurons of these layers are connected with all the neurons of previous layer.

2.4 Activation Function

According to the application, different activation functions can be used in CNN [9]. The different activation functions are as follows.

Sigmoid function: For this activation function, the output range is 0–1. Here, saturation means that the gradient with respect to argument goes to zero. Not zero-centered outputs are always positive updates of all weights connected to it in the same direction [10].

Tanh function: For this activation, function output lies in the range of −1 to 1. Saturation problem exists here. It acts like identity near origin [11].

ReLU (Rectified Linear Unit): In this activation function, saturation problem does not exist for large input. But near zero point, neurons will die and will get no updates. It initializes with positive bias [12].

2.5 Cost Function

There are different cost functions that may be used. For regression problem, least square cost function may be used but that is not so good for classification problem. So, for classification problem, Binary cross-entropy cost function may be used [13].

$$J = -\left\{y \log \hat{y} + (1 - y) \log(1 - \hat{y})\right\}$$

where J is the cost function, y is the actual output, and $\hat{y}$ is the estimated output.

The attractive properties of this cost function are as follows:

J is zero if $y = \hat{y}$, and J is very high for misclassification. And J is always greater than zero.

3 Popular CNN Architectures

The first convolutional neural network that made a breakthrough for image classification task is LeNet. After that, a lot of development had occurred and different
CNN models have been proposed. Different CNN models are discussed below.

3.1 LeNet

This model was proposed by Lecun [14], where handwritten digit was recognized
using this CNN technique. This network is the base of many advanced networks
found today. It took an input of 32 × 32 handwritten digit as input. The input was
followed by a 5 × 5 convolutional layer and a 2 × 2 pooling layer and it repeated,
and finally it went through two fully connected layers as shown below. It classified
the appropriate digit among ten different classes (Fig. 3).

3.2 AlexNet

This network was proposed by Krizhevsky [15] in ImageNet LSVRC-2010 challenge.
In this work, a deep convolutional network was trained to classify 1.2 million images
into 1000 different classes. Top 1 and top 5 test error rate of 37.5 and 17% were
achieved by this model. This model consists of 60 million parameters, overlapped
max pooling layers, five convolution layers, and three fully connected layers. In
this model, "Dropout" method was used to reduce overfitting due to fully connected
layers. A variant of this model was also used in ILSVRC-2012, and it achieved
winning top 5 test error rate of 15.3%.

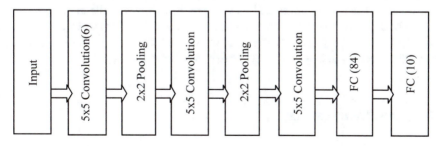

Fig. 3 LeNet network

3.3   VggNet

VggNet was developed by Simonyan and Zissermanin ILSVRC-2014 competition. The network consists of 138 million parameters and 16 convolutional layers. In this model, the number of features increases as the depth of the network increases. And during the whole network the kernel size is fixed. Only 3×3 kernels are used throughout the network. Due to more nonlinearity in this network, it is advantageous for classification task. Computational gain was also achieved in this model by reducing the number of parameters as it used successive convolutional layers.

3.4   GoogleNet

The basic building block of GoogleNet was inception model which was introduced in ImageNet competition-2014 by Christian Szegedy [16]. GoogleNet consists of 22 layers. In this model, the input is followed by sequence convolution layer, max pooling layer, and a sequence of inception layer, and it repeats. Finally, it passes through fully connected network to classify among 1000 different classes. It used very few parameters, that is, only 5 million and it won the ImageNet challenge-2014 with a winning top 5 test error rate of 7%. So, this network can be concluded as very deep network but with fewer parameters.

3.5   ResNet

ResNet model was proposed by He [17] at ILSVRC-2015 and won first place. The network achieved 3.75% error on ImageNet test data set. The model followed a residual learning to ease training of comparatively deeper network. Residual networks are easy to optimize and its accuracy increases with the increase of the network depth. In ImageNet competition, the model used 152 layers which is eight times than previously used VggNet. But the complexity of the model is less than VggNet. Not only that, it was able to achieve a better result.

4   Data Preprocessing

Before feeding the network, data should be formatted appropriately. Most of our image sources are in the form of jpg, jpeg, tiff, etc. After collecting those data, it should be decoded to RGB images and then it should be converted to grids of pixels. Next, it should be converted to floating-point tensors. Then the pixel values (between

0 and 255) should be rescaled to [0, 1] interval as neural networks prefer to deal with small input values [18].

5 Data Augmentation

Overfitting is a major problem to be solved in any network. Overfitting may occur due to the lack of training samples. Infinite training data may help to overcome this problem, but this is impossible in real world. In real world, we mostly find the lack of sufficient data. Besides dropout and weight decay, data augmentation is another technique by which overfitting can be removed. Data augmentation generates more training data from the existing samples by augmenting the sample via a number of random transformations keeping in mind that the model should not see the same image twice. This technique helps the network to take more data and to take better decision [18].

6 Conclusion

Different CNN models have been discussed in this paper. Depth of the network, number of parameters, and computational complexity are the major concerns for this purpose. In image classification, the described models performed well. Even ResNet achieved only 3.75% error on ImageNet test data set, which is great. But when it has to deal with object detection, finding the region of the object, described models do not perform well. To fulfill this objective, RCNN, Fast RCNN, Faster RCNN, and Mask RCNN may be studied.

References

1. Y. LeCun, Y. Bengio, G. Hinton, *Deep Learning* (Macmillan Publishers Limited, 2015). https://doi.org/10.1038/nature14539
2. S. Angra, S. Ahuja, Machine learning and its applications: a review, in *International Conference on Big Data Analytics and Computational Intelligence (ICBDAC)* (2017). https://doi.org/10.1109/ICBDACI.2017.8070809
3. D.H. Hubel, T.N. Wiesel, Receptive fields and functional architecture of monkey striate cortex. J. Physiol. (1968)
4. Y. Guo, L. Bai, S. Lao, S. Wu, M.S. Lew, *A comparison between Artificial Neural Network and Cascade-Correlation Neural Network in Concept Classification* (Springer, Cham, 2014)
5. S. Albawi, T. Abed Mohammed, S. Ai-Zawi, Understanding of a convolutional neural network. Int. Conf. Eng. Technol. (ICET) (2017)
6. N. Sharma, V. Jain, A. Mishra, An analysis of convolutional neural networks for image classification. Int. Res. J. Eng. Technol. **05**(07) (2018)

7. D. Scherer, A. Muller, S. Behnke, Evaluation of pooling operations in convolutional architectures for object recognition, in *20th International Conference on Artificial Neural Networks (ICANN)* (Thessaloniki, Greece, 2010)

8. M. Mishra, M. Srivastava, A view of artificial neural network, in *International Conference on Advances in Engineering & Technology Research* (2014). https://doi.org/10.1109/ICAETR.2014.7012785

9. V. Sharma, S. Rai, A. Dev, Int. J. Adv. Res. Comput. Sci. Softw. Eng. **2**(10) (2012)

10. T.M. Jamel, B.M. Khammas, Implementation of a sigmoid activation function for neural network using FPGA, in *13th Scientific Conference of Al-Ma'moon University College*, vol. 13, April 2012

11. S. Gomar, M. Mirhassani, M. Ahmadi, Precise digital implementations of hyperbolic tanh and sigmoid function, in *50th Asilomar Conference on Signals, Systems and Computers* (2016)

12. X. Glorot, A. Bordes, Y. Bengio, Deep sparse rectifier neural networks, in *14th International Conference on Artificial Intelligence and Statistics* (2011)

13. O.I. Abiodun, A. Jantan, A.E. Omolara, K.V. Dada, N.A. Mohamed, H. Arshad, State-of-the-art in artificial neural network applications: a survey. Heliyon **4**(11) (2018)

14. Y. LeCun, L. Bottou, Y. Bengio, P. Haffner, Gradient-based learning applied to document recognition. Proc. IEEE (1998)

15. A. Krizhevsky, I. Sutskever, G.E. Hinton, ImageNet classification with deep convolutional neural networks, in *NIPS* (2012)

16. C. Szegedy, W. Liu, Y. Jia, P. Sermanet, S. Reed, Going deeper with convolutions. CoRR (2014)

17. K. He, X. Zhang, S. Ren, J. Sun, Deep residual learning for image recognition. arXiv preprint arXiv:1512.03385 (2015)

18. F. Chollet, *Deep Learning with Python* (Manning Publications Co., 2018). ISBN:9781617294433

Design and Development of IoT-Based Low-Cost Robust Home Security System

Shauvick Das, Mayank Budia, Ratul Ghosh, Sourav Dhar, and H. S. Mruthyunjaya

Abstract Lower cost and lower power consumption are two emerging issues in the deployment of household IoT systems. In this paper, design and development of a prototype IoT-based home security system has been proposed. The low-cost sensors, Arduino board, and Raspberry Pi board are used to provide 24 × 7 home securities. In this system development, event-driven strategy is adopted to conserve energy.

Keywords Internet of things (IoT) · Wireless sensor network (WSN) · Home security system

1 Introduction

Theft activities are very common these days in a metropolitan city when the inhabitants are not present in the house. Also, accidents like LPG or any other harmful gas leakage are also very common. There are many security and surveillance systems in the market but most of them are very costly and everyone cannot afford them.

S. Das · M. Budia · R. Ghosh · S. Dhar (✉)
Department of Electronics and Communication Engineering, Sikkim Manipal Institute of Technology, Sikkim Manipal University, Majitar, Sikkim, India
e-mail: sourav.dhar80@gmail.com

S. Das
e-mail: shauvick_201600603@smit.smu.edu.in

M. Budia
e-mail: mayank_201600025@smit.smu.edu.in

R. Ghosh
e-mail: ratul_201600110@smit.smu.edu.in

H. S. Mruthyunjaya
Department of Electronics and Communication Engineering, Manipal Institute of Technology, Manipal, Karnataka, India
e-mail: mruthyu.hs@manipal.edu

© Springer Nature Singapore Pte Ltd. 2020
R. Bera et al. (eds.), *Advances in Communication, Devices and Networking*,
Lecture Notes in Electrical Engineering 662,
https://doi.org/10.1007/978-981-15-4932-8_51

Therefore, it is the need of the hour to develop a low-cost, robust, and user-friendly home security that can be used by everyone.

One of the major aspects of an IoT system is the seamless connection of heterogeneous devices and to combine the collected data to yield new services [1]. Rich services are the type of service-oriented architecture for the integration of services in large-scale systems. Achieving a highly scalable and on-the-fly integration of devices in an IoT environment is one of the major challenges in IoT development, which is proposed to be solved by using OASIS model [1]. An energy-efficient system is proposed by Abedin et al. [2]. They have introduced an expert algorithm to make the flow of energy reduced and more efficient. This algorithm has three modes by which the limitation and the energy consumption of a sensor device can be managed depending upon the remaining battery life. But the deployed sensors also need to be taken care of from getting damaged. Considering the importance of reduced power consumption in an IoT system, Electric Power and Energy Systems (EPES), considered in [3], can provide clean distributed energy for maintaining a sustainable environment. The role of IoT is very important in EPESs. Energy wastage can be reduced by digitizing the EPESs using IoT. It will also improve the reliability, efficiency, security, etc. An IOT system, described in [4], is used to monitor a lab environment to manage the optimal usage of computers, and controls air conditioning system of the lab in an account of how many persons are in it. The overall system is not efficient since the computers shut down automatically if found idle for some time. The proposed IoT-based safety expert system (SES) [5] is established in a coal mine which includes different monitoring and control systems. It monitors real-time status of different parts of a coal mine, analyzes risks, and determines the thresholds of risks and according to that it takes necessary actions. There is no power management algorithm applied. An IoT system, introduced in [6], is a platform where all the everyday electronic devices become smarter, intelligent, and provide more informative communication. Humans are considered as a layer of the IoT architecture proposed in [6] to enable things to interact and interconnect with homogeneous as well as heterogeneous objects. The narrowband (NB)-IoT and Long Range (LoRa) are the two principal technologies for enabling the communication in a low-power wide-area (LPWA) technology [7]. Both the technologies have their own advantages and disadvantages. Primary focus of LoRa is more on low-power applications, while NB-IoT is eyeing toward high QoS and low latency applications. Atzori et al. [8] surveyed some very important characteristics of the IoT with the focus on what is currently being done and what are the problems that require further research. This paper states the IoT as a promising model which is the combination of a number of technologies and communication mediums. In this paper, several wired and wireless sensor networks, actuator networks, and distributed intelligence for smart objects are identified. The Green IoT for a smart world, identified in [9], have various issues regarding the reduction of the energy consumption of IoT devices to be used in recent generation. In this work, green information and communication technologies (ICT) are reviewed and general principles for green ICT have been summarized. There are lots of challenges in green IoT that are faced because of too much usage of energy inefficient end devices [10]. Thus, the literature survey points out the necessity of low-cost, energy-efficient IoT system development for strategic smart world development. In this paper, we have shown the design and development of a low-cost IoT system for home security.

2 Problem Definition

The primary objectives of this work are as follows:

1. Development of WSN for the detection of LPG/toxic gases and motion.
2. Real-time data acquisition.
3. Development of communication system for real-time surveillance.
4. Development of an automated lock system that can be operated locally as well as remotely.

Power consumption is also one of the main concerns in these systems. There are lots of sensors and actuators in this system which will remain active 24×7, hence, cumulatively consuming a lot of energy. Therefore, a green IoT network needed to be developed which will reduce the amount of energy consumption.

3 Solution Strategy

The concept of master and slave is used in this system where a Raspberry Pi is used as a master and Arduino is used as a slave. The Arduino nodes are detecting the toxic/LPG gases and motion and transmitting the generated data to the Raspberry Pi. The automated door lock system which includes a camera is connected to the Raspberry Pi. The camera is used for the purpose of monitoring and capturing image of any intruder, in case any suspicious activity is detected by the motion sensor when the door is locked, and nobody is inside or if someone tries to break the lock system. The captured image will be sent to the user with a warning message via WhatsApp. Similarly, a warning message will also be sent to the user in case any toxic gas

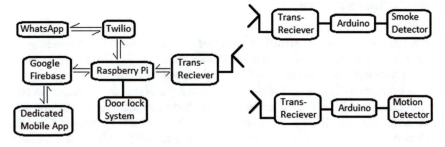

Fig. 1 System design

is detected by the sensor. A database is created in the Raspberry Pi using Google Firebase which will contain all the sensor data and captured images. And the whole system can be accessed by a mobile application (Fig. 1).

4 System Design

4.1 Hardware Design

For the detection of LPG/toxic gases, MQ2 sensor module has been used. This module is useful for detecting any gas leakage in home or industry. This module is suitable for detecting H2, LPG, CH4, and other toxic gases. It has high sensitivity and less response time, which makes it suitable for real-time applications. The same is operated using an Arduino nano at the end node. Also, for the detection of motion PIR sensor has been used in this work. By using this we can detect if anyone is present inside the house, when no one supposed to be inside. PIR sensor is also operated by an Arduino nano at another end node. Both the Arduino nodes and the Raspberry Pi are connected to a WSN by using a wireless transceiver nRF24L01. This nRF24L01 module is operated in 2.4 GHz worldwide ISM band and uses GFSK modulation for the transmission of data. All the data that are generated by the MQ2 gas sensor and PIR sensor are transmitted to the Raspberry Pi through nRF24L01 transceiver module. And based on the received data from the end nodes the Raspberry Pi will be able to decide whether to send a warning message to the user or not (Figs. 2 and 3).

For the automated lock system, two servo motors can be used, one will lock the hatch and the other will open/close the door flap. The lock can be accessed locally by entering correct password on the keypad. On wrong password entry, the user will receive a warning message. The lock system is also connected with a camera that takes a snapshot at every wrong password attempt. The whole lock system is operated by a Raspberry Pi. For the remote access of the lock system, a mobile application of the also developed (Fig. 4).

4.2 Software Design

All the data that are generated at the sensor nodes are transmitted to the Raspberry Pi in real time through a wireless transceiver module nRF24L01. These data are saved in a separate database present at the Raspberry Pi, and the same data are uploaded to the cloud storage of Google Firebase. These data are analyzed by a program present at the Raspberry Pi and based on the analysis it decides whether to send an alert message (toxic gas/LPG leakage or unusual motion detected) or not (Figs. 5 and 6).

For the surveillance of the house, a camera is connected to the automated lock system. This camera clicks a snapshot of the person whenever there is a wrong

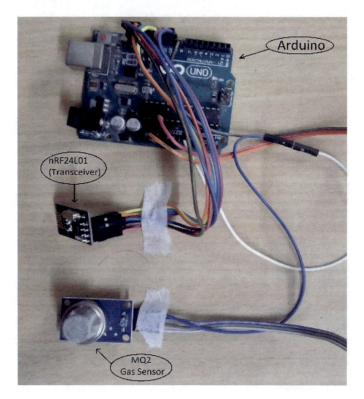

Fig. 2 Gas sensor node

password attempt, and this photo is sent to the user with a warning message through WhatsApp. The captured images are also uploaded to Google Firebase cloud storage by the Raspberry Pi. The same images can be retrieved by a simple mobile application designed to monitor and control the lock system remotely (Fig. 7).

The WhatsApp alert messages are sent by the Raspberry Pi through Twilio, which is a cloud communication platform. This alert message is directly sent to the end user automatically.

5   Results and Discussion

A complete automated security system has been developed using Raspberry Pi and Arduino. The sensor nodes work independently and transmit the collected data to the Raspberry Pi over a wireless network. The data are then stored and analyzed by an algorithm in the Raspberry Pi. Based on the analysis, it is decided whether to send an alert message or not. The system can also be operated and monitored remotely by a mobile application (Fig. 8).

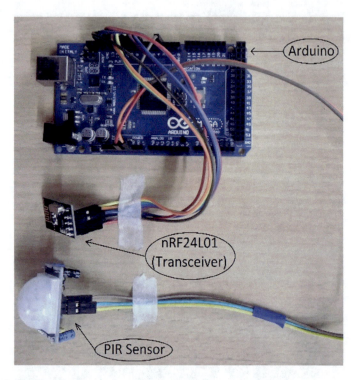

Fig. 3 Motion sensor node

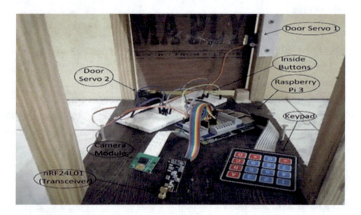

Fig. 4 Automated lock system

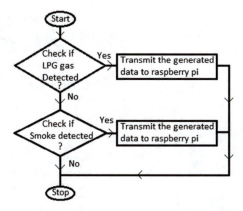

Fig. 5 Flowchart of gas sensor

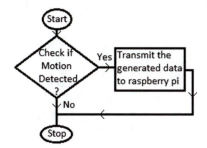

Fig. 6 Flowchart of motion sensor

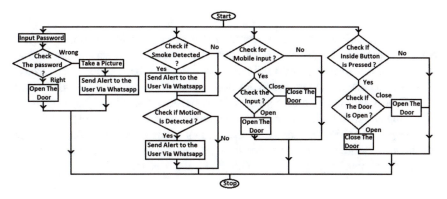

Fig. 7 Flowchart of entire system

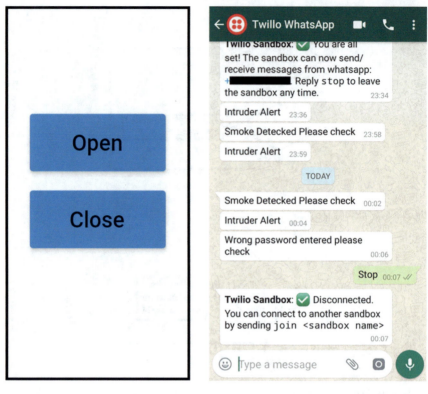

Fig. 8 Mobile application and WhatsApp alert feature of the system

Whenever a harmful smoke or any motion is detected inside the house when no one is supposed to be present inside, then alert messages are sent to the user via WhatsApp. For this purpose, Twilio API services have been used in this project. Also, a basic mobile application is developed using flutter which is an open-source UI software. This application consists of two switches "open" and "close" through which the automated lock can be accessed remotely using Internet (Fig. 9).

The photos which are taken by the camera module during wrong password attempts in the lock system are stored in the Google Firebase and are accessible to the user by just logging into their Google Firebase account.

6 Conclusion

The holistic development of IoT-based prototype home security system has been tested successfully. Automated WhatsApp alert messaging is novel feature of this system. The sensors, actuators, and the development boards used in this project are low-power devices, thus ensuring nominal energy consumption. Further, the camera

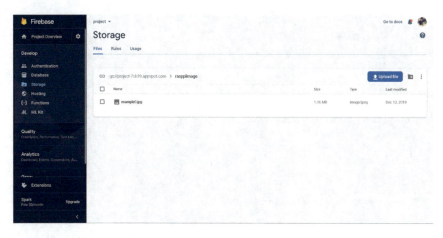

Fig. 9 Images stored in Google Firebase

used here is event-driven which makes sure that the system is not becoming an energy-hungry system.

References

1. S. Shokrollahi, F. Shams, Rich Device-Services (RDS): a service-oriented approach to the internet of things (IoT). Wirel. Pers. Commun. **97**(2), 3183–3201 (2017)
2. S.F. Abedin, M.G.R. Alam, R. Haw, C.S. Hong, A system model for energy efficient green-IoT network, in *2015 International Conference on Information Networking (ICOIN)* (Cambodia, 2015), pp. 177–182
3. G. Bedi, G.K. Venayagamoorthy, R. Singh, R.R. Brooks, K. Wang, Review of Internet of Things (IoT) in electric power and energy systems. IEEE Internet of Things J **5**(2), 847–870 (2018)
4. H. Wang, Toward a green Campus with the internet of Things—the application of lab management, in *Proceeding of the World Congress on Engineering 2013*, vol. 2 (London, UK, WCE, 2013). ISBN:978-988-19252-8-2
5. W. Liang, F. Xue, B.S. Nie, C.L. Wang. Study on safety expert system based on Internet of Things. Adv. Mater. Res. **962–965**, 2716–2720 (2014)
6. P.P. Ray, A survey on Internet of Things architectures. J. King Saud Univ. Comput. Inf. Sci. **30**(3) (2018)
7. R.S.Sinha, Y. Wei, S.H. Hwang, A survey on LPWA technology: LoRa and NB-IoT. ICT Express **3**(1) (2017)
8. L. Atzori, A. Iera, G. Morabito, The Internet of Things: a survey. Comput. Netw. **54**(15) (2010)
9. C. Zhu, V.C. M. Leung, L. Shu, E.C. Ngai, Green Internet of Things for Smart World. IEEE Access **3**, 2151–216 (2015). https://doi.org/10.1109/access.2015.2497312
10. R. Arshad, S. Zahoor, M.A. Shah, A. Wahid, H. Yu, Green IoT: an investigation on energy saving practices for 2020 and beyond. IEEE Access **5**, 15667–15681 (2017). https://doi.org/10.1109/ACCESS.2017.2686092

Development of a Holistic Prototype Hadoop System for Big Data Handling

Manisha K. Gupta, Md. Nadeem Akhtar Hasid, Sourav Dhar, and H. S. Mruthyunjaya

Abstract In the upcoming smart world where billions of sensors, actuators, communication, and networking devices are expected to be connected with each other, these devices will be of various types. The volume of the data expected to be generated will be huge and due to the heterogeneity of the connected devices, there will be various types of data as well as the rate of data generation will be variable. This leads to a challenge of storage and processing these "Big data". In this paper, development of a Hadoop prototype system is presented. The system is developed using HDFS and MapReduce techniques. Red Hat Linux, Windows 10 with Oracle VM VirtualBox are used as the platform to develop this system.

Keywords Big data · Hadoop distributed file system (HDFS) · MapReduce cluster

1 Introduction

In recent time, big data handling is one of the most emerging challenges technical world. Data that is too large, fast, distinct coming from variety of sources in different formats normally termed as big data [1]. Data can be unstructured, semi-structured, or structured which is not managed by traditional database management or traditional data processing applications. The characteristics of big data are mainly defined by 3Vs [2, 3]. They are Volume, Velocity, and Variety. Also, big data analytics offers

M. K. Gupta · Md. N. A. Hasid · S. Dhar (✉)
Department of Electronics and Communication Engineering, Sikkim Manipal Institute of Technology, Sikkim Manipal University, Majitar, Sikkim, India
e-mail: sourav.dhar80@gmail.com

M. K. Gupta
e-mail: guptamanisha3845@gmail.com

H. S. Mruthyunjaya
Department of Electronics and Communication Engineering, Manipal Institute of Technology, Manipal, Karnataka, India
e-mail: mruthyu.hs@manipal.edu

© Springer Nature Singapore Pte Ltd. 2020
R. Bera et al. (eds.), *Advances in Communication, Devices and Networking*, Lecture Notes in Electrical Engineering 662, https://doi.org/10.1007/978-981-15-4932-8_52

three models, namely, descriptive, predictive, and prescriptive [4]. With the growth of Internet of things (IoT), the volume of big data gradually expanding beyond terabytes to petabytes and even to exabytes [5–7]. Velocity of the data, another key parameter of big data, refers to the rate of data being generated or processed [8]. The third key parameter, variety, refers to the different types of data in different formats which are generated due to the use of heterogeneous sensors in IoT. The primary challenges include in big data the storage, search, capture, transfer, analysis, etc. [9]. Considering the challenges of big data, this paper proposed a solution to manage big data using Hadoop Distributed File System (HDFS) and MapReduce cluster.

2 Hadoop Architecture

Big data Hadoop is not a problem; it's a solution to overcome the problem faced in today's scenario related to storage and computation. Big data Hadoop is an open-source platform which works on the principle of distributed and parallel storage. This framework mostly works on Java Programming Language. It is designed to handle very large files with streaming data access patterns. For storage, it consists of HDFS and for processing of big data it utilizes MapReduce technique. Hadoop HDFS cluster can consist of thousands of hardware nodes. It uses blocks to store a file or parts of a file. Further these blocks, referred to as packaged codes, are sent to nodes for processing in a parallel mode. On an advantage, these nodes have access to modify the dataset which results in fast and a much efficient processing.

In Fig. 1, Hadoop architecture is shown. Hadoop works on master/slave model. In the above HDFS layer, we have a single master which is known as NameNode and three slaves known as DataNode. These DataNodes provide their storage to the cluster, so that users can have access to the storage accordingly. In MapReduce layer, we have a JobTracker and three TaskTrackers. Here JobTracker acts as a master which manages the metadata of TaskTrackers. The TaskTracker acts as a slave which

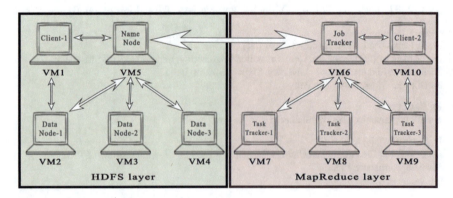

Fig. 1 Hadoop architecture

provides the platform for computation purpose. To implement MapReduce cluster, we should ensure that HDFS cluster is implemented first with a port number of "9002".

3 System Development and Validation

Here we are using RedHat Linux to implement the work. In windows 10, we use Oracle VM VirtualBox.

3.1 Hadoop Distributed File System (HDFS)

Hadoop distributed file system (HDFS) is a platform written in Java that stores data in high-ended machines which offers high bandwidth for data storage. It is highly scalable and portable in nature. Its main components are NameNode and DataNode corresponding to Master and Slave.

NameNode has the whole information regarding the entire data existing in the Cluster. The Hadoop Cluster developed here is consisting of only one NameNode. The primary duty of the NameNode is to get the data stored in the Cluster by utilizing the DataNodes. The NameNode divides the new data into smaller blocks and submit in the Hadoop Cluster.

Though the default block size is set to 64 MB, however, the user will have the option to change the block size according to their needs. Then it identifies the DataNodes, which actually stores the partitioned data. Then it transfers the blocks in each DataNodes and maintains a table of data allocation called metadata.

From the Client perspective, for retrieving the data from HDFS Cluster, the client needs to contact the NameNode for the locational information of the targeted data. The NameNode fetches the locational information from its allocation table and delivers the location information to the client. The NameNode intermittently updates the status of stored data as the periodical information received from DataNodes. The DataNodes, which are also called Slave in HDFS Cluster, are the real storage location in Hadoop Cluster. There may be "n" numbers of DataNodes in the Hadoop Cluster as per the storage requirement. These store the data as directed by the NameNode. Whenever a DataNode goes down, it will not affect the data of the file.

In Fig. 2, HDFS Cluster is implemented in RHEL 7.5 (RedHat Linux Enterprise). Here three live nodes provide their storage to the Cluster. The total storage provided by the DataNodes is an aggregate of 18.56 GB, out of which Distributed File System (DFS) used is 803.54 MB and DFS remaining is 2.26 GB. In Hadoop to run a process, it requires a daemon program which runs at the back end. A host file is configured in every node/VM, which consists of each IP address with their respected hostname. To identify the Master in the cluster, we have to configure "core-site.xml" in each

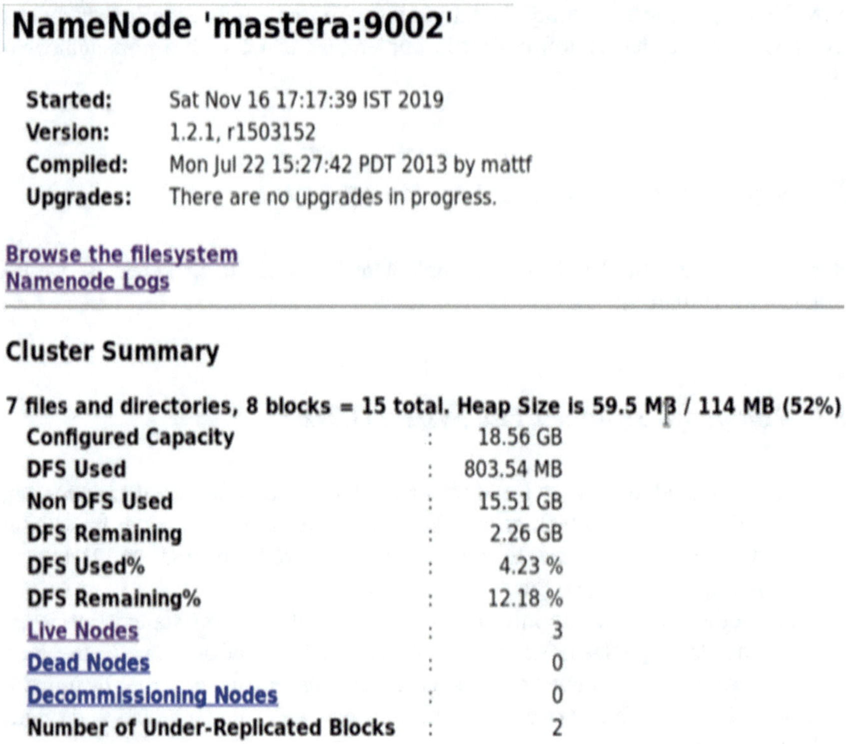

NameNode 'mastera:9002'

Started: Sat Nov 16 17:17:39 IST 2019
Version: 1.2.1, r1503152
Compiled: Mon Jul 22 15:27:42 PDT 2013 by mattf
Upgrades: There are no upgrades in progress.

Browse the filesystem
Namenode Logs

Cluster Summary

7 files and directories, 8 blocks = 15 total. Heap Size is 59.5 MB / 114 MB (52%)

Configured Capacity	:	18.56 GB
DFS Used	:	803.54 MB
Non DFS Used	:	15.51 GB
DFS Remaining	:	2.26 GB
DFS Used%	:	4.23 %
DFS Remaining%	:	12.18 %
Live Nodes	:	3
Dead Nodes	:	0
Decommissioning Nodes	:	0
Number of Under-Replicated Blocks	:	2

Fig. 2 HDFS cluster

node assigning IP of the master with port number "9001/9002". In the DataNodes "hdfs-core.xml" file needs to be configured, so that it provides the storage to the cluster.

Considering a client uploads a file in Hadoop cluster; firstly the file is divided into chunks of blocks and simultaneously distributed among DataNodes. For each block created, three replicas are formed and stored in different locations. In the above file "ak.txt", the file size is 143.72 MB. When it stores the file in the cluster, it will create three blocks. Out of these three blocks, two blocks are in size 64 MB and one block of 15.72 MB. Only the client has the root access over the file to maintain the security.

3.2 MapReduce

Hadoop offers not only storage but also a program processing tool which helps in computation tasks. The program works in two phases, one is for mapping and other is for reducing. In mapping phase, each input of a file is assigned to a tag

Name	Type	Size	Replication	Block Size	Modification Time	Permission	Owner	Group
ak.txt	file	143.72 MB	3	64 MB	2019-11-16 17:36	rw-r--r--	root	supergroup
bk.txt	file	143.72 MB	3	64 MB	2019-11-16 17:48	rw-r--r--	root	supergroup
ck.txt	file	143.72 MB	3	64 MB	2019-11-16 17:56	rw-r--r--	root	supergroup
dir	dir				2019-11-16 16:14	rwxr-xr-x	root	supergroup
dk.txt	file	143.72 MB	3	64 MB	2019-11-16 17:59	rw-r--r--	root	supergroup
ek.txt	file	62.24 MB	3	64 MB	2019-11-16 18:04	rw-r--r--	root	supergroup
fk.txt	file	22.86 MB	3	64 MB	2019-11-16 18:05	rw-r--r--	root	supergroup
gk.txt	file	143.72 MB	3	64 MB	2019-11-16 18:06	rw-r--r--	root	supergroup
m.txt	file	2.29 MB	3	64 MB	2019-11-19 12:30	rw-r--r--	root	supergroup
my.txt	file	0 KB	3	64 MB	2019-11-16 16:15	rw-r--r--	root	supergroup
o1	dir				2019-11-19 12:43	rwxr-xr-x	root	supergroup
output	dir				2019-11-19 12:32	rwxr-xr-x	root	supergroup
passwd	file	2.87 KB	3	64 MB	2019-11-19 12:43	rw-r--r--	root	supergroup
tmp	dir				2019-11-19 12:16	rwxr-xr-x	root	supergroup
www.txt	file	10.49 MB	3	64 MB	2019-11-16 16:24	rw-r--r--	root	supergroup
xxx.txt	file	10.49 MB	3	64 MB	2019-11-16 16:27	rw-r--r--	root	supergroup

Fig. 3 HDFS cluster file

(key) and in reducing phase all the similar tags are assigned to a new tag. MapReduce consists of JobTracker and TaskTracker. JobTracker mainly emphasizes on monitoring, managing, and identifying the tasks. When Client assigns jobs to the JobTracker, it will contact NameNode for the location of data. JobTracker divides the tasks and distributes to the available TaskTrackers. Here JobTracker acts as a single point of failure.

In Fig. 4, MapReduce Cluster is shown. The cluster consists of three TaskTracker. Here no tasks are assigned to the Cluster, so running map tasks and running reduce tasks shows zero. The default task capacity is 6. Whenever task is assigned, SafeMode should be kept off.

Considering a log file, which consists of users' login name and password. From this file, we have to find the users' name who have the root access. Here, each word assigned to a key (tag) in the mapping phase and in the reducing phase all similar keys are assigned to a unique key. In the Mapping phase, it divides each input and then we have to use the reducing program that is available in MapReduce called "wordcount". This program search throughout tags consists of bash word.

When a file is uploaded to the Cluster for some processing tasks, it goes through four phases of execution, namely, splitting, mapping, shuffling, and reducing. When a task starts, it will start mapping. After completion of 100% mapping, only then it starts reducing.

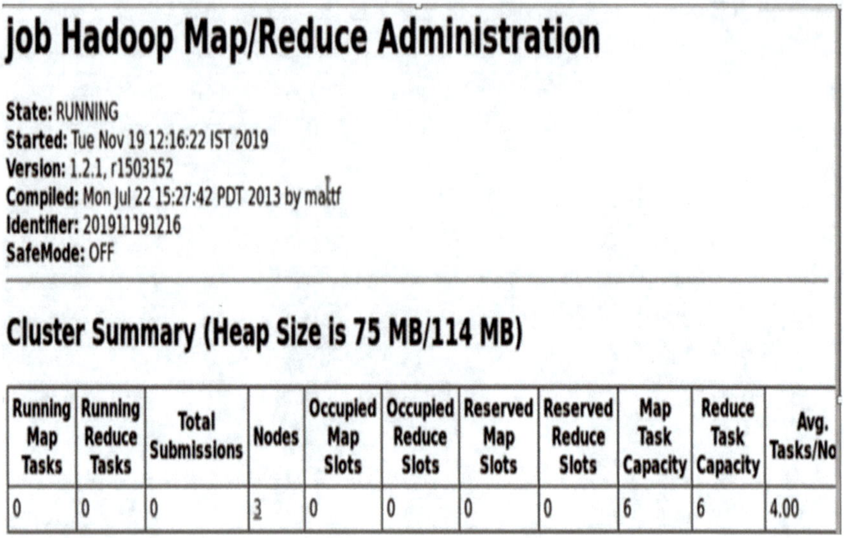

job Hadoop Map/Reduce Administration

State: RUNNING
Started: Tue Nov 19 12:16:22 IST 2019
Version: 1.2.1, r1503152
Compiled: Mon Jul 22 15:27:42 PDT 2013 by mattf
Identifier: 201911191216
SafeMode: OFF

Cluster Summary (Heap Size is 75 MB/114 MB)

Running Map Tasks	Running Reduce Tasks	Total Submissions	Nodes	Occupied Map Slots	Occupied Reduce Slots	Reserved Map Slots	Reserved Reduce Slots	Map Task Capacity	Reduce Task Capacity	Avg. Tasks/No
0	0	0	3	0	0	0	0	6	6	4.00

Fig. 4 MapReduce cluster

```
lp:x:4:7:lp:/var/spool/lpd:/sbin/nologin
sync:x:5:0:sync:/sbin:/bin/sync
shutdown:x:6:0:shutdown:/sbin:/sbin/shutdown
halt:x:7:0:halt:/sbin:/sbin/halt
mail:x:8:12:mail:/var/spool/mail:/sbin/nologin
operator:x:11:0:operator:/root:/sbin/nologin
games:x:12:100:games:/usr/games:/sbin/nologin
ftp:x:14:50:FTP User:/var/ftp:/sbin/nologin
nobody:x:99:99:Nobody:/:/sbin/nologin
systemd-network:x:192:192:systemd Network Management:/:/sbin/nologin
dbus:x:81:81:System message bus:/:/sbin/nologin
polkitd:x:999:998:User for polkitd:/:/sbin/nologin
libstoragemgmt:x:998:996:daemon account for libstoragemgmt:/var/run/lsm:/sbin/nologin
rpc:x:32:32:Rpcbind Daemon:/var/lib/rpcbind:/sbin/nologin
colord:x:997:995:User for colord:/var/lib/colord:/sbin/nologin
saslauth:x:996:76:Saslauthd user:/run/saslauthd:/sbin/nologin
abrt:x:173:173::/etc/abrt:/sbin/nologin
setroubleshoot:x:995:992::/var/lib/setroubleshoot:/sbin/nologin
```

Fig. 5 Log file

```
-rw-r--r--  3 root supergroup        0 2019-11-16 05:45 /my.txt
drwxr-xr-x  - root supergroup        0 2019-11-19 01:46 /tmp
-rw-r--r--  3 root supergroup 11000011 2019-11-16 05:54 /www.txt
-rw-r--r--  3 root supergroup 11000011 2019-11-16 05:57 /xxx.txt
[root@client ~]# hadoop jar /usr/share/hadoop/hadoop-examples-1.2.1.jar wordcount /m.txt /outpu
19/11/19 02:02:21 INFO input.FileInputFormat: Total input paths to process : 1
19/11/19 02:02:21 INFO util.NativeCodeLoader: Loaded the native-hadoop library
19/11/19 02:02:21 WARN snappy.LoadSnappy: Snappy native library not loaded
19/11/19 02:02:22 INFO mapred.JobClient: Running job: job 201911191216_0001
19/11/19 02:02:23 INFO mapred.JobClient: map 0% reduce 0%
19/11/19 02:02:30 INFO mapred.JobClient: map 100% reduce 0%
19/11/19 02:02:38 INFO mapred.JobClient: map 100% reduce 33%
19/11/19 02:02:39 INFO mapred.JobClient: map 100% reduce 100%
19/11/19 02:02:40 INFO mapred.JobClient: Job complete: job 201911191216_0001
19/11/19 02:02:40 INFO mapred.JobClient: Counters: 29
19/11/19 02:02:40 INFO mapred.JobClient:   Map-Reduce Framework
19/11/19 02:02:40 INFO mapred.JobClient:     Spilled Records=30
19/11/19 02:02:40 INFO mapred.JobClient:     Map output materialized bytes=114
19/11/19 02:02:40 INFO mapred.JobClient:     Reduce input records=10
19/11/19 02:02:40 INFO mapred.JobClient:     Virtual memory (bytes) snapshot=3939291136
19/11/19 02:02:40 INFO mapred.JobClient:     Map input records=100001
19/11/19 02:02:40 INFO mapred.JobClient:     SPLIT_RAW_BYTES=96
19/11/19 02:02:40 INFO mapred.JobClient:     Map output bytes=4400044
19/11/19 02:02:40 INFO mapred.JobClient:     Reduce shuffle bytes=114
19/11/19 02:02:40 INFO mapred.JobClient:     Physical memory (bytes) snapshot=334475264
19/11/19 02:02:40 INFO mapred.JobClient:     Reduce input groups=5
19/11/19 02:02:40 INFO mapred.JobClient:     Combine output records=10
19/11/19 02:02:40 INFO mapred.JobClient:     Reduce output records=5
```

Fig. 6 MapReduce phase

After completion of the task, it shows that there are 17 users who have the root access and can login in the root account.

4 Conclusion

The prototype Hadoop system developed has three live nodes to provide their storage to the Cluster. The aggregate of 18.56 GB total storage has been provided by the DataNodes. There is the provision to increase the number of DataNodes, and hence the total storage can be increased to a large extent.

After enhancing the storage, the system aims to enhance computation efficiency by introducing proper programmed processing tool. The complete system has been successfully tested in real time.

```
hdfs:x:201:123:Hadoop HDFS:/tmp:/bin/bash
nad:x:1001:1001::/home/nad:/bin/bash
mani:x:1002:1002::/home/mani:/bin/bash
q:x:1004:1004::/home/q:/bin/bash
myuser:x:1005:1005::/home/myuser:/bin/bash
nadd:x:1028:1028::/home/nadd:/bin/bash
nadddd:x:1029:1029::/home/nadddd:/bin/bash
user1:x:1049:1049::/home/user1:/bin/bash
yesss:x:1050:1050::/home/yesss:/bin/bash
nadim10:x:1058:1058::/home/nadim10:/bin/bash
username1:x:1059:1059::/home/username1:/bin/bash
nadim12:x:1060:1060::/home/nadim12:/bin/bash
user:x:1061:1061::/home/user:/bin/bash
user20:x:1063:1063::/home/user20:/bin/bash
user25:x:1064:1064::/home/user25:/bin/bash
[root@client ~]# cat /etc/passwd | grep bash | wc -l
17
[root@client ~]# vim /etc/passwd
[root@client ~]# hadoop fs -ls /o1
Found 3 items
-rw-r--r--   3 root supergroup          0 2019-11-19 02:13 /o1/_SUCCESS
drwxr-xr-x   - root supergroup          0 2019-11-19 02:13 /o1/_logs
-rw-r--r--   3 root supergroup          4 2019-11-19 02:13 /o1/part-00000
[root@client ~]# hadoop fs -cat /o1/part-00000
17
[root@client ~]# cat /etc/passwd | grep bash | wc -l
17
[root@client ~]#
```

Fig. 7 MapReduce computation

References

1. A.B. Patel, M. Birla, U. Nair, Addressing big data problem using hadoop and map reduce, in *2012 Nirma University International Conference on Engineering (NUiCONE)* (Ahmedabad, 2012), pp. 1–5. https://doi.org/10.1109/nuicone.2012.6493198
2. A. Verma, A.H. Mansuri, N. Jain, Big data management processing with Hadoop MapReduce and spark technology: a comparison, in *2016 Symposium on Colossal Data Analysis and Networking (CDAN)* (Indore, 2016), pp. 1–4. https://doi.org/10.1109/cdan.2016.7570891
3. K. Singh, R. Kaur, Hadoop: addressing challenges of big data, in *2014 IEEE International Advance Computing Conference (IACC)* (Gurgaon, 2014), pp. 686–689. https://doi.org/10.1109/iadcc.2014.6779407
4. S. Sharma, Rise of big data and related issues, in *2015 Annual IEEE India Conference (INDICON)* (NewDelhi, 2015), pp. 1–6. https://doi.org/10.1109/indicon.2015.7443346
5. A. Cuzzocrea, Big data mining or turning data mining into predictive analytics from large-scale 3Vs data: the future challenge for knowledge discovery, in *Model and Data Engineering. MEDI 2014*, ed. by Y. Ait Ameur, L. Bellatreche, GA Papadopoulos. Lecture Notes in Computer Science, vol 8748 (Springer, Cham, 2014)
6. A. Katal, M. Wazid, R.H. Goudar, Big data: issues, challenges, tools and Good practices, in *2013 Sixth International Conference on Contemporary Computing (IC3)* (Noida, 2013), pp. 404–409. https://doi.org/10.1109/ic3.2013.6612229
7. A. Rabkin, R.H. Katz, How hadoop clusters creak, in *IEEE Software*, vol. 30, no. 4, pp. 88–94, July–Aug 2013. https://doi.org/10.1109/ms.2012.73

8. A. Saldhi, D. Yadav, D. Saksena, A. Goel, A. Saldhi, S. Indu, Big data analysis using Hadoop cluster, in *2014 IEEE International Conference on Computational Intelligence and Computing Research* (Coimbatore, 2014), pp. 1–6. https://doi.org/10.1109/iccic.2014.7238418

9. J. Liu, F. Liu, N. Ansari, Monitoring and analyzing big traffic data of a large-scale cellular network with Hadoop, in *IEEE Network*, vol. 28, no. 4, pp. 32–39, July–Aug 2014. https://doi.org/10.1109/mnet.2014.6863129

C-Shaped Radiator Structure Antenna for Wireless Communication

Pooja Verma and Om Prakash

Abstract In this research outcome, a C-shaped microstrip antenna is developed for the applications of Worldwide Interoperability for Microwave Access (WiMAX) and Wireless Local Area Network (WLAN). The proposed C-shaped patch antenna exhibits two operating frequencies at 3.7 and 5.2 GHz with bandwidths 110 MHz (3.62–3.73 GHz) and 160 MHz (5.06–5.22 GHz), respectively. The antenna structure provides a peak gain of 5.53 dBi and 5.39 dBi at 3.7 GHz and 5.2 GHz, respectively. Antenna efficiency obtained at 3.7 GHz and 5.2 GHz are 34.5% and 31.5%, respectively. Key performance metrics of the antenna are evaluated by two different software.

Keywords C-shaped patch antenna · WiMAX applications · WLAN applications

1 Introduction

Wireless communication industries are growing rapidly, so nowadays a lot of research is going on the antenna with different shapes, different designs, and different specifications [1–6]. It is expected that these antennas have proper antenna gain, precise broadband matching, and useful radiation patterns throughout the frequency bands. The main motto of all the research is to achieve a high data rate with a small device size. For this requirement, several articles [7–9] have considered the development of alphabetical shaped patch antenna for WiMAX and WLAN applications.

P. Verma (✉)
Department of Electronics and Communication Engineering, Sikkim Manipal Institute of Technology, Sikkim Manipal University, Majitar, Sikkim, India
e-mail: puzavermaparhwara@gmail.com

O. Prakash
Department of Electronics and Communication Engineering, St. Mary's Engineering College, Hyderabad, Telangana, India
e-mail: om4096@gmail.com

© Springer Nature Singapore Pte Ltd. 2020
R. Bera et al. (eds.), *Advances in Communication, Devices and Networking*,
Lecture Notes in Electrical Engineering 662,
https://doi.org/10.1007/978-981-15-4932-8_53

473

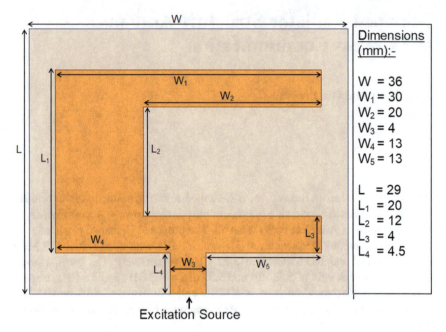

Excitation Source

Fig. 1 C-shaped patch antenna

In this paper, a compact C-shaped patch antenna is proposed for WiMAX and WLAN applications. Both CST-MWS and HFSS software platforms are explored for antenna simulation. Highly correlated results are obtained for different antenna metrics from two different computational electromagnetic software.

2 Antenna Construction

The construction of the proposed C-shaped microstrip antenna for 3.7 and 5.2 GHz frequency bands is displayed in Fig. 1. The presented antenna is developed on an FR4 substrate ($\varepsilon_r = 4.4$, loss tangent $\tan\delta = 0.002$) with height 1.59 mm. The overall dimension of this C-shaped antenna is $(36 \times 29 \times 1.59)$ mm^3. A conductive ground of thickness 0.05 mm is considered below the substrate. The detailed dimension of each segment is mentioned in Fig. 1.

3 Results and Discussion

The above-mentioned geometry is simulated by using HFSS and CST-MWS software. The variation of return loss with the frequency of C-shaped antenna is shown in Fig. 2, and the variation of VSWR is shown in Fig. 3 for the proposed C-shaped

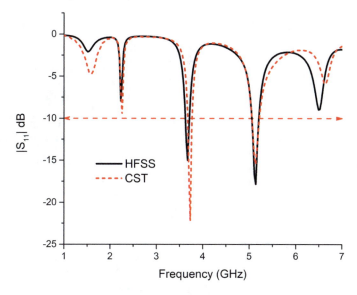

Fig. 2 S_{11} of C-shaped patch antenna

antenna. Simulated return loss confirms a value of more than 10 dB for both 3.7 and 5.2 GHz. The VSWR also confirms its good matching ability at both the desired frequencies. The peak gains obtained are 5.53 dBi and 5.39 dBi at 3.7 GHz and 5.2 GHz, respectively, as shown in Fig. 4 and Fig. 5. Antenna efficiencies are observed as 34.5% and 31.5% for 3.7 GHz and 5.2 GHz, respectively. Poor antenna efficiency may be due to the use of FR4 material as an antenna substrate. 3.7 GHz is a preferred frequency for WiMAX application, whereas 5.2 is popular for WLAN applications.

4 Conclusion

A simple structure of the C-shaped patch dual-band antenna is proposed in this paper for the application of 3.7 GHz WiMAX and 5.2 GHz WLAN band. The antenna structure provides a peak gain of 5.53 dBi and 5.39 dBi at 3.7 GHz and 5.2 GHz, respectively. Antenna efficiencies obtained at 3.7 GHz and 5.2 GHz are 34.5% and 31.5%, respectively. Key performance metrics of the antenna are evaluated by two different software.

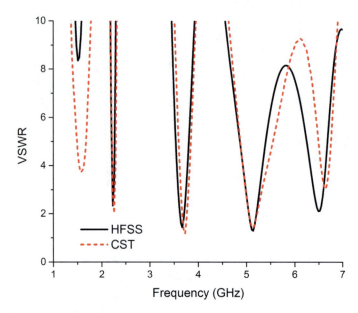

Fig. 3 VSWR of C-shaped antenna

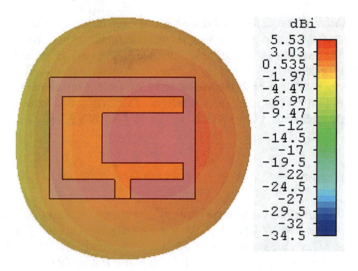

Fig. 4 Gain at 3.7 GHz

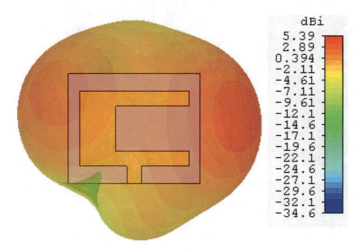

Fig. 5 Gain at 5.2 GHz

References

1. S. Das, T. Bose, H. Islam, Design and acceptance test of compact planar monopole antenna for lTE smartphone considering SAR, TRP, and HAC values. Int. J. Commun. Syst. e4155 (2019)
2. P. Liu, Y. Zou, B. Xie, X. Liu, B. Sun, Compact cpw-fed tri-band printed antenna with meandering split-ring slot for wlan/wimax applications. IEEE Antennas Wirel. Propag. Lett. **11**, 1242–1244 (2012)
3. C.Y. Huang, E.Z. Yu, A slot-monopole antenna for dual-band wlan applications. IEEE Antennas Wirel. Propag. Lett. **10**, 500–502 (2011)
4. B. Gautam, P. Verma, A. Singha, H. Islam, O. Prakash, S. Das, Design of multiple collar stay antennas for wireless wearable compact devices. Microw. Opt. Technol. Lett. 1–7 (2019)
5. W. Hu, Y.Z. Yin, P. Fei, X. Yang, Compact triband square-slot antenna with symmetrical l-strips for wlan/wimax applications. IEEE Antennas Wirel. Propag. Lett. **10**, 462–465 (2011)
6. S. Das, H. Islam, T. Bose, N. Gupta, Ultra wide band cpw-fed circularly polarized microstrip antenna for wearable applications. Wirel. Pers. Commun. **108**(1), 87–106 (2019)
7. P. Xu, Z.H. Yan, C. Wang, Multi-band modified fork-shaped monopole antenna with dual l-shaped parasitic plane. Electron. Lett. **47**(6), 364–365 (2011)
8. Y. Xu, Y.C. Jiao, Y.C. Luan, Compact cpw-fed printed monopole antenna with triple-band characteristics for wlan/wimax applications. Electron. Lett. **48**(24), 1519–1520 (2012)
9. W. Hu, Y.Z. Yin, X. Yang, P. Fei, Compact multiresonator-loaded planar antenna for multiband operation. IEEE Trans. Antennas Propag. **61**(5), 2838–2841 (2013)

Phased MIMO Radar Simulation for Moving Object Detection

Md Nadeem Akhtar Hasid, Subhankar Shome, Manisha Kumari Gupta, Bansibadan Maji, and Rabindranath Bera

Abstract Phased MIMO radar is a recent multi-antenna technique which is becoming popular because of its superior performance toward target detection. Multiple beams over space using different subarrays are the key advantage of the system. The array antenna available in a system is divided into subgroups which are coherently creating multiple channels over space which is basically MIMO operation. These multiple beams created by different arrays will improve the target detection in low SNR condition. This multi-beam can also be utilized for multi-target detection and tracking. All over the world researchers are planning to use multiple beams in favor of communication. The same can be applicable for radar also; multiple-vehicle tracking using multiple beams of phased MIMO radar can open a new direction for autonomous industry.

M. N. A. Hasid · S. Shome (✉) · M. K. Gupta · R. Bera
ECE Department, St. Mary's Technical Campus Kolkata, Barasat, West Bengal, India
e-mail: subho.ddj@gmail.com

M. N. A. Hasid
e-mail: nadim70.na@gmail.com

M. K. Gupta
e-mail: guptamanisha3845@gmail.com

R. Bera
e-mail: bansibadan.maji@ece.nitdgp.ac.in

B. Maji
ECE Department, National Institute of Technology Durgapur, Durgapur, West Bengal, India
e-mail: rbera50@gmail.com

© Springer Nature Singapore Pte Ltd. 2020   479
R. Bera et al. (eds.), *Advances in Communication, Devices and Networking*,
Lecture Notes in Electrical Engineering 662,
https://doi.org/10.1007/978-981-15-4932-8_54

1   Introduction

Almost in every 10 years, mobile communication steps forward into the next generation. Today's mobile communication is entering toward 5th generation; it comes into
reality because of advance baseband technology as well as advance antenna technologies. Technologies like CDMA, OFDM, and MIMO array antenna all together
make a strong platform for advance mobile communication. Application of all these
technologies in favor of multimedia transmission takes mobile communication into
a next height. Now researcher starts experimenting all these technologies in favor
of radar [1] which can take target detection, target imaging, and target tracking into
the next level. MIMO is a very popular technique in mobile communication. Array
antenna is also very popular for beam formation. 5th generation mobile is thinking
to create multiple beams using a group of array antennas which are basically called
subarray; these multiple beams can serve multiple users for mobile communication.
This concept is very recent and knows as phased MIMO [2]. Now radar researcher is
also thinking to borrow the same concept in favor of radar. Multiple beams created
by different subarrays can be utilized to track multiple vehicles in road scenario and
can be very much useful for autonomous industry. In view of the above context,
in this article author tried an end-to-end simulation of phased MIMO radar [3] for
autonomous industry, and also created moving target scenario, which is detected by
the radar, and few target parameter is tested.

2   Phased MIMO Radar Simulation

2.1   MIMO Tx Modeling

The complete phased MIMO radar end-to-end model is simulated for target detection,
velocity, and range measurement in a multi-target environment. Phased MIMO radar
simulation is divided into four parts: first is MIMO transmitter design, and second
is radar channel and target simulation. In this part transmit and receive antennas are
designed which is basically working in phased MIMO configuration. Third part is
MIMO receiver, and last part is designed for radar signal processing.

Frequency-Modulated Continuous-Wave (FMCW) signal generator is the first
block of the transmitter, and then TDM-MIMO signal generator block is used to
generate MIMO signal in TDM mode. TX up-convertor block is used to up-convert
the baseband signal into the RF frequency. Here the main lobe synthesization is done
using theta and phi toward target direction. This whole simulation is shown in Fig. 1.

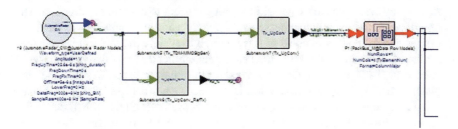

Fig. 1 Radar baseband design

2.2 Antenna Layer

The Tx and RX antenna setup block is used to input the platform location and target scatter location; these are the X, Y, Z values in the ECI frame. Automotive Radar_Platform and Automotive Radar_TargetScatter location are the blocks where simulated values can be put. Antenna plane type can be XYZ plane, which means array antenna positioned in XY-plane and radiation of the beam is in Z-plane. This portion of this model is also calculating the elevation and azimuth angles of various targets over the radar antenna frame. The simulation model is shown in Fig. 2.

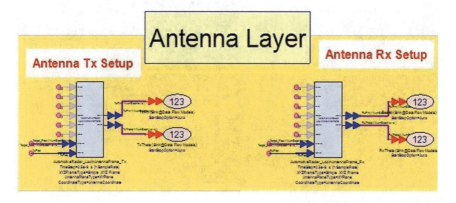

Fig. 2 Antenna layer

2.3 *Phased MIMO Antenna Modeling*

The main research area under this topic comes under in this section phased MIMO antenna modeling [4–6]. MIMO radar modeling or array radar modeling is a common topic these days but integrating both antenna techniques in favor of radar is a new area of work. In this work, 4-MIMO channel is used in the transmitter side; now [4 × 4] antenna elements are used in uniform rectangular array for each transmit channel. In this way, each transmit channel is forming a beam toward target direction. This type of configuration is achieving special diversity using array antenna [7]. Antennas can receive from all directions, so it is required to specify the incident wave angles for the input envelope signals. Target echo generation block simulating multi-target environment and reflecting the echo signal for the receiver. In receiver side, the same phased MIMO antenna configuration is used to create multiple beams over space using array antenna [8]. Four-target echo block is used as if all the MIMO channel or beam is reflected from the same targets, and the same echo is present at the receiver side. The simulation modeling is shown in Fig. 3.

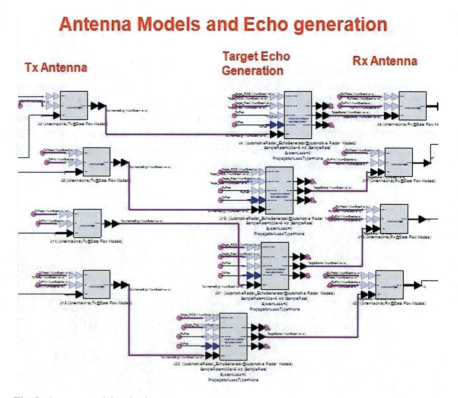

Fig. 3 Antenna model and echo generator

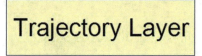

Radar Tx Platform   **Targtes' Location**   **Radar Rx Platform**

Fig. 4 Trajectory layer

2.4 Trajectory Layer

In Fig. 4, the simulation parameters are set. TX and RX platform position, target positions, is set in this section which has to be estimated in the receiver section. In this simulation radar transmitter and receiver, the platform is Static; only the Target is moving. This platform is simulated using TX, Rx platform block. Two-target location block is used to simulate two different targets, which is a multi-target environment. Target 1 range values in X, Y, Z coordinates are set to [10 0 250], and target 2 range value is set to [0 0 300]. Velocities for targets 1 and 2 are the same, which is [0 0 5]. These values are estimated in the receiver, and outputs are shown in the result section.

2.5 MIMO Receiver

Four-MIMO receive channel (Fig. 5) is simulated here; each RF component chain consists of Low-Noise Amplifier (LNA) [9] and transmit–receive code correlator, which mainly used as target detector; in this block an original transmitted FMCW reference is used for correlation. Then, range and velocity of the target are measured using range and doppler FFT block.

2.6 Phased MIMO Radar DSP

This processed data now passed through the Doppler Constant false alarm rate (CRAF) detection [10]. This is a very important block for radar detection (Fig. 6). In general, the detection is done based on a threshold value which is a function of

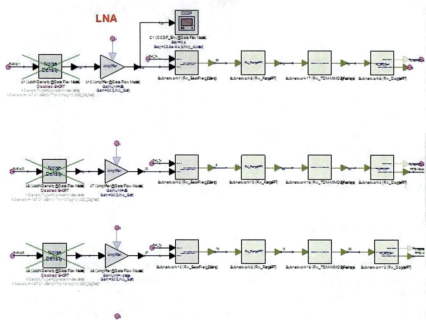

Fig. 5 MIMO Rx modeling

both the probability of detection and the probability of false alarm. This block is very much important to detect the original target out of false alarm. Another Matlab script is implemented to find out the angle and range measurement for multi-target.

2.7 Clutter Modeling

Sea clutter and ground clutter both conditions are simulated under this modeling. Sea state 3 is considered. Clutter code is written in Matlab script (Fig. 7).

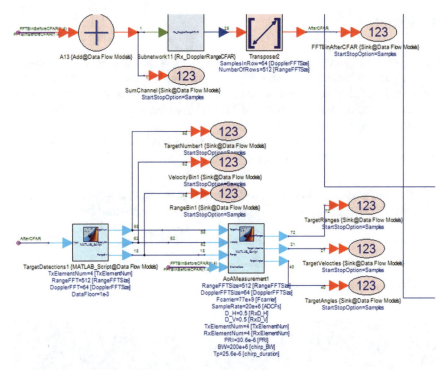

Fig. 6 Receiver signal processing for parameter estimation

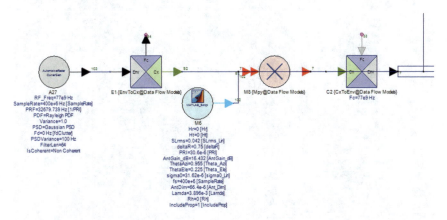

Fig. 7 Added clutter with channel

TargetVelocties_Index	TargetVelocties[1,1]	TargetVelocties[1,2]
0	4.974	4.974

Fig. 8 Measured target velocity for multi-target

3 Results

The measured target velocity is 4.974 as shown in Fig. 8, which is set to 5 in simulation model. The measured range is showing 250.5 for target 1 and 300.75 for target 2 as shown in Fig. 9. So far, the measurements of range and velocity for multi-target are very accurate using designed phased MIMO radar, which may be very much effective for moving target like a car.

Figure 10 shows the multi-target detection after CRAF processing; the targets are very well defined in the target range.

TargetRanges_Index	TargetRanges[1,1]	TargetRanges[1,2]
0	250.5	300.75

Fig. 9 Measured target range for multi-target

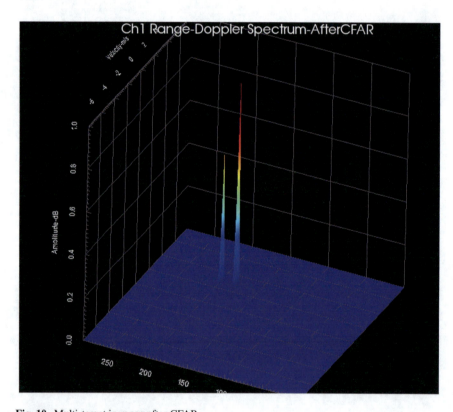

Fig. 10 Multi-target in range after CFAR

4   Conclusion

A new antenna technique is simulated for radar application which is basically the integration of two renowned antenna techniques which are MIMO and array antenna. Array antenna is divided into subarray to form multiple beams toward a certain direction. Array antenna will provide high gain within the beam, and MIMO beam will provide special diversity. Integrating both the advantages radar system will perform superior in every SNR condition. The system is tested for moving object detection even in the worst clutter condition. Sea clutter and ground clutter both are tested and phased MIMO radar is performed satisfactorily in both conditions.

References

1. M.I. Skolnik, *Introduction to Radar Systems*, 3rd edn. (Mc-Graw-Hill, New York, 2001)
2. A. Hassanien, S.A. Vorobyov, Why the phased- MIMO radar outperforms the phased array and MIMO radars, in *18 European Signal Processing Conference(EUSIPCO-2010)*, Aalborg, Denmark, August 23–27 (2010)
3. A. Hassanien, S.A. Vorobyov, Transmit/receive beamforming for MIMO radar with colocated antennas, in *Proceedings of the IEEE International Conference on Acoustic, Speech, and Signal Processing*, Taipei, Taiwan (2009), pp. 2089–2092
4. R. Fa, R.C. de Lamare, An adaptive LCMV beamforming algorithm based on dynamic selection of constraints, communications research group (Department of Electronics, University of York, YO10 5DD, United Kingdom)
5. B.G. Ferguson, Minimum variance distortionless response beamforming of acoustic array data. J. Acous. Soc. Am. **104**, 947 (1998). https://doi.org/10.1121/1.423311
6. Q. Zou, Z.L. Yu, Z. Lin, A Robust algorithm for linearly constrained adaptive beamforming. IEEE Sig. Process. Lett. **11**(1), (2004)
7. S.K. Tiong, B. Salem, S.P. Koh, K.P. Sankar, S. Darzi, Minimum variance distortionless response (MVDR) beamformer with enhanced nulling level control via dynamic mutated artificial immune system. Sci. World J. (2014)
8. http://www.labbookpages.co.uk/audio/beamforming/frost.html
9. H.L. Van Trees, *Optimum Array Processing* (Wiley, NY, 2002)
10. A. Hassanien, S.A. Vorobyov, K.M. Wong, Robust adaptive beamforming using sequential programming: an iterative solution to the mismatch problem. IEEE Sign. Process. Lett. **15**, 733–736 (2008)

Wireless Mobile Charger Using RF Energy (Green Charger)

Arun Kumar Singh and Pankaj Kumar Tiwari

Abstract This atmosphere is completely surrounded by radio frequency and microwave radiation, so a method is providing the power to charge a device or battery by the use of radio frequency. And we are representing a type of wireless charging system. So in this system, we can absorb the radio waves of certain range by the use of multiple receiver antennas, and after this we can convert this energy into DC power by the RF to DC converter circuit, as the input of wireless mobile charger. And this supply gives the required output for the device to charge.

Keywords Mobile phone · Wireless battery charger · RF energy · Batteries

1 Introduction

In this modern time of technology, we are surrounded by the radio frequency and the microwave radiation. And the sources of this energy generate highly electromagnetic fields such as mobile phones, television, and Wi-Fi.

In this method and system [1–3], we are providing electrical input without any external source but by the RF energy. By the help of receiver antenna, we will absorb the radio frequencies of a certain range/up to 915 MHz. This range of frequency produces the approximated 4 w power by the use of voltage multiplier which is enough to charge up the mobile batteries. We will convert the RF energy into electrical energy by the direct connection of rectifier. And the work of this is to convert the RF energy to DC voltage. And the wireless charging circuit also comprises the wireless controller. This circuit is used to compare the supplied power in the DC voltage charging signal and the required power by the battery [4] (Fig. 1).

A. K. Singh (✉) · P. K. Tiwari
Department of Electronics and Communication Engineering, Sikkim Manipal Institute of Technology, Sikkim Manipal University, Gangtok, India
e-mail: arunsingh.smit@gmail.com

P. K. Tiwari
e-mail: pankajt85597@gmail.com

© Springer Nature Singapore Pte Ltd. 2020
R. Bera et al. (eds.), *Advances in Communication, Devices and Networking*,
Lecture Notes in Electrical Engineering 662,
https://doi.org/10.1007/978-981-15-4932-8_55

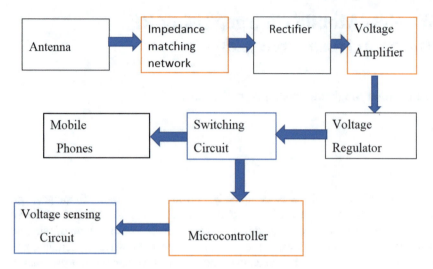

Fig. 1 Block diagram of proposed charger

2   Description of the Work

Antenna from the environment has the tendency to receive electromagnetic waves from the environment. Waveforms of different frequencies are available in the environment. In our proposed work, we will be receiving the waveform of cellular network (2G/3G/LTE). The received waveform will now be amplified, converted, and regulated to the appropriate level so that it can charge the electronic devices like mobile phones. The brief description of components used is shown below.

2.1   *Antenna*

Antenna is a bi-directional transducer which is used to transmit and receive the radio frequencies, and in this green charger it is used first as a receiver antenna because first it will receive the radiated waves which are present in the environment and after this it will convert that received power into the voltage, which will be the input of the green charger.

2.2   *Voltage Amplifier*

Voltage amplifier is used to produce the high voltage output to the respect of given input voltage, so the input of the voltage amplifier of first stage will be the output of the antenna. Normally, the output of antenna will be around 10 uv and by the use of this first-stage amplifier it will give the output of maximum 2 v. In the amplifier, we use the type of op-amp so the gain also will be high and the input impedance of amplifier also matches with the impedance of antenna.

2.3 Voltage Amplifier

It is a type of system which is used to design and maintain a fixed level of voltage. And it is also a monolithic switching regulator and used for DC–DC converter. And simply it will be used for implementing high efficiency and also for the simple supply power switching. The use of switching regulator is to pronounce for the linear regulator because of its size reduction. And another thing is that it will increase the device flexibility of output voltage. And the output can vary and or change in polarity.

2.4 Microcontroller

We will use a low-power high-performance 8-bit microcontroller with 4 k bytes of in-built programmable flash memory. It is designed by the use of non-volatile, high-density memory technology. And it is in standard of 8051 instruction set and pin diagram. And the flash memory will allow to program memory to re-programmed in the system or by a non-volatile memory programmer. It will help in the section of wireless to start the communication between the device and the green charger output.

2.5 Batteries

Batteries are classified into many ways, which will use according to work of that embedded system. All batteries have different types of advantage and disadvantage.

2.6 Rectifier

A rectifier is a device which is used to convert the alternating current which comes from the side of antenna transmitter and it will convert that current into direct current and that current will be supplied to the voltage amplifier and after this it will regulate the input voltage and gives the required output.

3 Social Impact of the Design

The design will have a great impact on the society as this will be a device charger where no cost is involved. Neither electricity is required nor any charge storage device. There will be no need of connecting cellular phone with the power adapter in order to charge the battery. If there is a cellular network, the device's battery will get charged automatically. This will be a charger which can charge the device anywhere, anytime.

4 Conclusion

According to this paper, it is the idea to charge mobile phones using RF energy without any use of external electricity and wiring. The idea of unwired charger and without the external electricity will comfort for the mobile users. The users can charge mobile phone at any place and anytime.

References

1. Drayson Technologies, Press Release, RF Energy Harvester: Freevolt™. London 30 Sept. (2015)
2. P. Surendar, S. Pramodh, M.R. Kannan, *Wireless Charging Using RF Energy*, vol. 03, issue: 05 (IJRET, 2016), pp. 1870–1872
3. D.W. Harrist, Wireless battery charging system using radio frequency energy harvesting. A project report (Master of Science, University of Pittsburgh, 2004)
4. Krishnan, Method and Apparatus for Wireless Powering and Recharging. U.S. Patent No. 6,127,799, Date of Patent: Oct. 3 (2000)

A Review on 5G Channel Characterization

Battu Phalguna, Subhajit Paul, Samarendra Nath Sur, Arun Kumar Singh, Rabindranath Bera, and H. Srikanth Kamath

Abstract The tremendous development of 5G enables its utilization for various applications over different environmental conditions. But successful deployment of the 5G system depends on the acquit understanding of the behavior of the wireless channel. Depending on the environment, the channel model can be formulated. And this paper highlights some of the key aspects of the channel modeling and the measurement techniques.

Keywords MIMO · mmWave · 5G · AOA · AOD · HST · V2V

1 Introduction

The 5G technology system is predicted to provide very good quality connectivity among various devices in various conditions of diverse network topologies [1]. It will enable to share information over the wireless scenario with less latency and high data rates [2, 3]. There are several approaches of technologies in 5G systems which are compatible and useful; one of them is the millimeter-wave communication which has a very huge bandwidth but sometimes when we use large frequency we get a severe path loss but in order to overcome this loss, massive multiple-input multiple-output (MIMO) communication can be used [4]. Massive MINO can be reliable as it increases the efficiency of the system. And the development of 5G systems is focused on its variant application over a large range of scenarios like rural area, suburban area, urban area, etc. [5, 6]. But the efficient and successful development of any wireless communication system depends on the meaningful characterization of the channel.

B. Phalguna (✉) · S. Paul · S. N. Sur · A. K. Singh · R. Bera
Electronics and Communication Engineering Department, Sikkim Manipal Institute of Technology, Sikkim Manipal University, Majitar 73736, Sikkim, India
e-mail: phalgunabattu007@gmail.com

H. S. Kamath
Electronics and Communication Engineering Department, Manipal Institute of Technology, Manipal, India

© Springer Nature Singapore Pte Ltd. 2020
R. Bera et al. (eds.), *Advances in Communication, Devices and Networking*,
Lecture Notes in Electrical Engineering 662,
https://doi.org/10.1007/978-981-15-4932-8_56

The 5G channel modeling sets new requirements for all kinds of technologies; some of the requirements are as follows:

- The frequency range of 5G channel models should be wide, e.g., 500 MHz to 100 GHz. But the system should have a compatibility with both the lower and higher frequency bands.
- The 5G channel model should support on a wide range of scenarios like rural, urban, suburban, indoor, high-speed train (HST), and many more.
- The 5G channel model should provide double-directional three-dimensional modeling, where 3D propagation modeling and 3D antenna modeling should have accuracy.
- The 5G channel model must develop with time in order to support bean tracking and mobility for communication system.
- The 5G channel model should support spatial consistency where the transmitter and receiver should have the same channel characteristics. The channel model should include some parameters, in continuous- and real-time as a function of position, like large-scale/small-scale fading parameters, line of sight (LOS), and non-line of sight (NLOS) parameters and indoor and outdoor conditions. And also, it should have a very good frequency dependency and frequency consistency. It should have a strong correlation with the adjacent frequencies of the channel parameter.
- The 5G channel should support massive MIMO channeling. There are two parts, which must be properly modeled and those are spherical wave front and array non-stationarity. Spherical wave front is considered in place of plane wave front because of the shorter distance between transceiver and cluster in comparison to the Rayleigh distance. Array non-stationarity refers to the aggregate that may either appear or disappear from the visual range of the antenna.
- The 5G channel model should include high mobility conditions such as HST where the speed is 500 km/h or even above. This model should be able to capture the Doppler frequency and non-stationarity characteristics and many more. It should work over various HST scenarios such as hilly areas, open space, cutting, tunnel, etc. [3–11].

2 Channel Measurements

5G system is going to operate over a wide scope of situations, notwithstanding, some unique or outrageous situations, for example, environment tunnel, underground situations, etc. And those are the current research topics for the channel modeling. In general, modeling of these situations emphatically relies upon channel measurements because of their unique surroundings and distinct scattering properties. There is a lot of research occurring in the advancement to give suggestions. In this area, we will briefly survey few channel measurements and characterization techniques in the light of various 5G communication advancements.

2.1 MmWave Channel Measurements

High data rate under high mobility is the main challenge for the researchers under current circumstances. To address these challenges, we must move into millimeter-wave bands (mmWave), where here huge data transmissions are accessible. MmWave technology can accomplish a lot of higher information rate by utilizing basic air interfaces because of the enormous bandwidth [12]. Recently, this band got utmost attention as a prospective band for 5G systems [13]. The mmWave carrier frequencies support bigger data transfer capacity, which convert into the higher speed transmission [14]. The advantage of higher frequency operation is not just too only increase the bandwidth as well as have the ability of radio antennas to achieve higher directivity. This will enable to improve the signal quality and spectral efficiency of the system [13]. Because of little wavelengths at millimeter waveband, diffraction turns into the weakest and least solid engendering mechanism millimeter wave for communication frameworks [15].

For the most part of mmWave channel, measurements were centered on a severe path loss, which is the greatest test for the utilization of mmWave communication. Measurements in [4, 39, 40] were campaigns at 38 and 28 GHz to pick up insight on RMS delay spread, angle of departure (AOD), path loss, angle of arrival (AOA), and reflection characteristics for the structure of their future mmWave channel measurement. By watching their experimentation for the path loss and delay spread for the urban regions and the light urban regions, they found generous contrasts in proliferation parameters. Multipath delay spread is seen as a lot higher in the substantial urban zones than the light urban territories, where the outcomes in [16] were analyzed for single-frequency, directional and omnidirectional, and multifrequency models.

2.2 Massive MIMO Channel Measurement

MIMO techniques have come in attention in the wireless communications for so many years because it offers multiple advantages such as high data rate and reliable link without any power boosting of signals. Orthogonal frequency-division multiplexing (OFDM) with MIMO has been accepted and is being used in the fourth-generation long-term evolution (LTE) cellular networks.

In [17], 2.6 GHz frequency channel measurements have experimented with a 50 MHz bandwidth, and 128-element linear virtual arrays were used at the base station (BS). In this paper, authors have observed significant variations of the angular power spectrum (APS), K-factor, and channel gain over the linear array. This paper presents the fact that there is no validation of the plane wave assumption or far-field assumption in case of massive MIMO systems. Other than the linear array, different types of large antennas were likewise announced in massive MIMO channel. In [18], 15 GHz frequency band where the measurements were conducted with a 5 GHz bandwidth. 1600 elements of virtual planar antenna array were used. The overall antenna structure is segmented into multiple sub-arrays to observe the nonstationary behavior of massive MIMO channels over the antenna elements. As presented in the paper, it highlights some key parameters like delay spread, azimuth angular spread of arrival (AASA), elevation angular spread of arrival (EASA), and K-factor varied in blocks with clear boundaries over the array plane. Recently, MIMO channel measurement was conducted with frequency bands of 16, 11, 38, and 28 GHz over the indoor environment [19, 42–45].

2.3 HST Channel Measurement

The high-speed railway/train (HST) propagation channel has a significant impact on the structure and execution investigation of remote railway route control system. High-speed railways are minimal effort and earth benevolent methods for mass transportation over the huge separation. Therefore, in railways wireless connectivity is fundamental for the trade control and security application and for the GSM-R [20] have been created. However, the propagation channel in the cellular system is significantly different from the high-speed railway because of the different antenna positions and the propagation environment. The utilization of directional transmitting antenna is the main important feature of high-speed railway communication. Here attaining an accurate characterization is a challenging task for the propagation channels.

Authors in [21] and [22] demonstrate a position-based radio propagation channel model for high-speed railway based on the measurements at 2.35 GHz. This paper explores the small-scale fading properties, for example, K-factor, Doppler frequency feature, and time delay spread. The statistical position-based channel models have been developed to characterize the channel, which significantly advances the assessment and verification of wireless communication in relative scenarios.

In [23] the conducted measurement campaign for HST channels measurement was presented where the LTE-A system has been utilized. The channel impulse responses (CIRs) are extricated from the got cell-specific reference signals (CRSs). The outcomes show that most of the channels contain a defer direction comparing to a LOS path. The geometry parameters, for example, the train speed and the negligible separation between the transmitter and the beneficiary are extricated from the defer directions. Besides, path loss characteristics tributes along the directions are explored and contrasted and the model separated for the Universal Mobile Telecommunications System (UMTS). In [24, 25], a ray-tracing approach was applied for producing the engendering models in the burrow and open-air situations. These models rely upon specific areas and neglect to create channel acknowledge for arbitrary situations [24, 26].

2.4 V2V Channel Measurement

The vehicle-to-vehicle (V2V) communication is an empowering influence for improved traffic safety and clogs control [25]. For any wireless framework, a definitive performance limit is controlled by the propagation channel. As in [27], the V2V channel measurement includes both city and highway situations over 5 GHz band. For the experimental propose, authors have used 50 MHz test signal and omnidirectional antennas, both inside and outside the vehicles. Here, the channel modeling has been expressed in terms of frequency correlation functions and RMS delay spread. The estimated RMS delay spreads are of the order of a microsecond, in urban areas with dense vehicle traffic, and it corresponds to a coherence bandwidths of 1–2 MHz. Some researchers are also looking for the 2.4 GHz band. As in [28], which reported the variation of Doppler spread, and [29], which characterized attenuations in short-range indoor environment for mobile-to-mobile (M2M) communication system, similar type of analytical studies has been accomplished under the same type of channel condition, and the analytical models have been included in [30, 31]. Some narrowband estimations in the 5 GHz band were presented in [32]. In [33], the author analyzed the nonstationary fading procedure of vehicular channels. For that analysis, they utilized radio channel measurement data which is collected from the traffic scenarios. They estimated the LSF from the measured sample of time-varying frequency response. They also reported the second-order statistical model for the Doppler spread and the time–frequency-varying RMS delay spread. The empirical distribution of the Doppler spread, and the RMS delay spread for different traffic scenarios have been modeled with the help of bimodal Gaussian model.

3 Channel Modeling for Indoor Scenario

This section presents some review on the channel modeling for indoor scenario. In [34], the measurement system utilized a vector network analyzer (VNA)-based set up in an indoor environment, i.e., office at 60 GHz frequency band. The basic approach for the mmWave indoor channel measurements revolves around rotated directional antenna (RDA)-based method and uniform virtual array (UVA)-based method. Under indoor channel condition, parameters including power, delay, azimuth point, and rise edge were extracted using the pace-alternating generalized expectation–maximization (SAGE) algorithm. Two IEEE standards, 802.15.3c and 802.11ad, have been proposed for 60-GHz indoor wireless communications. And particularly for those two standards, the channel models fully relies on the extended Saleh–Valenzuela (S-V) model. The prime focus of the paper [34] is to characterize 60-GHz indoor channel and also to parameterize the environment based on the angular extended S-V model and temporal and spatial clustering properties. Similarly, some 3D S-V-based millimeter-wave channel models with between and intra-cluster parameters for indoor environment conditions were exhibited in the channel model [35–38, 41].

4 Conclusion

As represented in this paper, channel can be mathematically formulated based upon the clear understanding of the behavior or the statistical distribution of the channel defining parameters. And in order to model it correctly, channel measurements play an important role. As pointed out in this paper, for different sets of characterization, the approach for the measurements should be different. And 5G channel measurement and characterization is of utmost important as it is the future of the wireless communication evolution.

References

1. F. Boccardi, R.W. Heath, A. Lozano, T.L. Marzetta, P. Popovski, Five disruptive technology directions for 5G. IEEE Commun. Mag. **52**(2), 74–80 (2014)
2. C.-X. Wang, F. Haider, X. Gao, X.H. You, Y. Yang, D. Yuan, H.M. Aggoune, H. Haas, S. Fletcher, E. Hepsaydir, Cellular architecture and key technologies for 5G wireless communication networks. IEEE Commun. Mag. **52**(2), 122–130 (2014)
3. P. Popovski et al., METIS, ICT-317669/D1.1, Scenarios, requirements and KPIs for 5G mobile and wireless system (2013)
4. T.S. Rappaport, S. Sun, R. Mayzus, H. Zhao, Y. Azar, K. Wang, G.N. Wong, J.K. Schulz, M. Samimi, F. Gutierrez, Millimeter wave mobile communications for 5G cellular: it will work! IEEE Access **1**, 335–349 (2013)
5. S. Kutty, D. Sen, Beamforming for millimeter wave communications: an inclusive survey. IEEE Commun. Surv. Tut. **18**(2), 949–973 (2016)
6. E.G. Larsson, O. Edfors, F. Tufvesson, T.L. Marzetta, Massive MIMO for next generation wireless systems. IEEE Commun. Mag. **52**(2), 186–195 (2014)
7. F. Rusek, D. Persson, B.K. Lau, E.G. Larsson, T.L. Marzetta, O. Edfors, F. Tufvesson, Scaling up MIMO: opportunities and challenges with very large arrays. IEEE Sign. Process. Mag. **30**(1), 40–60 (2013)
8. A. Maltsev et al., MiWEBA, FP7-ICT-608637/D5.1 Channel modeling and characterization, V1.0 (2014)
9. V. Nurmela et al., METIS, ICT-317669/D1.4, METIS Channel Models (2015)
10. I. Tan, W. Tang, K. Laberteaux, A. Bahai, Measurement and analysis of wireless channel impairments in DSRC vehicular communications, in *Proceedings of the IEEE ICC'08* (Beijing, China, 2008), pp. 4882–4888
11. O. Renaudin, V.M. Kolmonen, P. Vainikainen, C. Oestges, Wideband measurement-based modeling of inter-vehicle channels in the 5 GHz band. IEEE Trans. Veh. Technol. **62**(8), 3531–3540 (2013)
12. L. Wei, R. Hu, Y. Qian, G. Wu, Key elements to enable millimeter wave communications for 5G wireless systems. IEEE Wireless Commun. Mag. **21**(6), 136–143 (2014)
13. O.H. Koymen, A. Partyka, S. Subramanian, J. Li, *Indoor mm-Wave Channel Measurements: Comparative Study of 2.9 GHz and 29 GHz* (Qualcomm R&D, Bridgewater, New Jersey, 2015)
14. M. Elkashlan, T.Q. Duong, H. Chen, Millimeter-wave communications for 5G: fundamentals: part I. IEEE Commun. Mag. **52**, 52–54 (2014). https://doi.org/10.1109/MCOM.2014.6894452
15. T.S. Rappaport, R.W. Heath, R.C. Daniels, J.N. Murdock, *Millimeter Wave Wireless Communications* (Pearson Education, 2014)
16. Indoor Office Wideband Millimeter-Wave Propagation Measurements and Channel Models at 28 GHz and 73 GHz for Ultra-Dense 5G Wireless Networks

17. S. Payami, F. Tufvesson, Channel measurements and analysis for very large array systems at 2.6 GHz, in *Proceedings of the EUCAP'12*, (Prague 2012), pp. 433–437
18. J. Chen, X. Yin, X. Cai, S. Wang, Measurement-based massive MIMO channel modeling for outdoor LoS and NLoS environments. IEEE Access **5**(99), 2126–2140 (2017)
19. J. Huang, C.-X. Wang, R. Feng, J. Sun, W. Zhang, Y. Yang, Multi-frequency mmWave massive MIMO channel measurements and characterization for 5G wireless communication systems. IEEE J. Sel. Areas Commun. **35**(7), 1591–1605 (2017)
20. Z. Zhong, X. Li, W. Jiang, *Principles and Foundation of Integrated Digital Mobile Communicationsystem for Railway* (2003)
21. Measurement Based Channel Modeling with Directional Antennas for High-Speed Railways
22. L. Liu, C. Tao, Ji. Qiu, H. Chen, L. Yu, W. Dong, Y. Yuan, Position-based modeling for wireless channel on high-speed railway under a viaduct at 2.35 GHz
23. X. Ye, X. Cai, Y. Shen, X. Yin, X. Cheng, School of Electronics and Information Engineering, Tongji University, Shanghai, China, Institute of Automation, Chinese Academy of Science, Beijing, China, School of Electronics Engineering and Computing Sciences, Peking University, Beijing, China
24. J. Lu, G. Zhu, C. Briso-Rodríguez, State key laboratory of rail traffic control and safety fading characteristics in the railway terrain cuttings
25. R. He, A.F. Molisch, F. Tufvesson, Z. Zhong, B. Ai, T. Zhang, Vehicle-to-vehicle propagation models with large vehicle obstructions
26. K. Guan, Z. Zhong, B. Ai, and T. Kurner, "Deterministic propagation modeling fortherealistic high-speed railway environment," inIEEE77th Vehicular Technology Conference (VTC Spring), 2013, pp. 1–5
27. D.W. Matolak, I. Sen, W. Xiong, N.T. Yaskoff, GHz Wireless Channel Characterization for Vehicle to Vehicle Communications. School of Electrical Engineering & Computer Science Avionics Engineering Center Ohio University Athens, OH 45701
28. G. Acosta, K. Tokuda, M. Ingram, Measured joint doppler-delay power profiles for vehicle-to-vehicle communications at 2.4 GHz, in *Proceedings of the GLOBECOM' 04*, Dallas, TX, vol. 6, (2004), pp. 3813–3817
29. T.J. Harrold, A.R. Nix, M.A. Beach, Propagation studies for mobile-to-mobile communications, in *Proceedings of the IEEE 54th Veh. Tech. Conf (Fall), Atlantic City, NJ*, vol. 3, (2001), pp. 1251–1255
30. A. Akki, Statistical properties of mobile-to-mobile land communication channels. IEEE Trans. Veh. Tech. **43**(4), 826–831 (1994)
31. C.S. Patel, G.L. Stuber, T.G. Pratt, Simulation of rayleigh faded mobile-to-mobile communication channels. Proc. IEEE Veh. Tech. Conf. **1**, 163–167 (2003)
32. J. Maurer, T. Fugenm, W. Wiesbeck, Narrowband measurement and analysis of the intervehicle transmission channel at 5.2 GHz. Proc. IEEE Veh. Tech. Conf **3**, 1274–1278 (2002)
33. L. Bernadó, T. Zemen, F. Tufvesson, A.F. Molisch, C.F. Mecklenbräuker, Delay and doppler spreads of nonstationary vehicular channels for safety-relevant scenarios
34. X. Wu, C.-X. Wang, J. Sun, J. Huang, R. Feng, Y. Yang, X. Ge, 60-GHz millimeter-wave channel measurements and modelling for indoor office environments. IEEE Trans. Antennas Propag. **65**(4), 1912–1924 (2017)
35. Maltsev et al., IEEE doc. 802.11ad 09/0334r8,2014, Channel models for 60 GHz WLAN systems, (2009)
36. A. Maltsev, R. Maslennikov, A. Sevastyanov, A. Khoryaev, A. Lomayev, Experimental investigations of 60 GHz WLAN systems in office environment, IEEE J. Sel. Areas Commun. **27**(8), 1488–1499 (2009)
37. A. Maltsev, R. Maslennikov, A. Sevastyanov, A. Lomayev, A. Khoryaev, Statistical channel model for 60 GHz WLAN systems in conference room environment, in *Proceedings of the EuCAP'10* (2010), pp. 1–5
38. A. Maltsev et al., IEEE doc. 802.11-08/1044r0, 60 GHz WLAN Experimental Investigations (20080

39. Z. Lin, X. Du, H.-H. Chen, B. Ai, Z. Chen, D. Wu, *Millimeter-Wave Propagation Modeling and Measurements For 5 g Mobile Networks*
40. M.B. Majed, T.A. Rahman, O.A. Aziz, M.N. Hindia, E. Hanafi, *Channel Characterization and Path Loss Modeling in Indoor Environment at 4.5, 28, and 38 GHz for 5G Cellular Networks*
41. A.M. Al-Samman, T.A. Rahman, M.H. Azmi, M.N. Hindia, I. Khan, E. Hanafi, *Statistical Modelling and Characterization of Experimental mm-Wave Indoor Channels for Future 5G Wireless Communication Networks*
42. C.-X. Wang, S. Wu, L. Bai, X. You, J. Wang, and C.-L. I, Recent advances and future challenges for massive MIMO channel measurements and models. Sci. China Inf. Sci, Invited Paper **59**(2), 1–16 (2016)
43. X. Gao, F. Tufvesson, O. Edfors, F. Rusek, Measured propagation characteristics for very-large MIMO at 2.6 GHz, in *Proceedings of the ASILOMAR'12* (Pacific Grove, CA, 2012), pp. 295–299
44. X. Gao, O. Edfors, F. Rusek, F. Tufvesson, Massive MIMOperformance evaluation based on measured propagation data. IEEE Trans. Wireless Commun. **14**(7), 3899–3911 (2015)
45. À.O. Martínez, E. De Carvalho, J.Ø. Nielsen, Towards very large aperture massive MIMO: a measurement based study, in *Proceedings of the IEEE GC Wkshps'14* (Austin, TX, 2014), pp. 281–286

Estimation of Gain Enhancement of Circular Sector Microstrip Antenna with Air Substrate

Sudip Kumar Ghosh, Abhijyoti Ghosh, Subhradeep Chakraborty, L. Lolit Kumar Singh, and Sudipta Chattopadhyay

Abstract A quick hand estimation of gain enhancement with air substrate in a circular sector microstrip antenna (CSMA) has been presented in paper. The standard gain estimation process has been carried out with 90° CSMA (air substrate) in comparison with conventional CSMA (PTFE substrate). Around 2.8 dB gain improvement has been achieved.

Keywords Circular sector microstrip antennas · Air substrate · Gain enhancement · Radiation patterns

1 Introduction

In the recent past, CSMA has grown into popularity in different wireless applications due to its small size and easy to fabricate in the printed board. Although the CSMA attains miniaturization with respect to other regular geometry, the gain of such antenna is not adequate [1]. However, gain improvement is one of the primary requirements of modern antenna technology. Therefore, the gain can be enhanced in such antennas if the conventional dielectric substrate (PTFE) is replaced by simple air substrate. Although several attempts have been made to use CSMA in modern wireless communications for different applications, the issue of gain enhancement has not been studied yet [2, 3]. In [4], the application of CSMA in Ku band was reported. The calculation of accurate resonance frequency of CSMA has been studied in detail in [5]. The radiation characteristics of semicircular patch have been reported in [6, 7]. The CSMAs of 90° and 180° sectors have been found to be the suitable

S. K. Ghosh
Department of ECE, Siliguri Institute of Technology, Siliguri, West Bengal, India

A. Ghosh (✉) · L. L. K. Singh · S. Chattopadhyay
Department of ECE, Mizoram University, Aizawl, Mizoram, India
e-mail: abhijyoti_engineer@yahoo.co.in

S. Chakraborty
3 Microwave Devices Area, CSIR-CEERI, Pilani, India

© Springer Nature Singapore Pte Ltd. 2020
R. Bera et al. (eds.), *Advances in Communication, Devices and Networking*,
Lecture Notes in Electrical Engineering 662,
https://doi.org/10.1007/978-981-15-4932-8_57

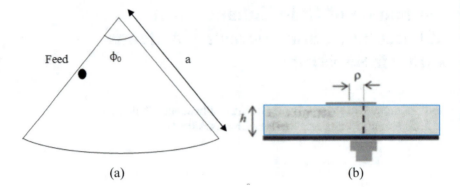

Fig. 1 Schematic diagram of 90° CSMA **a** top view **b** cross-sectional views

replacement of conventional circular patch and are reported in [8, 9] recently. Never-theless, the issue of gain in such small antennas has not been properly addressed in the above-mentioned papers. The use of air substrate in microstrip antennas of common geometries is the common and popular technique to enhance the gain.

In this paper, the air substrate has been utilized instead of PTFE substrate, and an improvement of gain is observed like other geometries. Further, a quick hand formulation has been established to estimate the gain enhancement of CSMA with air substrate compared to standard CSMA on PTFE substrate. In the present investigation, 90° CSMA has been considered.

2 Antenna Structures

In this present study, a 90° CSMA ($\varphi_0 = 90°$) of radius $a = 30$ mm on PTFE ($\varepsilon_r = 2.33$) and air substrate ($\varepsilon_r = 1$) with $h = 1.575$ mm has been investi-gated on 60×60 mm^2 ground plane. The antenna is fed at proper position to excite its lowest order dominant mode following [10]. The schematic representation of the structure is shown in Fig. 1.

3 Theoretical Calculation

The relation between gain (G) and effective radiating area (A_{eff}) of a CSMA is as follows:

$$\Delta G[dB] = 10 \log_{10} \left[\frac{\left(\frac{A_{eff}}{\lambda_0^2}\right)_{air}}{\left(\frac{A_{eff}}{\lambda_0^2}\right)_{ref}} \right] \tag{1}$$

where $A_{eff} = \dfrac{(a_{eff}^2 \varphi_0)}{2}$

$a_{eff} = (a \times \sqrt{(1+q)}$,

q = fringing factor and can be obtained from [5]

λ_0 = Operating wavelength.

As soon as the conventional PTFE substrate is replaced by air substrate, the electric field becomes loosely bound. It results in wider effective area, and hence provides better gain.

4 Result and Discussions

The simulation study has been carried out using commercially available HFSS (v.14). The reflection coefficient profiles for both the CSMAs (with PTFE and air substrates) have been presented in Fig. 2. As expected, the resonant frequency of CSMA with air substrate has been shifted toward the higher side of the spectrum in comparison with CSMA on PTFE substrate. The resonant frequency of CSMA with air substrate is around 4.4. GHz, while the same with PTFE substrate is around 2.9 GHz. Moreover, the bandwidth in case of the air substrate is better (6% more) as compared to PTFE substrate.

The simulated radiation pattern of same CSMA with different substrates (PTFE and air) has been presented in Figs. 3 and 4.

It is observed that the co-pol gain of CSMA with air substrate is around 9.9 dBi, whereas the same with PTFE substrate is only 7.1 dBi. 2.8 dBi of gain improvement is revealed from the proposed technique. It is also found that the isolation between co-pol and cross-pol in both the principle planes has improved from 16 dB to 18 dB.

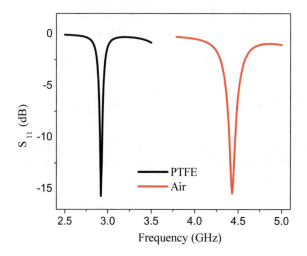

Fig. 2 Reflection coefficient profile of 90° CSMA with PTFE substrate and air substrate

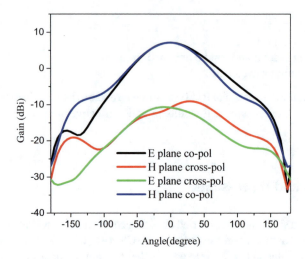

Fig. 3 Radiation pattern (co-pol and cross-pol) of CSMA with PTFE substrate

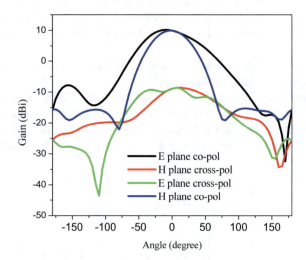

Fig. 4 Radiation pattern (co-pol and cross-pol) of CSMA with air substrate

The comparison of simulated and calculated gain enhancement of CSMA with air substrate has been presented in Table 1. A close agreement has been revealed between computed and simulated results.

Table 1 The calculated values of the effective patch dimensions and gain enhancement compared with the simulated results

Substrate	Resonance frequency (GHz)	Effective radial dimension (a_{eff})	Simulated gain (dBi)	$\Delta G[dB]$	
				Simulated	Calculated
PTFE	2.92	32.49	7.1	2.8	3.14
Air	4.43	33.94	9.9		

5 Conclusion

90° CSMA with air and PTFE substrate has been thoroughly investigated for gain enhancement. A quick hand formulation has been developed to estimate the gain enhancement which is in close agreement with simulated results. The proposed techniques to estimate the gain enhancement for CSMA with air substrate are very easy and simple which will be surely beneficial for scientist and practicing engineers looking for high-gain simple CSMA.

References

1. R. Garg, P. Bhartia, I. Bhal, A. Ittipibun, Microstrip antenna design handbook, in *Artech House*, (Norwood, 2001)
2. R.A. Dalli, L. Zenkouar, S. Bri, Theoretical analysis and optimization of circular sector microstrip antenna. Int. J. Comput. Sci. Info. Technol. Secur. **1**, (2011)
3. A. Deshmukh, A.R. Jain, A.A. Joshi, T.A. Tirodkar, K.P. Ray, Broadband proximity fed modified circular microstrip antenna, in *Proceedings of Advances in Computing and Communications (ICACC)*, (India, 2013)
4. G.K. Oğuz, Ş. Tahaİmeci, A compact size circular sector patch antenna for Ku-band applications, in *2017 International Applied Computational Electromagnetics Society Symposium (ACES)*, (Italy, 2017)
5. S.K. Ghosh, S. Chakraborty, L.L.K. Singh, S. Chattopadhyay, Modal analysis of probe fed circular sector microstrip antenna with and without variable air gap: investigations with modified cavity model. Int. J. RF Microwave Comput. Aided Eng. **28**(1), 1–14, (2018). https://doi.org/10.1002/mmce.21172
6. W.F. Richards, J.D. Ou, S.A. Long, A Theoretical and Experimental Investigation of Annular Sector, and Circular Sector Microstrip Antennas. IEEE Trans. Antennas Propag. **32**(8), 864–867 (1984)
7. T.T. Lo, D. Solomon, W.F. Richards, Theory and experiments on microstrip antennas. IEEE Trans. Antennas Propag. **27**, 137–145 (1979)
8. S.K. Ghosh, S.K. Varshney, S. Chakraborty, L.L.K. Singh, S. Chattopadhyay, Probe-fed semi circular microstrip antenna vis-à-vis circular microstrip antenna: a necessary revisit, in 3rd *International Conference on Communication Systems (ICCS)*, (India, 2017)

9. S.K. Ghosh, S. Chakraborty, S.k. Ghosh, A. Ghosh, L.L.K. Singh, S. Chattopadhyay, An investigation into the Equivalence of radiation characteristics of 90° Circular Sector microstrip antenna vis-à-vis circular microstrip antenna, in *IEEE Indian Conference on Antennas and Propagation (InCAP)*, 1–4, (India, 2018). https://doi.org/10.1109/incap.2018.8770757
10. S.K. Ghosh, A. Ghosh, S. Chakraborty, L.L.K. Singh, S. Chattopadhyay, The Influence of Feed Probes on the Modes of Circular Sector Microstrip Antennas: An Investigation. IEEE Antenna Propag. Mag. (2020). https://doi.org/10.1109/MAP.2019.2958520.

Design of an Electronic System for Analysis of Body Postural Sway Using Inertial Measurement Unit

Nirmal Rai, Saumen Gupta, and Rinkila Bhutia

Abstract Inertial measurement units are widely used for body movement analysis and balance assessment which helps in diagnosis of balance disorder in older people and people with movement disorder. This paper presents the design of a body postural sway analysis system built using Arduino and Inertial Measurement Unit (IMU).

Keywords MPU-9250 · Inertial measurement unit (IMU) · Sway analysis · Arduino Nano

1 Introduction

Old age as well as disease causes body sway to increase [1, 2]. This paper describes the design and development of a low-cost electronic system for analysis of body postural sway. Roll, pitch, and yaw angular displacement, which are indicators of sway [3, 4], are obtained using IMU and Arduino Nano.

Roll, pitch, and yaw angles are indicative of sway, respectively, in left–right, forward–backward plane, and horizontal planes, respectively [5]. Further, the system consists of a Python program for plotting the values in real time as well as records it for further processing.

N. Rai (✉)
Department of Electronics and Communication, Sikkim Manipal Institute of Technology, Majitar, India
e-mail: nirmal.r@smit.smu.edu.ins

S. Gupta
Department of Physiotherapy, Sikkim Manipal College of Physiotherapy, Gangtok, India
e-mail: saumen.g@smims.smu.edu.in

R. Bhutia
Department of Mathematics, Sikkim University, Gangtok, India
e-mail: rinkila.skm@gmail.com

© Springer Nature Singapore Pte Ltd. 2020
R. Bera et al. (eds.), *Advances in Communication, Devices and Networking*,
Lecture Notes in Electrical Engineering 662,
https://doi.org/10.1007/978-981-15-4932-8_58

507

2 System Design

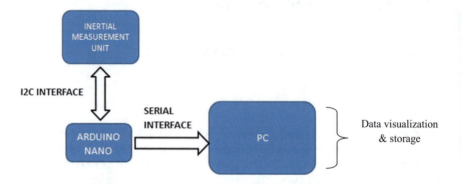

The Inertial measurement unit used in the device is MPU-9250. The MPU-9250 is a sensor module that combines two chips: the MPU-6500, which contains a 3-axis gyroscope, a 3-axis accelerometer, and an onboard Digital Motion Processor (DMP) capable of processing complex motion fusion algorithms; and the AK8963, the market-leading 3-axis digital compass. The raw data obtained from the MPU-9250 is relayed via I2C to Arduino Nano. The Arduino Nano is a small, complete, and breadboard-friendly board based on the ATmega328P microcontroller. It will process the raw data to produce output in the form of pitch, roll, and yaw angles. Processing involves calibration, filtering, and averaging of the raw data. These processed data are sent to computer through serial port for real-time visualization and storage. For these, a program has been developed in Python. The Python program receives the data from the serial port and stores it in csv format as well as plot it in real time (Fig. 1).

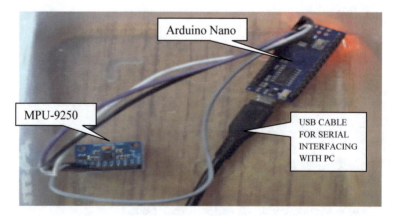

Fig. 1 Electronic system for body postural sway analysis

```
Roll: 50.67,Pitch: 40.57,Yaw: -139.79♦
Roll: 50.53,Pitch: 40.79,Yaw: -139.17♦
Roll: 50.67,Pitch: 40.58,Yaw: -139.11♦
Roll: 50.66,Pitch: 40.75,Yaw: -138.61♦
Roll: 50.98,Pitch: 40.55,Yaw: -138.93♦
Roll: 51.33,Pitch: 40.13,Yaw: -139.81♦
Roll: 50.67,Pitch: 40.77,Yaw: -139.31♦
Roll: 50.89,Pitch: 40.63,Yaw: -139.22♦
Roll: 51.07,Pitch: 40.49,Yaw: -139.50♦
Roll: 50.94,Pitch: 40.43,Yaw: -139.69♦
Roll: 50.89,Pitch: 40.35,Yaw: -140.10♦
Roll: 50.79,Pitch: 40.54,Yaw: -139.88♦
Roll: 50.87,Pitch: 40.45,Yaw: -140.11♦
Roll: 50.86,Pitch: 40.54,Yaw: -139.93♦
Roll: 50.98,Pitch: 40.37,Yaw: -140.25♦
Roll: 51.22,Pitch: 40.29,Yaw: -140.60♦
Roll: 50.98,Pitch: 40.51,Yaw: -140.40♦
Roll: 50.98,Pitch: 40.39,Yaw: -140.61♦
Roll: 50.76,Pitch: 40.48,Yaw: -140.58♦
```

Fig. 2 Processed output from Arduino Nano

3 Result and Discussion

Figure 2 shows the processed output of the Arduino Nano. The output consists of roll, pitch, and yaw angular rotation values in degrees. Figure 3, Fig. 4, and Fig. 5 are the real-time visualization of yaw, pitch, and roll angular rotation, respectively.

4 Conclusion

The paper has discussed the development of an electronic system for body postural sway analysis which can help in diagnosing movement disorder. The system makes use of IMU and Arduino Nano for obtaining roll, pitch, and yaw angular displacement values. The system also consists of Python program for data visualization and recording. The presented work has paved the way for application of machine learning algorithms for detection of movement disorder. Future work involves the real-world testing and validation of the sensor values.

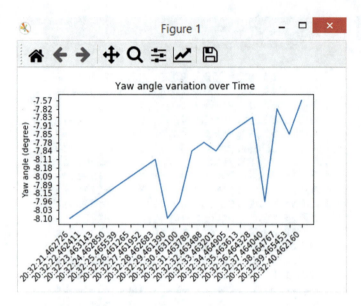

Fig. 3 Real-time variation of yaw angular rotation

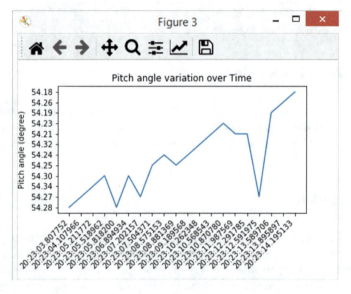

Fig. 4 Real-time variation of pitch angular rotation

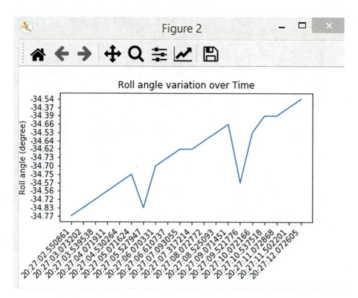

Fig. 5 Real-time variation of roll angular rotation

References

1. L.M. Nashner, M. Woollacott, The organization of rapid postural adjustments of standing humans: an experimental-conceptual model. Posture Move. **243**, 257 (1979)
2. M. Davidson, M.L. Madigan, M.A. Nussabaum, Effects of lum-bar extensor fatigue and fatigue rate on postural sway. Eur. J. Appl. Physiol. **93**(1–2), 183–184 (2004)
3. K. Taraldsen, S.F. Chastin, I.I. Riphagen, B. Vereijken, J.L. Helbostad, Physical activity monitoring by use of accelerometer-based body-worn sensors in older adults: a systematic literature review of current knowledge and application. Maturitas **71**(1), 13–19 (2012)
4. T. Shany, S.J. Redmond, M. Marschollek, N.H. Lovell, Assessing fall risk using wearable sensors: a practical discussion. Z Gerontol Geriatr **45**(8), 694–670 (2012)
5. M. Ghahramani, D. Stirling, Others, Body postural sway analysis in older people with different fall histories. Med. Biol. Eng. Comput. (2018)
6. J. Howcroft, J. Kofman, E.D. Lemaire, Review of fall risk assessment in geriatric populations using inertial sensors. J Neuroeng Rehabil. **10**(1), 91 (2013)
7. G.-A. Constantin, G. Ipate, *Evaluation of hand-arm vibration transmitted during soil tillage based on Arduino and MPU9250 motion sensor, symposium* (Actual Tasks on Agricultural Engineering, Opatija, Croatia, 2018)

Design of a Microwave Radiometer Using DTH Setup

Prashant Sharma, Hardick Saikia, Preman Chettri,
and Samarendra Nath Sur

Abstract In this paper, we have discussed the design of a full power radiometer to detect any object which radiates electromagnetic field above 0 °C. For this system, we have taken commercially available DTH setup for the development of the radiometer and also for the object detection. We have used components which are easily available in the market and are user friendly. Also, in this paper for the data acquisition and for the signal processing, we have utilized the LabView platform.

Keywords DTH · Radiometer · Detection · Noise · Amplifier

1 Introduction

In this paper, we described a radiometer designed to detect an object. As electromagnetic radiation is emitted by a warm object [1], research and development of microwave sensor is highly demanded due to its non-invasive detection or monitoring devices in the industry [2]. Microwave radiometer is a passive component which is used for measurement of electromagnetic field emitted by an object in a microwave frequency range [3]. Study of microwaves has been of interest for many years [4–6]. Radiometer has been used in many medical applications like treating cancer, detection and treatment of hyperthermia, and many other microwave thermographies [7–10]. As it is capable of monitoring object under sensitive environment, the system is placed on a roof; first of all, we will measure the roof environmental temperature. The baseline of a radiometer is given by the roof temperature. The system will detect a difference between roof environmental temperature and the temperature after placing any radiating object in antenna near field. The intensity of radiation is proportional to the temperature of the object. The aim of this work is research and development toward surface monitoring of absolute temperature from

P. Sharma (✉) · H. Saikia · P. Chettri · S. N. Sur
Department of Electronics and Communication Engineering, Sikkim Manipal Institute of
Technology, Sikkim Manipal University, Majitar, Rangpo, East Sikkim 737136, India
e-mail: prashant.dhungana1999@gmail.com

© Springer Nature Singapore Pte Ltd. 2020
R. Bera et al. (eds.), *Advances in Communication, Devices and Networking*,
Lecture Notes in Electrical Engineering 662,
https://doi.org/10.1007/978-981-15-4932-8_59

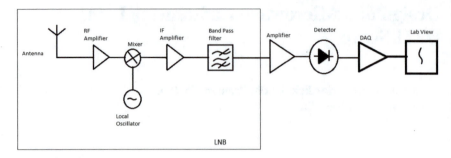

Fig. 1 A block diagram of a total power radiometer for object detection

near field of the object. We are particularly interested in measuring thermal emission emitted from an object.

As in Fig. 1, all the components inside the dashed line are the component present in the LNB which sense the incoming signal. At the output of an LNB, we connect a square-law detector followed by a low-pass filter and the integrator.

The main objective of our project is to find the change in the electromagnetic field emission when we bring any radiating object in the antenna near-field region, and to find the difference in temperature between the normal temperature condition and after the arrival of any object or body above 0 °C in antenna real field.

Some of the problems of radiometer are discussed in [11, 12]. The sensitivity of radiometer has been discussed in [13, 14].

2 Theoretical Background

A microwave radiometer is a handheld device that can measure the temperature of an object. The measuring principle depends on the thermally emitted power over the given band of frequency. The amount of spectral emitted by the body and received by the receiver is given by Plank's radiation law. The total power P over a frequency range Δf can be written as [11]

$$P = kTB \tag{1}$$

T: change in temperature of the system (K), k: 1.38e−23 (J/K), Boltzmann constant, and B: IF bandwidth (Hz).

In practical cases, there is an introduction of noise at every stage in cascade system and it is given by

$$T_n = T_{n1} + T_{n2}/G_1 + T_{n2}/G_1 G_2 + \cdots + T_{nN}/(G_1 G_2 \ldots G_{N-1}) \tag{2}$$

T_n: noise temperature of the system, and G_i: gain of individual stage.

Equation (1) shows that the power depends on the temperature T and the bandwidth B and can be rewritten as the total power (P_t) received in the receiver side with some noise temperature is given by "Nyquist equation":

$$P_t = k(\eta T + T_n)B \text{ (W)} \tag{3}$$

η: antenna efficiency.

Now the total power available at the output of the receiver with antenna gain is given by

$$P_t = k G_t(\eta T + T_n)B \text{ (W)} \tag{4}$$

G_t: system total gain.

Square-law detector is connected at the output of the LNB. The output of the diode is given by (V_d):

$$V_d = S \cdot P_n \tag{5}$$

In quadratic region and in linear region, it is given by

$$V_d = \sqrt{S \cdot Pn} \tag{6}$$

where S is the detector sensitivity, and its unit is $(\text{mV}/\mu\text{W})$. We use low-pass filter at the output of the decoder to make the signal band limited.

The main problem we face in designing radiometer is due to temperature fluctuation, as different components behave differently in different temperatures. In our project we use LNB and dish antenna which is used in satellite TV communication. The system parameters are mentioned below:

Antenna reflector	Gain	40 dB
	Efficiency	60%
Low noise block (LNB)	Radio frequency band	10.7–11.7 GHZ
	Intermediate frequency	1–2 GHZ
	Local oscillator frequency	9.75 GHZ
	Gain	60 dB
	Noise figure	0.2 dB
Detector	Type	Schottky diode
IF amplifier	Gain	20 dB
	Noise figure	1.3 dB

3 Hardware and Software Description

Since the system is designed to detect an object which enters the antenna beam and the temperature of an object should be above 0 °C, we can find the difference between two signals by subtracting from the instantaneous value of the signal. The receiver is sensitive to the fast change. We have observed the graphical changes in software as the change in amplitude of the voltage level and also in the spectrum.

The main function of LNB, as in Fig. 2, is to amplify the incoming signal. It converts the high frequency into the low-frequency component. As the body emits low frequency due to which the frequency of output of LNB is very low, therefore we use high-gain amplifier at the output of the LNB. Our system should provide some indications of whether the object is present or not. So we use a linear detector in the circuit. The work of detector in the linear region is to avoid the low-frequency gain. We can achieve this by increasing the power level of the input frequency signal supplied to the detector input. After detector, we have connected DAQ, as in Fig. 3. The work of DAQ is to measure the voltage. It converts analog data into digital data and gives the readings to the computer. We see the change in signal on the computer.

Figures 4 and 5 show how the software work. First figure shows the connection diagram inside the software, and second figure shows the received signal by the software and its noise cancellation. It shows its respective DC value and the RMS value of the input signal.

Fig. 2 DTH antenna with LNB

Fig. 3 Hardware setup of radiometer system

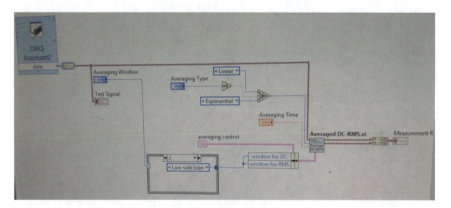

Fig. 4 Software program for data acquisition using LabView

4 Results and Discussion

We have observed the change in signal level both in DC voltage level and also in spectrum level. To detect the change in the spectrum level, we have used Keysight to make spectrum analyzer PXI M9010A. First of all, we measure the incoming signal without the object and after that we observe the change of the waveform after inserting the object.

From Figs. 6 and 7, we can see the difference in the spectrum of the received signal. First figure shows the field spectrum when there is no object in front of antenna near-field region, and second figure shows the change in the spectrum when any object is brought in front of antenna near-field region.

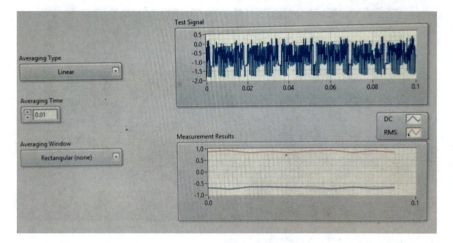

Fig. 5 DC/RMS voltage level of received signal (after windowing and averaging)

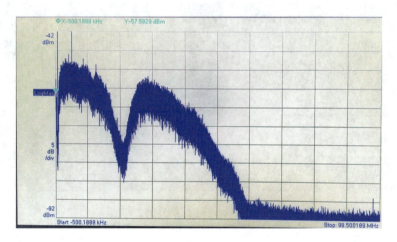

Fig. 6 Received signal spectrum without object

5 Conclusion

This paper deals with the development of the microwave radiometer using DTH setup in order to detect the object that is placed broadside of the antenna. Detection is quantified by observing the change in DC level change in the detected voltage level and by observing the change in the spectrum of the signal.

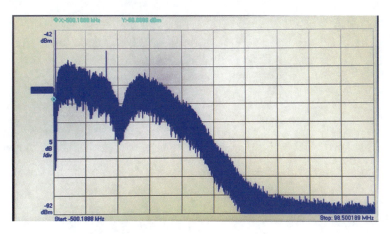

Fig. 7 Received signal spectrum with object that is brought in the antenna near field

References

1. L. Amaduzzi, M. Tinti, low cost components radiometer implementation for human microwave electromagnetic field emission detection. Prog. Electromagn. Res. Lett. **23**, 9–18 (2011)
2. Q. Bondst, T. Wellert, B Roeder, P. Herzig, A total power radiometer (TPR) and measurement test bed for non contact biomedical sensing applications. IEEE 978-1-4244-4565-3 (2009)
3. K.L. Carr, Antenna: the critical element is successful medical technology, in *IEEE MTT-S International Microwave Symposium Digest*, vol. I (1990), pp. 525–527
4. G. Konieczny, Z. Opilski, T. Pustelny, E. Maciak, T. Pustelny, Acta Phys. Pol. A **116**, 389 (2009)
5. A.-M.M. El-Sharkawy, P.P. Sotiriadis, P.A. Bottomley, E. Atalar, Absolute temperature monitoring using RF radiometry in the MRI scanner. IEEE Trans. Circuits Syst. I Reg. Papers **53**(11), 2396–2404 (2006)
6. E.A. Cheever, K.R. Foster, Microwave radiometry in living tissue: what does it measure? IEEE Trans. Biomed. Eng. **39**(6), 563–568
7. S.A. Kostopoulos, A.D. Savva, P.A. Asvestas, C.D. Nikolopoulos, C.N. Capsalis, D.A. Cavouras, Early breast cancer detection method based on a simulation study of single-channel passive microwave radiometry imaging. J. Phys. Conf. Ser. **633**(1), 012120 (2015)
8. F. Sterzer, Microwave medical devices. IEEE Microw. Mag. **3**, 6570 (2002)
9. E. Zampeli et al., Detection of subclinical synovial inflammation by microwave radiometry. PLoS ONE **8**(5), e64606 (2013)
10. E.C. Fear, S.C. Hagness, P.M. Meaney, M. Okoniewski, M.A. Stuchly, Enhancing breast tumor detection with near-field imaging. IEEE Microw. Mag. **3**, 48–56 (2002)
11. B. Przywara-Chowaniec, L. Poloński, M. Gawlikowski, T. Pustelny, Acta Phys. Pol. A **116**, 344 (2009)
12. W.J. Wilson, A.B. Tanner, F.A. Pellerano, K.A. Horgan, Ultra stable radiometers for future sea surface salinity missions. JPL Report D-31794 (2005)
13. A. Camps, J.M. Tarong'ı, Microwave radiometer resolution optimization using variable observation time. Remote Sens. **2**(7), 1826–1843 (2010)
14. G. Evans, C.W. McLeish, *R.F. Radiometer Handbook* (Artech House, Norwood, 1977)

Correction to: System Identification and Speed Control of a DC Motor Using Plug-In Repetitive Controller in the Presence of Periodic Disturbance

Rijhi Dey, Anirban Sengupta, Naiwrita Dey, and Ujjwal Mondal

Correction to:
Chapter "System Identification and Speed Control of a DC Motor Using Plug-In Repetitive Controller in the Presence of Periodic Disturbance" in:
R. Bera et al. (eds.), *Advances in Communication, Devices and Networking*, Lecture Notes in Electrical Engineering 662,
https://doi.org/10.1007/978-981-15-4932-8_48

The affiliation of the author "Anirban Sengupta" and "Ujjwal Mondal" in the original version of the book has been corrected in the chapter "System Identification and Speed Control of a DC Motor Using Plug-In Repetitive Controller in the Presence of Periodic Disturbance". The chapter and book have been updated with the changes

---

The updated version of this chapter can be found at
https://doi.org/10.1007/978-981-15-4932-8_48

© Springer Nature Singapore Pte Ltd. 2020
R. Bera et al. (eds.), *Advances in Communication, Devices and Networking*,
Lecture Notes in Electrical Engineering 662,
https://doi.org/10.1007/978-981-15-4932-8_60

Printed in the United States
by Baker & Taylor Publisher Services